# Related Books of

## DITA Best Practices
### A Roadmap for Writing, Editing, and Architecting in DITA

By Laura Bellamy, Michelle Carey, Jenifer Schlotfeldt
ISBN-13: 9780132480529

**The Start-to-Finish, Best-Practice Guide to Implementing and Using DITA**

Darwin Information Typing Architecture (DITA) is today's most powerful toolbox for constructing information. By implementing DITA, organizations can gain more value from their technical documentation than ever before. Now, three DITA pioneers offer the first complete roadmap for successful DITA adoption, implementation, and usage.

Drawing on years of experience helping large organizations adopt DITA, the authors answer crucial questions the "official" DITA documents ignore, including: *Where do you start? What should you know up front? What are the pitfalls in implementing DITA? How can you avoid those pitfalls?* If you're a writer, editor, information architect, manager, or consultant who evaluates, deploys, or uses DITA, this book will guide you all the way to success.

## The IBM Style Guide
### Conventions for Writers and Editors

By Francis DeRespinis, Peter Hayward, Jana Jenkins, Amy Laird, Leslie McDonald, Eric Radzinski
ISBN-13: 9780132101301

*The IBM Style Guide* distills IBM wisdom for developing superior content: information that is consistent, clear, concise, and easy to translate. This expert guide contains practical guidance on topic-based writing, writing content for different media types, and writing for global audiences. The guidelines are especially valuable for businesses that have not previously adopted a corporate style guide, for anyone who writes or edits for IBM as an employee or outside contractor, and for anyone who uses modern approaches to information architecture.

Filled with many examples of correct and incorrect usage, *The IBM Style Guide* can help any organization or individual create and manage content more effectively.

Sign up for the monthly IBM Press newsletter at
ibmpressbooks.com/newsletters

# Books of Interest

## Patterns of Information Management

By Mandy Chessell, Harald Smith
ISBN: 9780133155501

**Use Best Practice Patterns to Understand and Architect Manageable, Efficient Information Supply Chains That Help You Leverage All Your Data and Knowledge**

In the era of "Big Data," information pervades every aspect of the organization. Therefore, architecting and managing it is a multi-disciplinary task. Now, two pioneering IBM® architects present proven architecture patterns that fully reflect this reality. Using their pattern language, you can accurately characterize the information issues associated with your own systems, and design solutions that succeed over both the short- and long-term.

## Mobile Strategy
### How Your Company Can Win by Embracing Mobile Technologies

By Dirk Nicol
ISBN-13: 9780133094916

**Navigate the Mobile Landscape with Confidence and Create a Mobile Strategy That Wins in the Market Place**

*Mobile Strategy* gives IT leaders the ability to transform their business by offering all the guidance they need to navigate this complex landscape, leverage its opportunities, and protect their investments along the way. IBM's Dirk Nicol clearly explains key trends and issues across the entire mobile project lifecycle. He offers insights critical to evaluating mobile technologies, supporting BYOD, and integrating mobile, cloud, social, and big data. Throughout, you'll find proven best practices based on real-world case studies from his extensive experience with IBM's enterprise customers.

**IBM Press**

Visit ibmpressbooks.com for all product information

# Books of Interest

## Is Your Company Ready for Cloud?
Choosing the Best Cloud Adoption Strategy for Your Business

By Pamela K. Isom, Kerrie Everall Holley
ISBN-13: 9780132599849

**Make the Right Cloud Adoption and Deployment Decisions for Your Business**

This is the first complete guide to cloud decision making for senior executives in both technology and non-technology roles. IBM® Global Business Services® Executive Architect Pamela K. Isom and IBM Fellow Kerrie Holley present practical business cases, vignettes, and techniques to help you understand when cloud investments make sense and when they don't. You'll find decision models that are anchored with practical experiences and lessons to guide your decision making, best practices for leveraging investments you've already made, and expert assistance with every aspect of the cloud transition.

## Get Bold
Using Social Media to Create a New Type of Social Business
By Sandy Carter
ISBN-13: 9780132618311

## Decision Management Systems
A Practical Guide to Using Business Rules and Predictive Analytics
By James Taylor
ISBN-13: 9780132884389

## The New Era of Enterprise Business Intelligence
Using Analytics to Achieve a Global Competitive Advantage
By Mike Biere
ISBN-13: 9780137075423

## Opting In
Lessons in Social Business from a Fortune 500 Product Manager
By Ed Brill
ISBN-13: 9780133258936

## The Greening of IT
How Companies Can Make a Difference for the Environment
By John Lamb
ISBN-13: 9780137150830

Sign up for the monthly IBM Press newsletter at
ibmpressbooks.com/newsletters

# Developing Quality Technical Information

# Developing Quality Technical Information

## A Handbook for Writers and Editors

**Michelle Carey • Moira McFadden Lanyi • Deirdre Longo • Eric Radzinski • Shannon Rouiller • Elizabeth Wilde**

IBM Press
Pearson plc
Upper Saddle River, NJ • Boston • Indianapolis • San Francisco
New York • Toronto • Montreal • London • Munich • Paris • Madrid
Cape Town • Sydney • Tokyo • Singapore • Mexico City
ibmpressbooks.com

The authors and publisher have taken care in the preparation of this book, but make no expressed or implied warranty of any kind and assume no responsibility for errors or omissions. No liability is assumed for incidental or consequential damages in connection with or arising out of the use of the information or programs contained herein.

© Copyright 2014 by International Business Machines Corporation. All rights reserved.

Note to U.S. Government Users: Documentation related to restricted right. Use, duplication, or disclosure is subject to restrictions set forth in GSA ADP Schedule Contract with IBM Corporation.

IBM Press Program Managers: Steven M. Stansel, Ellice Uffer

Cover design: Michael Rouiller and Polly Hughes

Editor-in-Chief: Bernard Goodwin
Marketing Manager: Stephane Nakib
Acquisitions Editor: Bernard Goodwin
Publicist: Heather Fox
Managing Editor: Kristy Hart
Designer: Alan Clements
Project Editor: Andy Beaster
Full-service composition: codeMantra US LLC
Project Manager: Francesca Monaco, codeMantra US LLC
Manufacturing Buyer: Dan Uhrig

Published by Pearson plc

Publishing as IBM Press

For information about buying this title in bulk quantities, or for special sales opportunities (which may include electronic versions; custom cover designs; and content particular to your business, training goals, marketing focus, or branding interests), please contact our corporate sales department at corpsales@pearsoned.com or (800) 382-3419.

For government sales inquiries, please contact governmentsales@pearsoned.com.

For questions about sales outside the U.S., please contact international@pearsoned.com.

The following terms are trademarks or registered trademarks of International Business Machines Corporation in the United States, other countries, or both: IBM, the IBM Press logo, AIX, z/OS, DB2, InfoSphere, Optim, pureQuery, ibm.com, developerWorks, Maximo, WebSphere, Passport Advantage, FileNet, Cognos, Guardium, Redbooks, and IMS. Worklight is a trademark or registered trademark of Worklight, an IBM Company. A current list of IBM trademarks is available on the web at "copyright and trademark information" as www.ibm.com/legal/copytrade.shtml.

UNIX is a registered trademark of The Open Group in the United States and other countries. Linux is a registered trademark of Linus Torvalds in the United States, other countries, or both. Java and all Java-based trademarks and logos are trademarks or registered trademarks of Oracle and/or its affiliates. Microsoft, Windows, Windows Phone, Internet Explorer, SharePoint, Access, Excel, and the Windows logo are trademarks of Microsoft Corporation in the United States, other countries, or both.

Other company, product, or service names mentioned herein may be trademarks or service marks their respective owners.

Library of Congress Control Number: 2014931735

All rights reserved. This publication is protected by copyright, and permission must be obtained from the publisher prior to any prohibited reproduction, storage in a retrieval system, or transmission in any form or by any means, electronic, mechanical, photocopying, recording, or likewise. To obtain permission to use material from this work, please submit a written request to Pearson Education, Inc., Permissions Department, One Lake Street, Upper Saddle River, New Jersey 07458, or you may fax your request to (201) 236-3290.

ISBN-13: 978-0-13-311897-1

ISBN-10: 0-13-311897-5

Text printed in the United States on recycled paper at R.R. Donnelley in Crawfordsville, Indiana.
First printing: June 2014

# Contents

| | |
|---|---|
| Preface | xvii |
| Acknowledgments | xix |
| About the authors | xxiii |

## Part 1. Introduction — 1

### Chapter 1. Technical information continues to evolve — 3
- Embedded assistance — 4
  - Progressive disclosure of information — 9
- The technical writer's role today — 11
- Redefining quality technical information — 13

### Chapter 2. Developing quality technical information — 15
- Preparing to write: understanding users, goals, and product tasks — 16
- Writing and rewriting — 17
- Reviewing, testing, and evaluating technical information — 19

## Part 2. Easy to use — 21

### Chapter 3. Task orientation — 23
- Write for the intended audience — 25
- Present information from the users' point of view — 27
- Focus on users' goals — 32
  - Identify tasks that support users' goals — 33
  - Write user-oriented task topics, not function-oriented task topics — 35
  - Avoid an unnecessary focus on product features — 41

# Contents

    Indicate a practical reason for information    46
    Provide clear, step-by-step instructions    49
        Make each step a clear action for users to take    51
        Group steps for usability    53
        Clearly identify steps that are optional or conditional    58
    Task orientation checklist    64

## Chapter 4. Accuracy    67
    Research before you write    69
    Verify information that you write    74
    Maintain information currency    79
        Keep up with technical changes    79
        Avoid writing information that will become outdated    82
    Maintain consistency in all information about a subject    86
        Reuse information when possible    86
        Avoid introducing inconsistencies    88
    Use tools that automate checking for accuracy    93
    Accuracy checklist    96

## Chapter 5. Completeness    99
    Make user interfaces self-documenting    101
    Apply a pattern for disclosing information    107
    Cover all subjects that support users' goals and only those subjects    115
        Create an outline or topic model    115
        Include only information based on user goals    118
        Make sure concepts and reference topics support the goals    122
    Cover each subject in only as much detail as users need    123
        Provide appropriate detail for your users and their experience level    123
        Include enough information    130
        Include only necessary information    136
    Repeat information only when users will benefit from it    141
    Completeness checklist    148

# Part 3. Easy to understand    151

## Chapter 6. Clarity    153
    Focus on the meaning    155
    Eliminate wordiness    161

|   |   |
|---|---|
| Write coherently | 174 |
| Avoid ambiguity | 180 |
|     Use words as only one part of speech | 180 |
|     Avoid empty words | 183 |
|     Use words with a clear meaning | 187 |
|     Write positively | 189 |
|     Make the syntax of sentences clear | 194 |
|     Use pronouns correctly | 199 |
|     Place modifiers appropriately | 201 |
| Use technical terms consistently and appropriately | 205 |
|     Decide whether to use a term | 205 |
|     Use terms consistently | 207 |
|     Define each term that is new to the intended audience | 210 |
| Clarity checklist | 212 |

## Chapter 7. Concreteness — 215

|   |   |
|---|---|
| Consider the skill level and needs of users | 220 |
| Use concreteness elements that are appropriate for the information type | 223 |
| Use focused, realistic, and up-to-date concreteness elements | 240 |
| Use scenarios to illustrate tasks and to provide overviews | 243 |
| Make code examples and samples easy to use | 247 |
| Set the context for examples and scenarios | 251 |
| Use similes and analogies to relate unfamiliar information to familiar information | 253 |
| Use specific language | 256 |
| Concreteness checklist | 259 |

## Chapter 8. Style — 261

|   |   |
|---|---|
| Use active and passive voice appropriately | 263 |
| Convey the right tone | 267 |
| Avoid gender and cultural bias | 273 |
| Spell terms consistently and correctly | 276 |
| Use proper capitalization | 280 |
| Use consistent and correct punctuation | 284 |
| Apply consistent highlighting | 296 |
| Make elements parallel | 302 |
| Apply templates and reuse commonly used expressions | 305 |
| Use consistent markup tagging | 311 |
| Style checklist | 314 |

# Part 4. Easy to find     317

## Chapter 9. Organization     319

Put information where users expect it     322
    Separate contextual information from other types of information     324
    Separate contextual information into the appropriate type of embedded assistance     332
    Separate noncontextual information into discrete topics by type     337
Arrange elements to facilitate navigation     345
    Organize elements sequentially     350
    Organize elements consistently     354
Reveal how elements fit together     360
Emphasize main points; subordinate secondary points     366
Organization checklist     376

## Chapter 10. Retrievability     379

Optimize for searching and browsing     381
    Use clear, descriptive titles     381
    Use keywords effectively     384
    Optimize the table of contents for scanning     389
Guide users through the information     394
Link appropriately     399
    Link to essential information     400
    Avoid redundant links     405
    Use effective wording for links     409
Provide helpful entry points     413
Retrievability checklist     420

## Chapter 11. Visual effectiveness     421

Apply visual design practices to textual elements     424
Use graphics that are meaningful and appropriate     431
    Illustrate significant tasks and concepts     431
    Make information interactive     441
    Use screen captures judiciously     448
Apply a consistent visual style     460
Use visual elements to help users find what they need     467
Ensure that visual elements are accessible to all users     478
Visual effectiveness checklist     483

# Part 5. Putting it all together — 485

## Chapter 12. Applying more than one quality characteristic — 487
    Applying quality characteristics to progressively disclosed information — 488
    Applying quality characteristics to information for an international audience — 494
    Applying quality characteristics to topic-based information — 501

## Chapter 13. Reviewing, testing, and evaluating technical information — 515
    Reviewing technical information — 516
    Testing information for usability — 518
    Testing technical information — 524
    Editing and evaluating technical information — 527
    Reading and editing the information — 531
    Reviewing the visual elements — 536

# Part 6. Appendixes — 543

## Appendix A. Quality checklist — 545
## Appendix B. Who checks which characteristics? — 549

Glossary — 555
Resources and references — 565
Index — 573

# Preface

## About this book

Many books about technical writing tell you how to develop different elements of technical information, such as headings, lists, tables, and indexes. We took a different approach with this book; we organized it to show you how to apply quality characteristics that make technical information, including information embedded in user interfaces, easy to use, easy to understand, and easy to find. We hope you will find our approach useful and comprehensive—and we hope that you will find the information in this book easy to use, easy to understand, and easy to find!

## Is this book for you?

If you are a writer, editor, information architect, or reviewer of technical information and user interfaces, then yes, this book is for you. If you work on software information, this book will be of particular interest to you because most of the examples in it come from the domain of software. However, the quality characteristics and guidelines apply to all technical information.

In general, this book assumes that you know the basics of good grammar, punctuation, and spelling as they apply to writing. It does not assume that you are familiar with what makes technical information effective or ineffective.

## Changes in this edition

The organization of the book and the quality characteristics remain the same. However, within each quality characteristic, we made significant changes by replacing some guidelines with new ones, adding many new examples, and broadening the scope of the kinds of information that we discuss. If you are familiar with previous editions, you'll find a great deal of new content in this edition. For example, the following guidelines are among those that we added:

- "Apply a pattern for disclosing information" in the chapter about completeness
- "Guide users through the information" in the chapter about retrievability
- "Put information where users expect it" in the chapter about organization
- "Make information interactive" in the chapter about visual effectiveness

These changes resulted from several developments in technical communication:

- Greater emphasis on the embedded assistance in user interfaces
- The need to plan for information access from mobile devices
- The pervasiveness of Google and other search engines as users' preferred method for looking for information
- Video as a delivery medium for technical information

As with earlier developments in this field during the many years that these quality characteristics have been in use, the characteristics remain relevant while the definition of technical information expands in scope. This quality framework continues to apply to the information that we provide today. In addition, we have found that the characteristics apply well to user interfaces, which benefit from application of the guidelines much as other content does.

We hope that you find this book useful in improving the quality of the information that you develop.

# Acknowledgments

The predecessor of this book was an internal document called *Producing Quality Technical Information*. That document led to the first edition of *Developing Quality Technical Information*, which was published in 1998, followed by the second edition in 2004. And here we are 10 years later with the third edition.

After the second edition of *Developing Quality Technical Information* was published, its lead author and project manager, Gretchen Hargis, passed away.

Throughout the writing process for the first two editions of this book, Gretchen was vigorous in pushing the authors to do what was necessary to make the book as good as it could possibly be. We planned, we drafted, we edited, we haggled, we revised, we reedited, we proofread. Throughout the process, the concept of "good enough" never entered Gretchen's mind.

Sometimes, Gretchen's coauthors wished "good enough" had been just that, but in retrospect we are so glad that Gretchen persevered. Without Gretchen, neither the first nor second edition of the book would ever have been completed. Gretchen is sorely missed by all of her coauthors and colleagues.

We felt that she was with us every step of the way as we wrote the third edition, and we hope that this edition lives up to her standards.

Over the years, nearly a hundred talented people have in some way contributed to this latest edition. We thank all the people who helped with this book and its predecessors.

One of the biggest challenges to writing this book was providing the vast number of examples in each chapter. We were fortunate to get help identifying many excellent examples. Many thanks to Hassi Norlen, Richie Escarez, Ellen

# Acknowledgments

Livengood, Ann Hernandez (author from the second edition), Beth Hettich, Erin Jerison, Marcia Carey (Michelle's mom), and Gary Rodrigues for helping with the nearly 400 examples provided in this book.

And thanks to the talented visual designers and writers who provided some of the examples in the "Visual effectiveness" chapter: Tina Adams, Daiv Barrios, Jessy Chung, Caroline Law, Adam Locke, Challen Pride-Thorne, Rene Rodriguez, Shannon Thompson, and Jacob Warren. Thanks especially to visual designers Clark Gussin and Sean Lanyi for always being available to advise and help.

Many of the clarity and style issues we discuss come out of trying to do what's best for translators, so we'd like to thank Sabine Lehmann, Ph.D., for her guidance about machine translation and linguistics and for her translations of French, German, and English examples.

We thank Michael Rouiller and Polly Hughes (second edition author) for their help with the cover graphic.

We'd also like to thank the following folks who helped us find the quotations that introduce each quality characteristic chapter: Christopher Clunas, Paula Cross, Fran DeRespinis, Jasna Krmpotic, Yvonne Ma, and Leslie McDonald.

For technical support, we'd never have finished the book if it hadn't been for Dan Dionne, Kevin Cheung, and Simcha Gralla. Many thanks to these gentlemen for their help.

Thanks to Andrea Ames for the hours and laughs we continue to share while defining and building education for embedded assistance and progressive disclosure of information within and outside of IBM. Thanks also to Jennifer Fell who gave us a wonderful metaphor that describes how users should be able to use technical information: "As a guided journey instead of a scavenger hunt."

Thanks to Lori Fisher who created a space over the course of many years for all of us to contribute to the craft of information development and to develop a framework for information quality. Thanks to Eileen Jones for sponsoring this edition and for fostering the profession of information development at IBM.

We thank our families, friends, cats, and dogs for their incredible patience and support throughout the writing process. We stole countless late nights, weekends, and decent meals from them, and we can never pay those back.

Lastly, we must thank our talented editor, Julian Cantella, for the many long hours he spent editing our manuscript. It's never easy editing a book that's written by a team of editors. Julian's thoughtful and meticulous work helped us add that extra polish to the book.

Michelle Carey
Moira McFadden Lanyi
Deirdre Longo
Eric Radzinski
Shannon Rouiller
Elizabeth Wilde

# About the authors

The authors are all long-standing and respected members of the information development community at IBM. Although the authors have served in various roles throughout their careers, information quality has always been and continues to be their primary focus.

**Michelle Carey** is an information architect and technical editor at IBM and has taught technical communication at University of California Santa Cruz Extension. Michelle is the co-author of the book *DITA Best Practices: A Roadmap for Writing, Editing, and Architecting in DITA*. She is an expert on topic-based information systems, software product error messages, grammar, embedded assistance for user interfaces, and writing for international audiences. She also writes computational linguistic rules for a grammar, style, and terminology management tool. Michelle enjoys teaching, grammar, herding cats, and riding and driving anything with a lot of horsepower.

**Moira McFadden Lanyi** is an information architect and technical editor at IBM. She has experience with topic-based writing, DITA, embedded assistance, user interface design, and visual design. She created 99% of the artwork in this book. She is a co-author of the book *An Introduction to IMS*. Moira enjoys visiting San Francisco with her family as often as possible, cooking fresh, healthy meals, and watching her courageous son ride his unicycle and surf.

**Deirdre Longo** is an information architect and strategist at IBM. She has been a pioneer for embedded assistance in IBM: defining the scope of that term, developing standards for embedded assistance, and modeling how to work effectively in cross-disciplinary teams. She has taught webinars for the Society of Technical Communication (STC) and published articles on information architecture topics in STC's *Intercom*. She is an avid yoga practitioner.

## About the Authors

**Eric Radzinski** is a technical editor and information architect for industry-leading mainframe database software at IBM. He is a co-author of *The IBM Style Guide: Conventions for Writers and Editors* and is well versed in topic-based writing, embedded assistance, DITA, and writing for a global audience. Eric makes his home in San Jose, California, with his wife and their three children.

**Shannon Rouiller** is an information architect and technical editor at IBM. She has experience with quality metrics, topic-based information systems, DITA, videos, embedded assistance, and user interface design. She is a co-author of the book *Designing Effective Wizards*. Shannon dabbles in sports photography and likes to solve puzzles.

**Elizabeth Wilde** is an information quality strategist at IBM, developing strategies and education for developing high-quality content. She develops Acrolinx computational linguistic rules that enforce grammar, style, and DITA tagging rules. She teaches an extension course in technical writing at the University of California Santa Cruz. Her hobbies include growing cacti and succulents and collecting tattoos.

# PART 1

# Introduction

**Chapter 1. Technical information continues to evolve** — 3
Embedded assistance — 4
    Progressive disclosure of information — 9
The technical writer's role today — 11
Redefining quality technical information — 13

**Chapter 2. Developing quality technical information** — 15
Preparing to write: understanding users, goals, and product tasks — 16
Writing and rewriting — 17
    Planning what to write for different stages of product use — 17
    Deciding exactly what to write — 18
    Starting to write — 18
    Refining what you write — 19
Reviewing, testing, and evaluating technical information — 19

CHAPTER 1

# Technical information continues to evolve

The nature of our work as technical communicators continues to change, more rapidly than ever. The authors of this book can see it even over the short course, relatively speaking, of our own careers in technical communication. Some of us began our careers delivering camera-ready copy for a shelf of physical books and then began producing context-sensitive online help that was installed with the product. With the advent of the web, we used our online help-writing skills to rework books into online topic-based documentation.

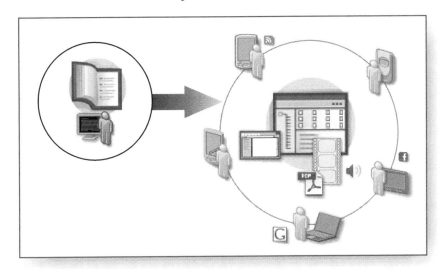

Today, writers sigh or laugh ruefully over the fact that users don't click help links. Testing with users validates this premise—that users don't want to ask for separate help—but that they *do* use all of the text they see in user interfaces to do their tasks. In surveys, users often say that their first response to trouble is to ask a colleague. In testing, when users were forced to seek additional assistance, a majority in our tests tried to search the Internet or visit a video site such as YouTube rather than reading the help. This finding is surprising at first, but on reflection, is the equivalent of asking a colleague.

One reason that users avoid help documentation is that we, as a profession, have taught them that, as one user told us, "There's nothing good there." For example, when we moved from command-oriented products to those with graphical interfaces, technical information was focused on helping users to understand how to manipulate the user interface. Although that focus made sense during the transition, many writers continue that focus today, 20 years after the transition. In spite of knowing better, we continue to produce huge amounts of help documentation.

As technical writers, we need to recognize this shift in our audience and move past it to address users where they are. A new generation of technology-savvy users is entering the workforce, existing workers are becoming more adept, and technology is becoming more sophisticated. Because of these changes, the emphasis is on more usable, intuitive, and appealing products. Now we need to expand our focus beyond topic-based information and onto the product user interfaces themselves, with input field labels, messages, and other embedded text, which collectively we refer to as *embedded assistance*.

We need to recognize that topics alone cannot address all needs. Topics work well in some contexts and for some types of documentation: planning, application programming, technical concepts, troubleshooting, and hardware diagrams. In many contexts, users expect to stay where they are and figure out how to do their tasks without reading separate documentation. But in many other contexts, especially mobile contexts, users want to watch a video introduction or a presentation by an expert. We need to write information for users where they are, focused on what they're trying to accomplish, instead of trying to make them read what they don't want to read.

## Embedded assistance

Our profession has different definitions for embedded assistance.

Some groups refer to static, descriptive inline text in a user interface as embedded assistance and differentiate it from the interface labels and messages. Others use the term to refer to the mechanism that displays a pane of online help text within the same window as the product. For the purposes of this book, we define embedded assistance as both of those and more—to define it more narrowly only reinforces the artificial separation between product and documentation that occurs because of the way most products and documentation are developed. When users buy or use a product, they don't differentiate between the interface, the documentation, and the functionality. To users, all of these are the product. We, with all members of our product development teams, must develop our products as a whole too.

Embedded assistance, therefore, encompasses all textual and graphical elements that users encounter in all types of products. In graphical user interfaces, embedded assistance includes (numbers refer to Figure 1.1 below):

- Labels for user interface controls such as fields, radio buttons, check boxes, push buttons, menus, window titles, and so on (1)
- Input hints in fields (2)
- Descriptive inline user interface text such as introductory text in a window (3)
- Messages that appear on fields, in sections of the interface, or in dialogs
- Tooltips, which are one- to two-word names for tools that do not have labels in the interface
- Hover help, which are one to two sentences of description for fields, check boxes, radio buttons, and so on (6)
- Wizards for simplifying complex interactions
- Embedded help panes (8)

The following illustration shows some of these elements in a user interface:

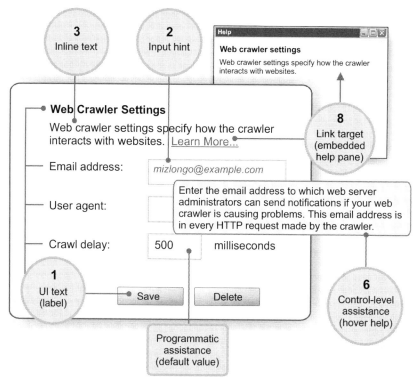

**Figure 1.1** Embedded assistance elements in a user interface.

Technical information continues to evolve

In nongraphical software contexts, such as ASCII-based interfaces, embedded assistance includes:

- Logged messages
- Command and parameter names
- Keyword names
- API names
- Utility or tool names

7

# Introduction

In hardware contexts, embedded assistance includes:

- Labels embedded on hardware wires, boards, or other equipment
- Labels attached on top of or around hardware, for example on an on/off switch
- Specifically sized slots for connectors
- Colors for wiring, for example, the color green indicates the ground wire in the US

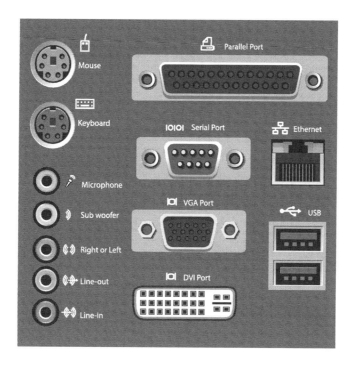

Embedded assistance also includes *programmatic assistance* that does a step or task for a user. Examples of programmatic assistance include:

- Default values
- Detected values
- Autocompleted values, as shown in the following user interface:

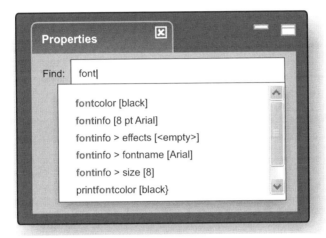

Although we writers usually don't have the programming skills to develop programmatic assistance, we do need to understand these types of assistance well enough to advocate for them when they'll be helpful to users.

Our skills with writing embedded assistance and our fluency with words and graphics will be crucial as software development shifts to focus on mobile devices. Most users don't follow help links in desktop and web applications; they are even less likely to do so in mobile environments. The lack of hover capabilities in the mobile environment removes an element of embedded assistance in an already small user interface, a user interface that makes web and desktop interfaces seem enormous by comparison. Because of the small screen size in a mobile interface, the small amount of text that is persistent gets even more attention.

## Progressive disclosure of information

Given the types of embedded assistance elements in the previous section, it's easy to see that writers can't work in isolation on each element, set of elements, or functional area of the user interface or piece of hardware. Instead, the entire set of embedded assistance, linked assistance, and separate documentation must tell a cohesive story.

## Introduction

The key to developing effective documentation is to apply and follow a pattern for progressively disclosing the information to the user. *Progressive disclosure* is not a new idea in the field of interaction design. Jakob Nielsen summarized it in 2006: "Progressive disclosure defers advanced or rarely used features to a secondary screen, making applications easier to learn and less error-prone." Applying such a progressive pattern to information ensures that you use available space in a user interface or on the hardware in the most effective way, consistently and without redundancy. Applying the pattern well also helps writing teams manage the complexity of information, providing clear paths to get to more complex or abstract information.

Information that is developed according to principles of progressive disclosure anticipates users' questions and provides a way for users to get additional contextual information when necessary. For example, in an installation wizard, a field might have the label **Application server version**, and a user might ask, "Is this the version I'm upgrading from or to?" Ideally, the label could be changed to clarify which version, or the field could be grouped under a heading **Upgrade from server**. If neither is possible, a hover help that explains which application server is being requested and how to find this information is helpful, but a hover help that says, "Enter the application server version" is not.

If you are used to writing books or help documentation, think about how you decide where (which book or help system) to deliver certain information today. Information that you put in an installation guide is not appropriate in an application development guide. A similar approach is true in user interfaces: different interface display mechanisms require different types of information.

Think of the available programmatic and textual assistance capabilities in a software or hardware interface as different delivery mechanisms. You can then use a pattern to map types of content to each mechanism. Because these delivery mechanisms are much smaller than a book or a web page, the pattern is also at a different scale. Instead of thinking about the type of content to deliver in a programming guide versus an installation guide, you think about the type of content to deliver in a field label versus hover help. Your pattern might look something like this:

Table 1.1  Sample pattern for progressively disclosing information in a web user interface

| User interface element | Content |
| --- | --- |
| Labels (for fields, windows, buttons, group boxes, and so on) | Succinct nouns based on a short, well-managed list of product terms. Repeatedly review these labels to ensure consistency and coherence of the interface as it is developed. |
| Messages | Full sentences that describe the situation. For error messages, provide an action so that users can solve the problem. |
| Static descriptive text at the top of windows | <ul><li>Describe the overall action that users accomplish on the window if it's not obvious.</li><li>Clarify anything users must do before completing this window.</li><li>Identify ramifications, if there are any, of the changes in this window.</li></ul> |
| Static descriptive text below fields | Examples for what to enter in a field. |
| Hover help | <ul><li>Syntax for what to enter in a field.</li><li>Ramifications of the field change.</li><li>Descriptive information for what to enter in the field.</li><li>Links to additional information if needed.</li></ul> |

The organization guideline "Separate contextual information into the appropriate type of embedded assistance" on page 332 describes the pattern in more depth, and the completeness guideline "Apply a pattern for disclosing information" on page 107 explains how to apply it to your information to ensure completeness.

When you become more adept at creating meaningful and effective embedded assistance and delivering it progressively, you create a better customer experience and become a more valuable member of your product team.

# The technical writer's role today

Our roles as technical writers are evolving as quickly as the products that we write information for. Because we develop embedded assistance, the timing and ways that we work with our extended teams have changed. We are more involved with product design and user interface development, which means that we must be involved earlier than ever in the development cycle.

As discussed in "Embedded assistance" on page 4, the separation between product and documentation is an artificial one, in large part a result of the historic waterfall development processes. The *waterfall development process* is made up of specific phases in which each participating team finishes its work and hands it to the next team. The problem with this process is that downstream teams have very little chance to change anything that happened upstream. Furthermore, because documentation is developed close to the end of the cycle, documentation often tries to describe poor design that can no longer be changed. Too often, technical writers who work in a waterfall development process must write comprehensive documentation that needs to atone for unwieldy design.

More and more development teams are using an *agile development process*, which depends on cross-functional teams working together throughout an iterative development cycle. Although members of these cross-functional teams all bring their own skills to the team from their unique disciplines, they are much more likely to look at and contribute to each others' deliverables so that products are a full team effort. Agile development, as the name implies, lends itself to making quick changes to product design when necessary.

In agile development, writers have a particularly effective role as the users' advocate. The Agile Manifesto (agilemanifesto.org) values "individuals and interactions over processes and tools" and "working software over comprehensive documentation." Writers who work on a project that follows the agile development process are critical members of the team throughout the entire process, from the earliest design phase, before a single line of code is written, to the final fit-and-finish stage. By participating in the design process in partnership with product developers, usability engineers, visual designers, and customers, writers can promote clear interaction and wise use of embedded assistance, thereby reducing or eliminating the likelihood of "papering the product" with unnecessary documentation.

The guidelines in this book describe the characteristics of quality technical information. However, your role in developing information and, indeed, in developing the product, is as important as any of the guidelines. Rather than trying to explain problems with the product design after the fact, focus on fixing real-world problems that users have.

When you develop quality technical information, you are responsible for:

- Knowing the user stories, which are the goals that users need to accomplish by using the product
- Being the users' advocate, ensuring that the product employs the necessary programmatic assistance and embedded assistance
- Owning the words, whether they are in labels in the user interface, error messages, or topics that are separate from the product

# Redefining quality technical information

Quality is ultimately determined by users. When users have questions and quickly find the exact information they need, they perceive the product (and the information, though they don't distinguish between the two) as being of high quality. In fact, an overwhelming majority of customers report that information quality both affects their view of the product quality and their overall product satisfaction. Information quality also has a significant impact on customers' buying decisions.

Almost always, users seek answers to specific questions and don't want to read a book from beginning to end to find those answers. Quality information addresses users where they are, for example, in the user interface. That quality helps them accomplish real goals rather than forcing them to figure out how to accomplish their goals in the product.

Content that focuses on *domain expertise*, provided by experts in the field based on their experience and judgement, is the most highly valued content today. We can already see the beginning of another technical communication transition toward artificial intelligence, and our role in gathering real domain expertise for users becomes critical. Think of voice-driven assistance that provides real-world information about proximity to gas stations with the lowest prices or guidance for how to choose the right app from an online store. In these situations, the writer is the trusted colleague or the concierge, directing users to exactly what they need at that moment. Domain expertise is described in more detail in the concreteness guideline, "Consider the skill level and needs of users" on page 220.

Technical writers must be the users' advocate throughout the product development process. Ideally, writers have access to users throughout that process, but user engagement alone cannot ensure information quality. Writers must apply their own skills and expertise based on solid research and proven methods.

Quality characteristics for technical information must reflect what users expect and want from the information. Based on comments from users and on experience in writing and editing technical information, the authors of this book have found that quality technical information has these characteristics:

| Easy to use | |
|---|---|
| Task orientation | In the context of a product, a focus on helping users do tasks that support their goals |
| Accuracy | Freedom from mistake or error; adherence to fact or truth |
| Completeness | The inclusion of all necessary parts—and only those parts |
| **Easy to understand** | |
| Clarity | Freedom from ambiguity or obscurity; using language in such a way that users understand it the first time that they read it |
| Concreteness | The inclusion of appropriate examples, scenarios, similes, analogies, specific language, and graphics |
| Style | Correctness and appropriateness of writing conventions and of words and phrases |
| **Easy to find** | |
| Organization | A coherent arrangement of parts that makes sense to the user |
| Retrievability | The presentation of information in a way that enables users to find specific items quickly and easily |
| Visual effectiveness | Attractiveness and enhanced meaning of information through the use of layout, illustrations, color, typography, icons, and other graphical devices |

You can apply the quality characteristics whether you're writing a book, a page, a paragraph, a sentence, or a single word in an interface. The quality technical information model of nine characteristics is flexible enough to support you as you develop ever smaller chunks of information to address the changing needs of users.

CHAPTER 2

# Developing quality technical information

The evolution in technical information isn't just about delivery formats. It's also about refining your ability to write meaningful, quality information.

In embedded contexts, when you're constrained to a few words, you need to use those words as effectively as possible. How do you develop quality information with so few words? In short, by following the same techniques you use to write separate documentation:

1. Investigating and understanding:
   - Users and their skills and knowledge
   - Users' goals, not to be confused with the tasks they do to achieve those goals
   - Scenarios for achieving goals
   - How the product helps users achieve goals
   - Terms and concepts used in the industry
2. Applying a pattern for progressively disclosing information
3. Iteratively writing, testing, and editing, using the quality guidelines described in this book

If you're sensing a theme in this summary, you've probably noticed how much of the work is background investigation and thinking and how little is writing. When this work is done well, you often spend more time thinking and investigating than writing and more time editing and revising than writing new content. As Blaise Pascal famously said, "I have made this [letter] longer than usual because I have not had time to make it shorter."

## Preparing to write: understanding users, goals, and product tasks

Before you write anything, you must understand your users: their skill levels, their challenges, and their environment.

When you write information, you need to build on the users' knowledge and speak the users' language while also considering the needs of people with disabilities (including limited vision, mobility, or attention), an aging population, nonnative language speakers, and others with learning challenges. You can observe users as they work, interview them, or survey a set of users to better understand them, their goals, and the tasks they undertake to achieve those goals.

The results of goal and task analysis can help you understand:

- What goals users are trying to achieve
- What tasks they do to achieve their goals
- Which tasks they complete outside of your product or tool
- Which tasks are most important
- Which tasks they spend the most time on
- Which tasks they struggle most with
- Which tasks are most tedious or most trivial
- When and where users typically need information
- What type of information users need

When you understand more about users and their goals, you can more easily contribute to the design of a meaningful, task-oriented product with well-crafted embedded assistance and supporting information.

Note the emphasis on goals, which are not to be confused with tasks. Goals are almost never product-focused, as they reflect a need in the real world. Tasks are the activities users do—some of which might be in different products or not in a product at all—to achieve their goals.

An example of this distinction is evident when you consider the chore of doing your taxes. Your goal is to accurately calculate your tax and submit your return on time to get the biggest possible refund and avoid penalties. You accomplish this goal by completing a series of tasks:

1. Calculate your taxes. You can use a variety of tools and methods to calculate your taxes, including a pencil and paper or a software program, or, if you are particularly averse to doing your own taxes, by hiring someone to do it for you.
2. Send your return. You can send your tax return by using different methods, including walking or driving a paper envelope to the post office or submitting electronically.
3. Pay your tax or receive a refund. You can pay your tax or get your refund electronically or as a paper check.

To work well, any software that aids in tax preparation must be oriented around the overall goal, recognize the different ways to do different tasks, and provide relevant information about the tasks that don't occur inside the software, such as taking the return physically to the post office. By grounding information in the real-world goals of users, writers can determine what information to write (tips for getting a bigger refund based on the data that users enter) and what information is extraneous and can be deleted (how to properly format currency values).

## Writing and rewriting

As with other kinds of writing, quality technical writing is the result of a process of refinement. When you first write about a subject, just getting your thoughts on the screen or paper might be enough of a challenge. Striving at that point for quality characteristics such as retrievability and style might be more inhibiting than helpful.

Eventually, as you develop the information, the guidelines for all of the quality characteristics should come into play. Many of them need frequent consideration.

### Planning what to write for different stages of product use

Just as you emphasize different characteristics at different times in your writing process, users benefit from different characteristics at different times during their stages of product use. If you think about where users of your planned documentation are in their product usage, you can see how much to emphasize different characteristics, for example:

- When they evaluate a product, users focus on visual effectiveness and style because these characteristics give the product appeal.

- When they learn about a new product, users lean on task orientation, organization, clarity, concreteness, and visual effectiveness because these characteristics help users conceptualize their goals within the product context.
- When they use the product, users value task orientation, retrievability, accuracy, concreteness, and completeness because these characteristics ensure that users get the right information at the right time.

## Deciding exactly what to write

You might not have the same expertise as many users, for example, IT administrators, but if you write about something that is new to you, pay close attention to your own questions as you learn about the subject. Your questions can help you identify the information to write. For example, in a user interface with a field labeled **Database name**, what questions might you ask? Perhaps: "Which database?" "Where can I find the name?" Validate your questions and answers with typical users or with those on your team who know more about the users than you do.

The information that you write might need to serve both novices and experienced users. (Remember, however, that users move from novice to some level of competency with most products fairly quickly, so you need to ensure that you write to the majority of your audience rather than only to novices.) Although all of the guidelines in this book are centered on users as the reference point for decisions about what to write and how to write it, the following three sections describe how to address multiple audiences:

- The task orientation guideline, "Write for the intended audience" on page 25
- The concreteness guideline, "Consider the skill level and needs of users" on page 220
- The completeness guideline, "Cover each subject in only as much detail as users need" on page 123

## Starting to write

When you write topics and other content that is delivered outside the product, you still need not concentrate on achieving all of the quality characteristics at the same time. Depending on your approach and where you are in the development cycle, you'll focus on some of the characteristics more than others:

- When you write an outline, you probably need to keep in mind the guidelines for task orientation, organization, and completeness.
- When you write first drafts of topics or individual pieces of embedded assistance, the guidelines for accuracy, concreteness, and clarity take precedence.

- When you write new topics for an existing deliverable:
  - Review the organization to ensure that your new topics fit well into the existing structure and if not, change the structure. You need to be willing to completely reorganize or discard material that no longer works. Each time you update information, it should still read as smoothly and make sense when progressively disclosed from the user interface, as if it were all written at the same time.
  - Review the existing retrievability structures (index, tables of contents, keywords) to see how to apply these to your new material.
- When you have decided where to insert new content and are ready to write it, you apply guidelines for accuracy, concreteness, clarity, style, and visual effectiveness.

## Refining what you write

Throughout the writing process, you refine what you've written by applying the characteristics, by incorporating review comments from users, developers, testers, and editors, by adding new material as product function changes, and by reading what you've written.

When you work on user interface labels and text, you refine your writing at a much lower level—on individual terms and how they fit together. Here, you must carefully select terms that make sense in the industry or to a specific set of users. You might select a metaphor with an associated set of terms. The desktop, for example, is a metaphor that has been used in most user interfaces since the early days of personal computers. The desktop metaphor incorporates the idea of documents in folders. Often in software development, you select a metaphor but discover that the metaphor won't work or that the terms are too broad or too narrow to work in all cases. As a result you might need to exchange the metaphor for a new one. Applying clarity guidelines is key to selecting good terminology.

Chapter 12, "Applying more than one quality characteristic," on page 487 follows several typical documentation scenarios and shows how you would apply the characteristics throughout the writing process.

# Reviewing, testing, and evaluating technical information

Users are your most essential reviewers. Agile development in particular incorporates frequent user testing to ensure that the product is meeting users' needs. Assistance, whether embedded or not, is a key part of this testing. Getting early feedback about terminology selections and where users need more assistance is essential to developing quality information. If customers aren't available to test, you can ask colleagues to test your information.

## Introduction

At the end of each of the quality characteristic chapters is a checklist of the problems to look for concerning each guideline. You can use the checklists to have users or your colleagues test what you have written. You can also use the checklists when you evaluate what someone else has written or to assess your own writing.

These individual checklists are gathered into one checklist in Appendix A, "Quality checklist," on page 545. You can put your findings from the individual checklists into this summary checklist to get an overall picture.

All of the checklists provide columns where you can give a numerical evaluation (from 1 to 5) based on the number and kinds of problems that you find. These numerical evaluations can help you see the areas that most need revising.

The major revision cycles in the process of developing technical information probably result from reviews by other people who are involved in the project, such as developers, designers, editors, usability engineers, users, and marketing professionals. You can use the checklist for the quality characteristics in Appendix B, "Who checks which characteristics?," on page 549 to help ensure that the different types of reviewers evaluate the appropriate quality characteristics.

Even after you publish information or release a product, you can continue to get feedback, including instant feedback. By incorporating social media commenting and rating capabilities into information delivery mechanisms, you can make it easy for your users to give you ongoing feedback.

Chapter 13, "Reviewing, testing, and evaluting technical information," on page 515 describes various methods for internal and external testing, reviewing, and evaluating information.

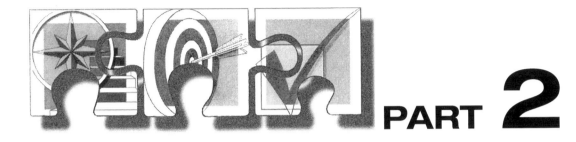

# PART 2

# Easy to use

Users access technical information to accomplish a goal, which typically involves doing one or more tasks. Technical information should make it easy for users to successfully achieve their goals. It should be trustworthy. Users should have all of the information they need without being overloaded with extraneous information.

| | |
|---|---|
| **Chapter 3. Task orientation** | **23** |
| Write for the intended audience | 25 |
| Present information from the users' point of view | 27 |
| Focus on users' goals | 32 |
|     Identify tasks that support users' goals | 33 |
|     Write user-oriented task topics, not function-oriented task topics | 35 |
|     Avoid an unnecessary focus on product features | 41 |
| Indicate a practical reason for information | 46 |
| Provide clear, step-by-step instructions | 49 |
|     Make each step a clear action for users to take | 51 |
|     Group steps for usability | 53 |
|     Clearly identify steps that are optional or conditional | 58 |
| Task orientation checklist | 64 |
| **Chapter 4. Accuracy** | **67** |
| Research before you write | 69 |
| Verify the information that you write | 74 |
| Maintain information currency | 79 |
|     Keep up with technical changes | 79 |
|     Avoid writing information that will become outdated | 82 |
| Maintain consistency in all information about a subject | 86 |

| | |
|---|---|
| Reuse information when possible | 86 |
| Avoid introducing inconsistencies | 88 |
| Use tools that automate checking for accuracy | 93 |
| Accuracy checklist | 96 |

## Chapter 5. Completeness — 99

| | |
|---|---|
| Make user interfaces self-documenting | 101 |
| Apply a pattern for disclosing information | 107 |
| Cover all subjects that support users' goals and only those subjects | 115 |
|     Create an outline or topic model | 115 |
|     Include only information based on user goals | 118 |
|     Make sure concepts and reference topics support the goals | 122 |
| Cover each subject in only as much detail as users need | 123 |
|     Provide appropriate detail for your users and their experience level | 123 |
|     Include enough information | 130 |
|     Include only necessary information | 136 |
| Repeat information only when users will benefit from it | 141 |
| Completeness checklist | 148 |

CHAPTER 3

# Task orientation

*Don't tell me how it works; tell me how to use it.* —An IBM customer

Task orientation is a focus on users' goals and the tasks that support those goals. You rarely help your users when you tell them only how a product works or how it is structured. Your users have a job to do, so they need practical information that helps them understand and complete their tasks.

## Easy to use

If your goal is to put your child's bicycle together the night before his birthday, you don't want to learn about how sturdy the tires are or what materials were used to make them. You want the steps to get from a box of pieces to an assembled bike in as little time as possible.

Task orientation applies well beyond step-based, procedural information. The text in user interfaces, such as in labels and error messages, and the text in concept and reference topics also need to be written from the perspective of your users' goals. Task-oriented writing keeps users "on task."

*[handwritten margin note: Who are you writing for?]*

To make information task oriented, follow these guidelines:

- ❑ **Write for the intended audience**
- ❑ **Present information from the users' point of view**
- ❑ **Focus on users' goals**
- ❑ **Indicate a practical reason for information**
- ❑ **Provide clear, step-by-step instructions**

# Write for the intended audience

When you plan what information to write, be sure that you have a clear understanding of your audience. For example, if you are writing for system architects, you might include only high-level tasks, such as evaluating and planning, or a high-level view of other tasks. Similarly, if you are writing for end users, avoid tasks that are appropriate for product administrators and focus on the tasks that end users do.

*[margin note: include what is relevant to the target user]*

Be sure that the information that you include in your topics and embedded assistance is relevant to the needs of your audience. For example, your product might have a powerful data analytics system, but a description of that system is of little interest to the person who is installing one of the product components on a client.

The following simple task is explained in detail. However, users who want to customize settings are experienced users and therefore do not need help performing simple tasks. This information will frustrate all but the most patient experienced users.

**Original**

1. Open the directory.
2. Click the `infodir` folder.
3. Right-click the `settings.def` file and select **Edit** from the menu.
4. Change the settings that you want in the file.
5. Click **File > Save** to save the file.
6. Click **File > Close** to close the editor.

**Revision**

To customize your settings, edit the `infodir/settings.def` file.

In the revision, the task is handled much more simply. The revision quickly provides users with the information that they need because it is written for the skill level of the audience.

# Easy to use

If you don't understand your audience, then you won't know what kind of assistance to provide for the user interface. Embedded assistance is useless if it is directed at the wrong audience, and in some cases it can be misleading. The text in the second paragraph of the following error message for a bank website is not meant for bank customers:

**Original**

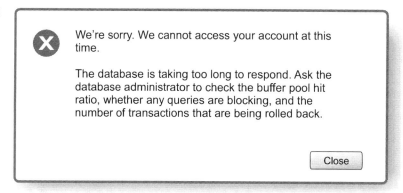

The second paragraph should never be viewed by the bank customers, who have no knowledge of a database and no interest in contacting a database administrator to troubleshoot an error with the website. Bank customers might also be concerned to read that transactions are rolling back if the customers assume that the transactions that are mentioned in the error message are bank transactions and not database transactions.

**Revision**

The revised error message tells the bank customers only what they need to know.

## Present information from the users' point of view

Writing information from the users' point of view brings users into the story so that they understand how information relates to them. Such information is predominantly written in second person and uses active voice. It also uses verbs that denote actions that users do as opposed to actions that the product does.

The following passage leaves users out of the story altogether:

**Original**

> Subsequent installation of the HIGS feature allows InfoProduct to run unattended.

The original passage is indefinite about who does the action and why. Users can't tell how the information in the original passage pertains to them.

**Revision**

> If you want to run InfoProduct unattended, you must install the HIGS feature. You can install the HIGS feature after you install InfoProduct.

The revision adds users to the story and explains what they need to do to run the product unattended. Notice that improving the task orientation also improves the clarity of the passage.

Easy to use

The following text is shown at the top of a list of recommendations that energy customers receive after they run tools to assess their home energy consumption:

Original

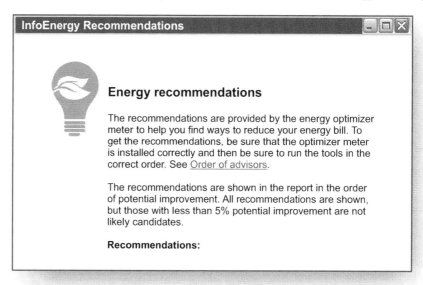

The original text wastes customers' time because it provides information about prerequisite steps that customers already completed. It even mentions checking whether the company technicians installed the meter correctly, which is hardly something energy customers are qualified to do.

Revision

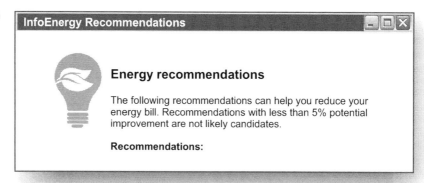

The revised text presents only information that makes sense based on the customers' position in the task. In addition, some descriptive text about the order of recommendations is removed because it describes what customers can already see.

# Task orientation

If you use the product, you will have a better idea of the point of view of your users, and you can present targeted embedded assistance. You might find places where it would help to add an example below a field, where an error message could include actions for users to complete, or where a label could be more specific. You might also find places in the product where programmatic assistance could reduce work for users.

In the following window, some of the text in the hover help is presented from the point of view of someone who is coding the product, not of someone who is using the product:

**Original**

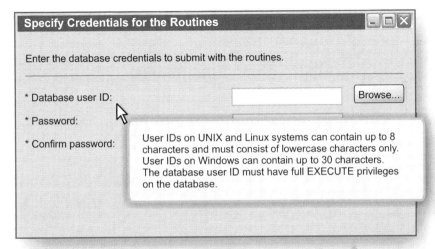

The original hover help for the **Database user ID** field contains information about how many characters user IDs can contain. However, users are being asked to enter an existing user ID for a database. Obviously, an existing user ID already follows the character restrictions for the operating system. The information about character restrictions is presented from the point of view of the product developer, who needs to know how many characters to expect in the field.

**Revision**

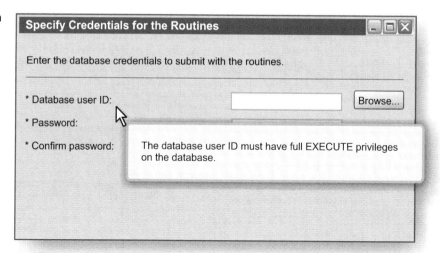

In the revision, the information that users do not need is removed. Users are no longer provided with the unnecessary reminder about character restrictions.

Error messages should be written specifically to accommodate users' situations. When you work with your product, make some mistakes on purpose to see whether the error messages are presented from the users' point of view. Not all error messages can be found by working with the product, so you also need to review your product's error strings and do some investigating.

The following error message is vague and is not written from the perspective of users:

**Original**

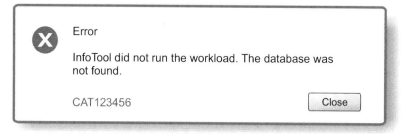

You can fix terse and unhelpful messages such as this one by doing some investigating. Determine what task users are trying to do, what reasons might cause the error, and how users can fix this problem.

# Task orientation

**Revision**

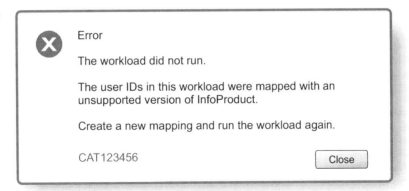

The revision looks like a completely different message. If you take the time to learn the real reason that the product is unable to find the database, you can provide a much more useful error message. In this case, users don't care that the product can't find the database. The mapping of the workload to the database is in the wrong format. From the users' perspective, the problem can be solved by updating the mappings for the workload, and that's all users need to know.

## Focus on users' goals

It's relatively easy to write about the details of a product, such as menu choices or report types, or to describe the simplest tasks for a product, such as logging in or changing preferences. However, if a task is easy to understand and describe, the documentation for that task is probably not very helpful to your users. You make the most impact if you ignore the easy details and simple tasks and focus on what matters most to your users—their goals.

Goals are not the same as tasks. A *goal* is an outcome that users want, and a *task* is an activity that users do. Achieving a goal generally involves doing multiple tasks. For example, users who log in to a dating website have a goal of finding a compatible person to date. These users might need to do many tasks on the website to reach their goal, such as creating a profile, reading through profiles of suggested matches, and making plans to meet. Users might also need to do some tasks outside the website, such as getting a haircut and finding the perfect photo to upload.

One way to ensure that your information is focused on goals is to base your information on scenarios. As explained in the concreteness guideline "Use scenarios to illustrate tasks and to provide overviews," a *scenario* is a story, often in the form of a high-level task, that shows how a product contributes to solving a business problem. Scenarios typically present a narrative of how a problem is solved in a real setting. They are commonly used in marketing or overview information but can also be used in tutorials or in place of examples.

*Scenario-based information* is information that is designed and written to support users' progress through scenarios to achieve goals. When you write scenario-based information, instead of thinking of a scenario as one topic that you write, you think of a scenario as the basis for everything that you write. You identify the key goals for users of the product, and then for each goal, you create a *scenario*, or path, for how users will accomplish the goal.

Know the goals of your users and follow the path through the products yourself. You might use multiple components or other products to complete the scenario. The path might involve several large, medium, and small tasks, and you need to learn how to do all of them. By following the path yourself, you can ensure that the embedded assistance is clear and supports the goal, and you can learn and document the necessary steps to get users from the beginning of the scenario to the end.

If you encounter any obstacles in the scenario, work with the design and development teams to address the problems before they become obstacles for users. You can improve the user experience through the scenario in these areas:

**Product design**
> Use your findings to ensure that the product design supports the path through the scenarios. For example, you might suggest wizards, welcome page content, or links to ensure that the path to start each key scenario is clear.

**Product integration**
> Look for ways to improve the integration between the products that are involved in the scenario. For example, because you know the point in the scenario when users will start to use another product, provide a link to open that product or provide a message to tell them how to start the other product and what to do. Look for opportunities to share information such as user names or account numbers between related products so that users don't need to provide details twice.

To focus on users' goals, follow these guidelines:

- Identify tasks that support users' goals
- Write user-oriented task topics, not function-oriented task topics
- Avoid an unnecessary focus on product features

## Identify tasks that support users' goals

Although users can probably do many tasks with your product, be sure that you provide sufficient embedded assistance and documentation for the tasks that support your users' goals. To identify the tasks that are associated with a goal, start with a high-level task that corresponds to the goal. For example, for the goal of securing company data, use "Securing company data" as the high-level task in your topic model. Divide the high-level task into groups of lower-level tasks that support the goal, such as setting up user access and setting up firewalls. Each of these tasks might also be divided into still lower-level tasks until you have an organized list of tasks in your model.

Each task in your model should be discrete and therefore make sense by itself to users. For example, creating an account for a retail website is a discrete task that makes sense for users to do. The action of selecting a user ID and password might be a step of a task, but it is not a task on its own.

Carefully consider the list of tasks in your model to determine which tasks users can do easily and which tasks you need to document in task topics. If you have good embedded assistance, many tasks will not need topics that provide step-level information. In addition, embedded assistance might replace the need to provide extra steps in the task topics that you do write.

If you are writing scenario-based information, you focus only on the tasks in the model that are part of the scenario path to achieve the goal. Although many tasks might support the goal or be related to the goal, you keep only the tasks that are part of the scenario in your model.

For the goal of creating a photo book, you might identify a scenario in which an existing user creates a project, uploads photos, plays with options for backgrounds, captions, and photo placement, and then orders a printed copy of the book. The following table shows some tasks that users might do when they create a photo book. Notice that a few of the tasks are not part of this particular scenario.

Table 3.1   Tasks associated with a goal of creating a photo book

|   | Task | Part of scenario | Documentation |
|---|---|---|---|
| 1 | Creating a photo book | Yes. | Need a task topic to show high-level steps. |
| 1.1 | Creating a user account | No. This scenario assumes the user has an account. | |
| 1.2 | Creating a project for the photo book | Yes. | No need for a topic. The web interface makes this task easy. |
| 1.3 | Gathering photos for the book | Yes. | No need for a topic. This task can be mentioned in a step in the high-level task. |
| 1.3.1 | Uploading photos to the project | Yes. | No need for a topic. The web interface makes this task easy. |
| 1.3.2 | Accessing photos from other projects | No. | |
| 1.4 | Choosing a background for the project | Yes. | No need for a topic. The web interface makes this task easy. |
| 1.5 | Organizing photos in the book | Yes. | No need for a topic. Users are more likely to experiment and find their own way than they are to read a task topic. |

Table 3.1 Tasks associated with a goal of creating a photo book *(continued)*

| | Task | Part of scenario | Documentation |
|---|---|---|---|
| 1.6 | Formatting photos | No. This scenario assumes the photos are preformatted. | |
| 1.7 | Adding captions to the photos in the book | Yes. | No need for a topic. This task can be mentioned in a step in the high-level task. |
| 1.8 | Sharing the book with other users | No. | |
| 1.9 | Ordering the printed book | Yes. | No need for a topic. Existing users know how to order, and the task is clear in the web interface. |
| 1.9.1 | Troubleshooting problems with the order | Yes. | No need for a topic. The web interface handles any problems. |

For a scenario-based model, the tasks that are not associated with the scenario are removed from the model. In this case, only one high-level task topic was needed. The high-level task topic will probably have a step for each row of the table that is part of the scenario. The tasks in these rows rely on effective embedded assistance and so can be done by users without the need to read additional topics. Therefore, the high-level task in this case will have steps but not child task topics.

## Write user-oriented task topics, not function-oriented task topics

Any task topic that you do write should be written from a perspective that users can relate to. A *user-oriented task* is a task that users want to perform regardless of whether they are using your product to do it. In contrast, tasks that are about using the product, service, or technology are *function-oriented tasks*.

Examples of task topics that are function oriented and corresponding task topics that are user oriented are shown in the following table:

| Function-oriented task topic | User-oriented task topic |
|---|---|
| Using the table editor | Editing a table |
| Using the CNTREC utility | Counting records in a file |
| Using business rules | Automating case management with business rules |
| Inserting a 5025 auxiliary chip | Improving server performance |

# Easy to use

| Function-oriented task topic | User-oriented task topic |
|---|---|
| Adjusting the pH balance of soil | Growing juicier tomatoes |
| Working with the diagram pane | Creating diagrams |
| Understanding the electrical system | Installing an electrical outlet |
| Using the MagicCook 5000 | Making pancakes |

When you work on a product for months, you might forget that users' tasks are not necessarily the same as product functions. Ensure that each task topic that you plan to write is user oriented, not function oriented. Tasks that start with "Using" or "Working with" are probably function oriented.

The following topic has a list of links to low-level task topics about all of the minor actions that users can take with an object in the product:

---

**Working with triggers**

You use triggers to automatically monitor the critical processes in your enterprise. The product polls for records that meet the criteria for the trigger. If a match is found, the appropriate action is taken.

Creating triggers
You create a trigger to monitor time-sensitive records that could require action.

Validating triggers
You must validate a trigger record before you can activate it.

Activating triggers
You can activate a trigger record that has at least one trigger point and one action defined for each trigger point.

Modifying triggers
You can change existing triggers.

Copying triggers
You can copy an existing trigger to create a similar trigger.

Deactivating triggers
You deactivate a trigger record to modify the trigger.

Deleting triggers
You can delete triggers with the **Delete** button.

**Parent topic:** Administering InfoTrigger

This list of task topics assumes that users want detailed steps for all the actions they might take for a trigger or that users are examining the interface and wondering what each option does. This type of information might be appropriate in embedded assistance, but it's not appropriate for task topics. Focusing the task topics on the triggers makes the topics function oriented. A user-oriented task topic would explain how to set up automatic monitoring for critical processes, and it would mention triggers but would not be about triggers.

Many products are designed or assembled one object or function at a time rather than around user scenarios. If the information is also written or assembled by object or function, then it's easy to end up writing tasks from the perspective of the function, widget, utility, or command and lose sight of users' goals. In addition, because function-oriented content can be much easier to write than user-oriented, goal-focused content, the user-oriented tasks are often not written at all, and users are left to figure out how to achieve their goals on their own.

In the following task topic, the task is described in product-specific terms:

**Original**

> **Using the InfoInstaller tool**
>
> InfoInstaller is a tool that you use to install InfoProduct. Using the InfoInstaller tool to install InfoProduct takes about 20 minutes.
>
> To use the InfoInstaller tool:
>
> 1. Open the InfoInstaller tool by entering `infoinst` at the command line.
> 2. Complete the InfoInstaller window by specifying the installation parameters. When you click **OK**, InfoInstaller installs InfoProduct.

The original topic title and content assume that users understand the task in terms of the tool that they need to use to do the task. Although some users might know what the InfoInstaller tool is, all users know what installation is.

# Easy to use

**Revision**

> **Installing InfoProduct**
>
> Installing InfoProduct takes about 20 minutes.
>
> To install InfoProduct:
>
> 1. Enter `infoinst` at the command line to open the InfoInstaller window.
>
> 2. Follow the prompts to specify the installation options for your computer and then click **OK** to install InfoProduct.

The revised introduction and steps separate the task from the tool so that users can relate to the real task of installing the product instead of to the tool that they are using.

You also need to be sure that the task topics that you plan to write really address the goals or needs of users. The title of the following topic makes it look like a user-oriented task, but the task is presented in terms of getting to the window that is used to do the task:

# Task orientation

**Original**

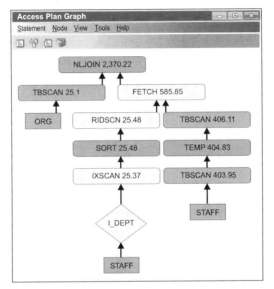

### Analyzing an access plan

You can view a visual representation of the access plan for your query to see how the database accesses the data that it needs for the query.

1. Select a query to analyze the access plan for.
2. Click **Show > Access plan**.
3. View the graph in the Access Plan Graph window that opens and optimize the access plan as needed.

If you want to view the plan for another query, go back and select another query.

The original topic has a title that sounds like a task, but the steps show users how to get to the product window and then stop as if the task ends as soon as users get to the window. The topic never explains how to do the task of analyzing the access plan. Imagine how frustrated users will be when they are faced with a big task but find that the only documentation tells them what they already know. The final sentence is especially useless. Why would users care about starting over if they haven't learned how to do their task in the first place?

# Easy to use

**Revision**

### Analyzing an access plan

You can use a visual representation of the access plan for your query to see how the database accesses the data that it needs for the query. You can look for ways to improve the access plan to shorten query response time.

1. Open the access plan for the query:

   a. Select a query to analyze the access plan for.

   b. Click **Show > Access plan**.

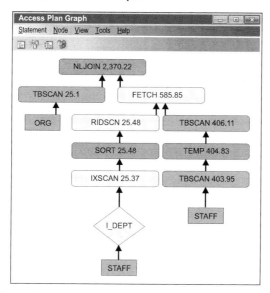

2. Look for ways to improve performance by reviewing the nodes in the access plan:

   - Double-click each table object in the graph to see if any tables have old or missing statistics. If you see the word "Default," then the statistics do not exist and need to be created. Tables are the rectangular objects at the end of each branch.

   - Look for tables that are accessed with table scans instead of index scans. Table scans are mauve, and index scans are light yellow. Table scans generally take longer than index scans and could indicate the need for an index or that any existing indexes are ineffective.

   - Identify other nodes where the query is consuming the most resources. Typically, you need to review *and so on...*

The revised task topic, while still explaining how to access the correct window, focuses more on the users' actual task. The title is the same, but now the title accurately reflects the content, which supports users' needs, not just their path in the interface.

By presenting the information from the perspective of tasks that users recognize and want to perform, you make the information more relevant to users.

## Avoid an unnecessary focus on product features

Content that is based on users' goals and tasks is about what users want to do, whereas content that is based on features is about what the product can do. Not all features need documentation, and further, many products have features that few or no users ever use.

Don't let unnecessary feature-focused content creep into your embedded assistance and topics. As you write, be careful not to "paper the product," which means be careful not to document everything a product can do or every object in a product. You will only end up with a lot of content that users don't want and don't need.

In the following welcome page, the descriptions under each key task include nearly everything the product can do:

Original

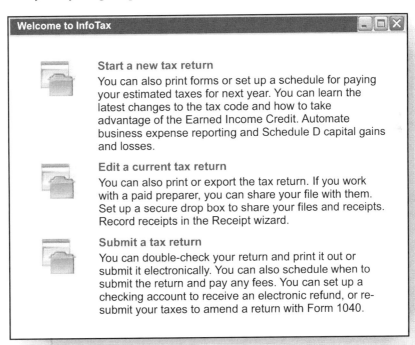

41

In the original welcome page, users who are getting started are seeing an overly comprehensive list of tasks and features that they don't want to think about right now.

**Revision**

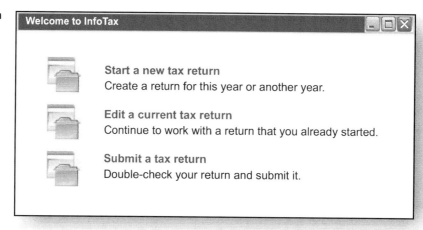

In the revised welcome page, users can get started down one of three paths, and they don't have to read about all of the features that they might, but probably won't, use in due course as they work on their tax returns.

Maintain your focus on goals and avoid the temptation to document every feature, component, and restriction of a product. If your product has basic features that are familiar to most of your users, such as print buttons, report options, or export options, then provide only necessary embedded assistance. These features are usually extraneous to users' goals, and users can get the information they need from the interface.

Suppose that a product introduces a new window for setting preferences so that users can change colors and defaults for their display. A common documentation mistake is to mention a new feature everywhere it might be of interest; in this case, that might be as a note in every task topic that mentions steps that are done in the user interface. However, common sense suggests that users do not want to stop in the middle of their tasks to go play with the colors of windows.

A better approach is to ensure that the feature is easy to use and that embedded assistance for the feature is available if necessary. Then, reference that feature in the documentation only when it is a required part of a user-oriented task. A feature such as a preferences window will likely not need to be referenced at all except perhaps in a "What's new" section that is separate from the goal-oriented task documentation.

When you add embedded assistance for a feature, make sure that the assistance is needed and supports the task that users are doing. The following hover help for a distribution bar chart explains how to change colors on the bar chart:

Original

| Workload Group | Response Time Distribution | Response Time Alerts | Average Time |
|---|---|---|---|
| ♦ cv105 | | | |
| ♦ sales | | You can change the colors for this bar in the Settings window. | |
| ♦ accounting | | — | 19.403500000 |
| ♦ engineering | | — | 18.425700000 |

The hover help is not wrong, but it is about a feature rather than the users' task.

Revision

| Workload Group | Response Time Distribution | Response Time Alerts | Average Time |
|---|---|---|---|
| ♦ cv105 | | | |
| ♦ sales | | The colors show the proportion of the response time spent by applications, data servers, and the network. Workloads with a large middle bar spend the longest time in the data server. | |
| ♦ accounting | | | |
| ♦ engineering | | — | 18.425700000 |

In the revision, the hover help is focused on the task the users are doing when they see the bar chart. Users want to understand and troubleshoot their workload performance, and so the hover help explains what the bars show with respect to performance. In this case, the hover help does not tell users how to do the task, but it does provide information that supports the task.

For task topics, keep your steps focused on the actual user-oriented task. Tasks that try to mention features where they don't belong can be easy to spot because they often contain notes or tips or have more detail than is required for a straightforward step.

Avoid littering task steps with feature "advertisements" unless the features are especially helpful to users as part of the task. Apply a minimalist approach as discussed in the completeness guideline "Cover each subject in just as much detail as users need."

# Easy to use

The following task topic is about a user-oriented task, but unnecessary details about features are mixed in with the steps of the task:

**Original**

> **Recording a conference call**
>
> 1. Start the online meeting as you normally would with your conference ID and moderator PIN.
>
> 2. Turn on your computer speakers.
>
> 3. Click **Record** in the online meeting and then enter the phone conference number and passcode in the dialog that opens. Do not add a 1 to the beginning of the phone number.
>
>    **Tip:** After you record a conference successfully, this information persists in future sessions.
>
>    You will hear a beep when the recording line joins the audio conference. The beep is an mp3 file that you can change based on your preferences. Several additional beep sounds are available in the sounds directory of the main folder.
>
> 4. Turn off your computer speakers after you hear the beep.
>
> 5. Click **Start Recording**.
>
> 6. Share your screen. After you start sharing your own screen, you can hand over control of the presentation to one or more participants just like you normally do. The options are **Make Presenter** or **Stop Sharing**.
>
> 7. Optional: Click **Pause** as needed.
>
> 8. Optional: You can also restart the recording to delete all previous content by clicking **Reset**.
>
> 9. If you need to recover content that was reset, submit a request within 12 hours with the InfoCompany portal.
>
> 10. When the call ends, go to your InfoCompany portal and click **Recordings** for a list of your recordings. You can send the link to other team members. You can also delete the recording or edit it with the tools in the portal.

Because most users who are recording conference calls are trying to start a meeting, they are not interested in how to change the sounds of the beeps. In addition, steps 7, 8, and 9 in the original task topic are not part of the task. These steps only clutter the procedure by mentioning features that users might use. If you document every feature like this, your tasks will become unwieldy, and focus will shift away from the information that users really need.

**Revision**

**Recording a conference call**

1. Start the online meeting as you normally would with your conference ID and moderator PIN.

2. Turn on your computer speakers.

3. Click **Record** in the online meeting and then enter the phone conference number and passcode in the dialog that opens. Do not add a 1 to the beginning of the phone number.

4. After the recording line joins the audio conference (with an audible beep), turn off your computer speakers.

5. Click **Start Recording**.

6. Share your screen. After you start sharing your own screen, you can hand over control of the presentation to one or more participants just like you normally do.

7. When the call ends, go to your InfoCompany portal and click **Recordings** for a list of your recordings. You can send the link to other team members.

The revision shows the real steps of the task. Users can start their conference call and move on.

Placing too much focus on features can make your content unnecessarily long or, in some cases, can result in topics or embedded assistance that users don't need at all. Feature information that probably does not need to creep into your content includes:

- Explaining how to delete something that users just created
- Explaining how to edit something that users just created
- Explaining how to undo something that users just did
- Explaining additional ways to do something that users just did
- Detailed instructions for logging on
- Detailed instructions for changing settings
- Explaining how to use an obvious feature, like a print button

Although "Focus on user goals" is a task orientation guideline, the outcome is closely related to the completeness of your information. Focusing on goals and following scenarios helps you to ensure that your embedded assistance and user interface support users' goals and that your documentation is streamlined to give users what they need.

Easy to use

## Indicate a practical reason for information

Giving users the information that they need to reach their goals is only part of task-oriented writing. Users need to understand *why* you are giving the information to them—how it is relevant to their task.

Users should never wonder, "But why are you telling me this?" Users need a practical reason for the information. For example, to state that the records in a file have a certain size might leave users wondering, "So what?" However, if you tell them that they must build a collection to hold the file and that the record size affects the way that they do this task, then they can understand why you are telling them about record size.

At first glance, the following sentence appears to be only descriptive and to have no practical application:

**Original**

> The BW_Mapping table in the Data_LM directory contains warning messages that are issued by the web server when you submit a request.

**Revision**

> If the request does not process as expected, check the BW_Mapping table to see whether the web server issued any warning messages. The BW_Mapping table is in the Data_LM directory.

Users might glance at the original sentence and move on because it doesn't seem relevant. The revised sentence relates the information to the task of troubleshooting a request so that users understand why the information is significant.

The purpose that information serves must be apparent. In many cases, simply stating the facts can seem puzzling if you don't indicate what significance the facts have, as shown in the following passage:

**Original**

> If the NORES option is used, the routines are link-edited as part of the load module. If the RES option is used, the routines are loaded separately.

46      Developing Quality Technical Information

# Task orientation

**Revision**

> Use the NORES option when you have sufficient space for routines to be link-edited as part of your load module. Use the RES option to save space by loading the routines only when you need them.

The original passage explains what the options do, but it does not relate that information to the users' task of deciding which option to use. In the revision, the facts are restated so that users understand when to use which option.

The text for the following choices provides general facts about each data collection method:

**Original**

> **Data collection method**
>
> ● Collect in-memory metrics
>
>   Uses the mon_req_metrics configuration parameter. Data is aggregated by an agent and accumulated in memory.
>
> ○ Collect snapshots
>
>   Uses the mon_obj_metrics configuration parameter. Large volumes of data are stored in the repository. This option supports deprecated clients.

Details are provided about each method, but the details are not presented in a way that helps users make a choice. Because the details about the configuration parameters and agent might not help at all, they are of no use to users here. In addition, the details for the deprecated method do not clearly imply that the method is suboptimal.

**Revision**

**Data collection method**

 Collect in-memory metrics

This option reduces disk storage requirements and is recommended for best performance.

 Collect snapshots

Use only if you need to support any deprecated clients. This option requires more repository storage and increases overhead.

The revision provides details that help users with the task of choosing which method to use. The information about large volumes of data is now phrased in terms of storage needs so that users understand the implications of choosing the deprecated method.

Provide only the details that users care about. Details that help users complete a task or that explain why users want to do a task are especially valuable to users, but this kind of information can be difficult to obtain. Consult with users, technical experts, or domain experts to ensure that you are providing useful facts and explanations.

## Provide clear, step-by-step instructions

Most tasks contain steps. Occasionally, a task has only one step and can be described in a paragraph, but most tasks are performed as a series of ordered steps. Task orientation extends from the highest-level goal down to the lowest-level step. Any step that is not clear or not ordered correctly can cause your users to make mistakes and be unable to complete a task.

When you write task information, you are usually in a position to notice usability problems and suggest product improvements. For example, you might identify cumbersome steps that can be streamlined or avoided or steps where users are repeating actions that they've already performed (such as typing a long serial number in more than one window). If you have difficulty documenting a task, consider whether there might be a problem with the way that the product is designed. Keep the needs and interests of your users in mind as you write; there is no such thing as a product that is too usable. When products are designed for usability, users need to rely less on low-level step information.

*Low-level step information* refers to information that includes each minor action that users take. At the lowest level, each click requires its own step, and all fields, keystrokes, clicks, or taps are mentioned regardless of how obvious they are. If you spend a good deal of your writing time documenting which buttons to click, you are probably not providing much value to users.

Steps such as "Specify today's date in the field" are generally too low level for most users, whereas steps such as "Complete the form" are at a more sensible level. Lower step levels require more resources to write and test. In addition, steps at the lowest level generally become out of date quickly.

The following table shows some possible levels of step information from the highest to the lowest:

Table 3.2  Example progressions from highest-level to lowest-level steps

| Highest-level steps | Mid-level steps | Low-level steps | Lowest-level steps |
|---|---|---|---|
| 1. Install the product and activate the license. | 1. Import the license file. | 1. Open the Import window by clicking **File > Import**. | 1. Click **File**.<br>2. Click **Import**. |
| 1. Complete the three forms and send them back. | 1. Complete Form 6-26. | 1. Complete the fields on Form 6-26.<br>2. Click **OK** when done. | 1. In the **Event** field, specify a name for the wedding or party. |

# Easy to use

**Table 3.2** Example progressions from highest-level to lowest-level steps *(continued)*

| Highest-level steps | Mid-level steps | Low-level steps | Lowest-level steps |
|---|---|---|---|
| 1. Personalize your mobile phone with contacts, ring tones, and alarms. | 1. Add contacts to your phone. | 1. Click **Address book** and follow the prompts. | 1. Click **Address book**. <br> 2. Select from one of the following three choices: *and so on…* |
| 1. Get directions to the airport. | 1. Specify the address in the Map window. | 1. Click **Map**. <br> 2. Specify the address. | 1. Type a street number and click **OK**. <br> 2. Type a street name and click **OK**. <br> 3. Select the city. |
| 1. Wash your hair with shampoo and conditioner. | 1. Shampoo your hair. <br> 2. Condition your hair. | 1. Lather. <br> 2. Rinse. <br> 3. Repeat. | 1. Open the lid. <br> 2. Squeeze a tablespoon of shampoo into your hand. <br> 3. Put the shampoo on your head. |

In a task, some low-level steps can be useful to give users markers to verify that they are on track. In general, however, try to write steps at the highest level that will work for your users. If the product is designed well and has good embedded assistance, the low-level instructions about where to click won't be needed, and you can devote more time to writing task topics with higher-level steps. Higher-level steps give users a better understanding of what they are doing in the context of a task or goal and help make the flow of a task or set of tasks clear.

Organize the steps in each task topic from your users' perspective. For each task that you write, consider the following questions:

- Is every step really an action?
- Where does each step start and end?
- Are some steps subordinate to others?
- Could some substeps really be subtasks?
- Which steps are optional?
- Which steps are conditional?

The following guidelines can help you provide clear, step-by-step instructions:

❑ **Make each step a clear action for users to take**
❑ **Group steps for usability**
❑ **Clearly identify steps that are optional or conditional**

## Make each step a clear action for users to take

Each step should correspond to an action (high level or low level) that users perform. Tasks do not discuss interactions between users and the product; tasks list only the actions that users do to complete their task.

A step isn't really a step unless it has an action for users to do. In the following set of steps, one item has no user action:

**Original**

1. Submit your application.
2. Applications are evaluated in the order received.
3. After a few days, check your portal to view the status of the application.

In the original set of steps, step 2 describes how a process works but is not a user action.

**Revision**

1. Submit your application.
2. After a few days, check your portal to view the status of the application. Applications are evaluated in the order received, and responses can take from 3 to 10 days depending on the number of applications.

In the revision, step 2 is combined with step 3 because the original step 2 provided details to help users understand step 3, but it was not a separate step for users to take.

To give clear direction to users, include an imperative verb (a verb that instructs users to take an action) in the first sentence of every step.

When you make style decisions for your information, you might pick a specific way to phrase step-level information. For example, you might choose to set the context for your users before stating the action, as in "In the first column of the table, type the date." An alternative is to state the user action before setting the context, as in "Type the date in the first column of the table." Both approaches include the imperative verb in the first sentence of the step.

# Easy to use

Some actions are trickier than others to make into imperative steps.

**Original**

> 1. Run the update program.
> 2. Start the sample web service.
> 3. The error should no longer occur.
>    - If the error occurs, repeat step 1 for each module.
>    - If there are no errors, go to step 4.
> 4. Run the InfoVerify tool to check for viruses.

Step 3 of the original set of steps breaks the rule of including an imperative verb in the first sentence of every step. Step 3 is not a step.

**Revision**

> 1. Run the update program.
> 2. Start the sample web service and check for errors.
> 3. If an error occurs, repeat step 1 for each module.
> 4. Run the InfoVerify tool to check for viruses.

In the revision, step 3 includes an imperative verb and is easier to follow.

Ensure that the imperative verb that you use conveys a real action. The action should make sense with respect to the task and not be overly generic. For example, a step that says "Run the following three commands in order" is imperative, but it would be more meaningful if it related the action to the purpose of the task. A more helpful phrasing might be "Set up the server by running the following three commands in order." Both approaches use an imperative verb, but the second approach is more task oriented because it gives a purpose to the step.

# Task orientation

The following steps use imperative verbs, but one of them is not very task oriented:

**Original**

> 1. Configure the server to collect data from application clients.
>
> 2. For each application, follow these steps:
>
>    a. Install the extended client software.
>    b. Open two port numbers between 60000 and 65000.
>
> 3. On the server, run the Summary report to verify that data is being collected for each client.

Step 2 in the original task uses *follow* as the imperative verb, which gives no indication of what the purpose of the step is going to be.

**Revision**

> 1. Configure the server to collect data from application clients.
>
> 2. For each application, configure the client to send data to the server:
>
>    a. Install the extended client software.
>    b. Open two port numbers between 60000 and 65000.
>
> 3. On the server, run the Summary report to verify that data is being collected for each client.

In the revision, step 2 gives an action that indicates the purpose of the step.

## Group steps for usability

Group steps to help users relate to the task. If you instruct users to do one action or click after another, then another, then another, the task can become mind-numbingly obscure. If you can group minor steps together into a larger step, users can think of the steps in relation to the task that they are completing.

For example, instead of interpreting the steps as "First I click here, then I fill out that field, then I click over there, then I select an item from a list here," users can more easily grasp steps in terms such as "First I set my preferences, and then I specify the server information." In this way, you not only help users relate to the steps that they are doing, you also streamline the steps and make each step easier for users to find.

# Easy to use

The following steps are in the correct order, but they don't correspond to the way that the users think about the task:

**Original**

> To add a setting to your profile:
>
> 1. Select the profile object that you want.
> 2. Right-click and select **Properties** from the menu.
> 3. In the Properties window, find the name and path of the profile file.
> 4. Close the Properties window.
> 5. Open your profile in a text editor.
> 6. Add the setting to your profile file in the settings section.
> 7. Save the profile file.
> 8. Run the profile command with the `-file yourProfileName` option.
> 9. Restart the web browser.

The original set of steps gives each step the same weight. Trivial clicks, such as closing a window, are treated in the same way as steps that are more significant, such as running the command. The original set of steps ignores the relationship of each step to its surrounding steps.

# Task orientation

**Revision**

To add a setting to your profile:

1. Find the name of the profile that you need:
   a. Right-click and select **Properties** from the menu.
   b. In the Properties window, find the name and path of the profile file.
2. Update the profile with the new setting:
   a. Open your profile in a text editor.
   b. Add the setting to your profile file in the settings section.
3. Run the profile command with the `-file yourProfileName` option.
4. Restart the web browser.

The revised set of steps shows the relationships of some of the steps to each other and shows how they make up the two higher-level steps: finding the name of the file and updating the file. The revised set of steps also downplays some of the trivial steps by merging them or omitting them. Where the original set of steps shows a linear progression of one action or click after another, the revised set of steps shows what each step accomplishes toward completing the whole task. Also, by combining the steps into higher-level steps, the revision reduces the number of steps.

Unordered, or bulleted, lists provide another way to subordinate information or actions in steps. Use unordered lists only for steps that are not sequential, and be sure that unordered lists support the flow of the task.

Easy to use

The following steps use unordered lists to try to get both new users and returning users through the same path:

**Original**

> To change the amount of your loan:
>
> 1. Log in to your student portal and review your loan information.
>
> 2. Choose the option to modify your loan and complete the fields and submit the changes:
>    - If you have a signature PIN, click **Submit** and specify your PIN.
>    - If you do not yet have a signature PIN, click **Save Only**. You will need to return to this page when you have your PIN. If you do not yet have a signature on file, complete form 10235 and sign and fax it. A PIN will be sent to your student email account within two business days.
>
> 3. Review your student account. Changes to the loan will affect your other balances, and you might owe additional tuition, or you might be due a refund.

In the original set of steps, step 2 directs both new users and users who already have a PIN to fill out a form that only users who have a PIN can complete. The original steps are not focused on the flow of the task for the new users.

**Revision**

> To change the amount of your loan:
>
> 1. Log in to your student portal and review your loan information.
>
> 2. If you do not yet have a signature PIN on file, complete form 10235 and sign and fax it. A PIN will be sent to your student email account within two business days.
>
> 3. From your student portal, choose the option to modify your loan and complete the fields and submit the changes. You must have a signature PIN to submit the changes.
>
> 4. Review your student account. Changes to the loan will affect your other balances, and you might owe additional tuition, or you might be due a refund.

In the revised set of steps, step 2 is used to prepare the new users for the remaining steps. Thus both types of users can follow one clear set of steps. For information about addressing a primary and secondary audience, see the completeness guideline "Provide appropriate detail for your users and their experience level" on page 123.

When you use sublists to group steps, be sure to use the right type of sublist for the situation. The following example shows a step divided into substeps:

**Original**

> 2. Copy the contents of the file system from the source disk to the target disk. The steps that you use depend on the location of the target disk:
>
>     a. If the target disk is accessible from the source computer, run the `copyfilesystem` command.
>
>     b. If the target disk is not accessible from the source computer, run the `copyfilesystem -tmb` command to copy the contents of the source disk to tape, and then use the `copyfilesystem -tmb` command to copy the contents from the tape to the target disk.

In the original step, ordered substeps are used to show two choices that are mutually exclusive. Because the choices are not meant to be performed in order, the substeps are misleading.

**First revision**

> 2. Copy the contents of the file system from the source disk to the target disk. The steps that you use depend on the location of the target disk:
>
>     - If the target disk is accessible from the source computer, run the `copyfilesystem` command.
>
>     - If the target disk is not accessible from the source computer, run the `copyfilesystem -tmb` command to copy the contents of the source disk to tape, and then use the `copyfilesystem -tmb` command to copy the contents from the tape to the target disk.

In the revised step, the substeps are replaced with an unordered list. Users follow the instructions in one list item or the other.

**Second revision**

> 2. Copy the contents of the file system from the source disk to the target disk. The steps that you use depend on the location of the target disk:
>
> | Option | Description |
> | --- | --- |
> | If the target disk is accessible from the source computer | Run the `copyfilesystem` command. |
> | If the target disk is not accessible from the source computer | Run the `copyfilesystem -tmb` command to copy the contents of the source disk to tape, and then use the `copyfilesystem -tmb` command to copy the contents from the tape to the target disk. |

In the second revision, the unordered list is replaced by a table to make the options more retrievable and visually effective.

## Clearly identify steps that are optional or conditional

*Optional steps* are steps that all users can choose to skip. *Conditional steps* are steps that apply only if certain criteria are met, such as if users have a certain fix pack or service agreement. Optional steps and conditional steps are not the same. A common mistake is to label a step as optional when it is really conditional.

Users need to know at a glance if a step can be skipped or if a step does not apply to them. The best place to tell users whether a step is optional or conditional is at the very beginning of a step before users take any action.

## Optional steps

Users can skip optional steps and still complete the task successfully. As mentioned in the guideline "Focus on users' goals," try to keep your tasks free of feature clutter by eliminating steps that are superfluous to the task. However, in cases where optional steps support the task, include them in the task, but identify them as optional. For example, the following step supports the task but can be skipped:

> 3. Optional: Set default values for your startup parameters.

# Task orientation

Mislabeling optional steps can cause logic problems in your task. The following set of steps identifies step 4 as optional but not step 5:

**Original**

1. ...
2. ...
3. Put the large tires on the bike wheels to complete the main bike.
4. Optional: Attach the training wheel assembly to the rear axle.
5. Put the small tires on the training wheel assembly.

The original set of steps is confusing. Users who choose to skip step 4 are unable to do step 5. Because step 5 makes sense to do only if users follow step 4, step 5 must also be optional.

**First revision**

1. ...
2. ...
3. Put the large tires on the bike wheels to complete the main bike.
4. Optional: Attach the training wheel assembly to the rear axle.
5. Optional: Put the small tires on the training wheel assembly.

The first revision shows both step 4 and step 5 as optional. However, the first revision is still not logical because users who choose not to perform step 4 cannot perform step 5. So step 5 is not optional by itself.

**Second revision**

> 1. ...
> 
> 2. ...
> 
> 3. Put the large tires on the bike wheels to complete the main bike.
> 
> 4. Optional: Add training wheels to the main bike:
> 
>    a. Attach the training wheel assembly to the rear axle.
> 
>    b. Put the small tires on the training wheel assembly.

The second revision shows step 4 as an optional step that consists of two substeps. If users choose to follow step 4, they must perform both steps 4a and 4b, which is the only combination that makes sense.

Take care to clearly identify optional steps. Use the word *optional* for optional steps. However, do not follow the word *optional* with the phrase *if you want to* or *you can* because these phrases are redundant with the word *optional*.

## Conditional steps

Users follow conditional steps only if certain criteria apply. Conditional steps generally begin with the word *if*, as in, "If you are running version 7 of the client application, install fix pack 2." Users who meet the criteria for the step must follow the step. Always start conditional steps with the condition. That way, users who do not meet the criteria can skip the step after reading the condition.

Although the following steps are conditional, users might find themselves halfway through any one of these steps before they realize that they don't need to do the step:

**Original**

> 1. Register your computer as a client if you are not yet registered on the LAN.
> 
> 2. Specify your 12-digit serial number in the **Number** field if your software is not yet registered.
> 
> 3. Run the InfoExec program to reconfigure your settings (InfoExtended only).

# Task orientation

Because users rarely read ahead when following steps, the original steps might cause some users to take links that do not apply to them, start typing serial numbers, or try to run programs that they don't need.

**Revision**

> 1. If you are not yet registered on the LAN, register your computer as a client.
> 2. If your software is not yet registered, specify your 12-digit serial number in the **Number** field.
> 3. InfoExtended only: Run the InfoExec program to reconfigure your settings.

In the revised steps, the condition for each step is stated before the action. Users who are registered on the LAN can skip step 1, users who registered their software can skip step 2, and users who are not using InfoExtended can skip step 3.

The following step is introduced in a potentially confusing way:

**Original**

> 4. To install the limited license, run the `infolicense.exe` file on the application server.

Users might read the original step as required, optional, or conditional. The revised steps show clearer phrasing for all three situations.

**Revision**

> **Required step**
>
> 4. Install the limited license. Run the `infolicense.exe` file on the application server.
>
> **Optional step**
>
> 4. Optional: Install the limited license. Run the `infolicense.exe` file on the application server.
>
> **Conditional step**
>
> 4. If you purchased a limited license, run the `infolicense.exe` file on the application server.

Be sure to clearly identify conditional steps. Use the word *if* and state conditions early for steps that do not apply in all situations. Remember that *if you want* is not a condition because steps that users can choose not to do are optional, not conditional.

Conditional steps should be used only when needed and should not be overused. The following set of steps uses redundant conditional steps for a single action:

**Original**

1. If you are using Chrome, block the pop-ups from the **Options** menu for security.

2. If you are using Firefox, block the pop-ups from the **Preferences** menu for security.

3. If you are using Internet Explorer, block the pop-ups from the **Settings** menu for security.

Although the original set of steps does indicate the condition of the step at the beginning, three steps are shown for one action. This approach causes users to have to read through unnecessary steps.

**First revision**

1. Block all pop-ups for security:
   - For Chrome, use the **Options** menu.
   - For Firefox, use the **Preferences** menu.
   - For Internet Explorer, use the **Settings** menu.

The revision shows a better approach that is not conditional. In the revision, the one general action is stated with the details below.

**Second revision**

1. Block all pop-ups for security:

| Browser | Description |
|---|---|
| Chrome | Use the **Options** menu. |
| Firefox | Use the **Preferences** menu. |
| Internet Explorer | Use the **Settings** menu. |

In the second revision, the bulleted list is replaced by a table to make scanning easy and to reduce words.

When you use conditional steps for a task that is shared by different audiences, ensure that the steps are easy to navigate.

# Task orientation checklist

Task orientation is a focus on users' actual goals and the tasks that support those goals. Task-oriented information keeps users "on task."

Information is task-oriented if it's focused on what users want to do, not on what the product can do. Be sure that you know your audience, and provide practical information for their needs. The most useful task topics are those that support scenarios to accomplish key goals. Make sure that the steps that you write in task topics are clear and not too low level for your users.

You can use the following checklist in two ways:

- As a reminder of what to look for, to ensure a thorough review
- As an evaluation tool, to determine the quality of the information

| Guidelines for task orientation | Items to look for | Quality rating<br>1=very dissatisfied,<br>5=very satisfied |
|---|---|---|
| Write for the intended audience | • Audience is clearly defined.<br>• Tasks are appropriate for the intended audience.<br>• Level of detail is appropriate for the intended audience.<br>• Information in messages and user interfaces is appropriate for the intended audience. | 1 2 3 4 5 |
| Present information from the users' point of view | • Information is directed at users.<br>• Information explains what users need to know or do, not the status of the product.<br>• Information reflects the users' position. | 1 2 3 4 5 |
| Focus on users' goals | • Task topics are provided that support goal-oriented scenarios.<br>• Task topics are user oriented, not product oriented.<br>• Task topics address users' needs.<br>• Focus is on tasks, not features and interface elements. | 1 2 3 4 5 |

# Task orientation

| Guidelines for task orientation | Items to look for | Quality rating<br>*1=very dissatisfied,*<br>*5=very satisfied* |
|---|---|---|
| Indicate a practical reason for information | • Conceptual information, reference information, details, and facts support tasks.<br>• Information that is provided is useful to users. | 1 2 3 4 5 |
| Provide clear, step-by-step instructions | • Steps are at the appropriate level for the audience.<br>• The first sentence of each step includes an imperative action.<br>• Steps are weighted appropriately.<br>• Substeps and unordered lists are used appropriately.<br>• Optional steps are clearly marked.<br>• Conditional steps begin with the condition. | 1 2 3 4 5 |

To evaluate your information for all quality characteristics, you can add your findings from this checklist to Appendix A, "Quality checklist," on page 545.

**CHAPTER 4**

# Accuracy

*Fast is fine, but accuracy is everything.* —Wyatt Earp

As a writer of technical information, you have an important responsibility to provide information that is accurate and free of errors. Users depend on you for that accuracy; they stake their time and money on it. Achieving accuracy in technical information is an all-important goal, albeit a goal that isn't easy to achieve.

For technical information to be accurate, every piece of information must be accurate, including conceptual information, procedures, embedded assistance, legal disclaimers, screen captures, and illustrations.

Inaccuracies might or might not be obvious. Those that aren't obvious are often more significant because users might act based on inaccurate information.

For example, imagine someone gives you driving directions that instruct you to turn right at the dead end. You immediately know that the directions are inaccurate because you can't turn right at a dead end. Therefore, you take steps to resolve the inaccuracy, such as looking at a map or asking the person for clarification.

Now imagine that someone gives you directions that instruct you to turn right at Front Street and to go straight until you cross the railroad tracks. You could drive for quite some time before realizing that no railroad tracks intersect

Front Street in the direction that you're traveling, and that you were supposed to turn left instead of right. You'll probably be more upset by the second set of directions than the first.

In the article "Docs in the Real World," Carolyn Snyder and Jared Spool recognize the importance of information accuracy when they coin the term *trust-breakers*. These are errors in information that cause users to avoid the information because they don't trust it. Snyder and Spool make several observations about the effects of inaccuracies on users:

> Even small inaccuracies damage the confidence of some users, especially those users who don't know how to work around the problem. When they discover an error, they tend to give up on the information and call the help desk.

> The information often takes the blame for inaccuracies in the underlying product or in related products. Users don't care whose fault an error is; they are simply less willing to use the information in the future.

> After experiencing major problems in information, users are less likely to give the information another try, even in future releases.

Regardless of how much time users take to detect an inaccuracy, their confidence in the information is eroded; they might even choose not to use the information or the product at all.

To make information accurate, follow these guidelines:

- **Research before you write**
- **Verify the information that you write**
- **Maintain information currency**
- **Maintain consistency in all information about a subject**
- **Use tools that automate checking for accuracy**

# Research before you write

Before you can tell users how to complete a task, you must understand why users need to perform the task in the first place, and you must understand each step in that task. Before you can explain a complicated technical concept, you must understand that concept yourself. Without a proper understanding of your subject, you risk writing information that is based on assumptions and misunderstandings.

If you don't understand your subject, the writing process can become an exercise in frustration because you don't have all of the raw material that you need to complete your help topic, scenario, embedded assistance, or illustration. Technical reviews tend to be less productive because reviewers must spend valuable review time correcting fundamental mistakes and misstatements instead of making comments that can result in significant improvements. Information must be accurate before it can be effective.

Although you need to understand the technical subject that you write about, you don't need to understand it as well as the engineers who develop the product. In fact, if you understand the internal workings of the product too well, you risk losing the perspective of your users. A good balance is having enough understanding to be able to ask your technical experts relevant questions that users might ask, but not so much understanding that you think that you don't need input from technical experts.

In addition to researching technical subjects, you should spend time researching your audience. As explained in Chapter 3, "Task orientation," a solid understanding of your audience and their goals is critical to providing the information that your users really need.

To obtain the knowledge that you need to document a new or complex subject:

- Rely on hands-on experience to gain the knowledge that you need
- Create outlines, topic models, information plans, and rough drafts to help you identify what you know and what you don't know
- Interview technical experts and domain experts

## Hands-on experience and direct observation

The best way to understand a technical subject that you write about is to use the product as a user would. The more that you can use a product, tool, or interface, the more responsible you can be for the accuracy of the information.

For example, when you write a procedure that requires users to create a server profile so that applications can connect to a server, you can use the interface to create your own server profile. In doing so, you determine which parts of the profile need to be documented. After you finish writing the procedure, you can ask your colleagues to test it; if they can successfully create a server profile by using the documentation that you wrote, you know that the procedure is accurate.

When time and circumstances allow, use one or more of the following activities to ensure the accuracy and usefulness of your information:

- Participate in design review meetings. In addition to contributing embedded assistance where it is needed and ensuring that a progressive disclosure pattern is followed, you'll be able to gain insight into how new functions will be used.
- Use prototypes to familiarize yourself with how a product, tool, or interface works. As an early user, you're also likely to suggest changes that result in some design improvements.
- Set up an informal usability walkthrough in which you and your colleagues use the information to perform some or all of the important user tasks just as users will. Rely primarily on the embedded assistance, and consult topics only when necessary.
- Participate in formal usability walkthroughs in which users use the information to do certain tasks. Contribute questions that will help you to validate the information, such as:
    - Does a particular task topic include the correct prerequisite steps?
    - Do users need more detail about a particular set of input fields?
    - Will users understand a new term based on the current definition?
- Visit one or more of the users of your information to observe them as they do their day-to-day tasks.

These kinds of activities increase your knowledge of how your users do their jobs so that you can develop accurate information that makes their jobs easier.

## Information plan, outline, and rough-draft reviews

Perhaps a product is not available for you to experience first hand, or technical experts are unable to explain the subject to you at the beginning of the writing cycle. Perhaps the design of major pieces of the product is still in flux. You must start with what you do understand to create information plans, outlines or topic models, and rough first drafts of topic-based information. Creating these *early cycle* documents is a good way to validate your assumptions, right and wrong.

**Information plans**

Information plans are contracts between the people who write the information and the other groups who participate in the creation of the information, such as development, test, translation, and so on.

In traditional development environments, information plans typically cover all the deliverables that will be created for a release cycle, the processes that will be followed, the schedule for producing the information, and the people who will be involved in writing, reviewing, translating, and publishing the information. In an information plan, you can list specific topics and embedded assistance content that you think need to be written and then have that list validated by reviewers of your plan.

In agile development environments, there is no one-size-fits-all information plan. In fact, because of the ever-shifting nature of agile projects, an information plan might not even mention specific information deliverables. It might include sections such as a definition of the writing team's role in the process, agreements between the writing team and the rest of the development team about how information will be authored and reviewed, and the processes that will be followed during the development cycle. However, as the development cycle progresses, you will likely need to create one or more plans to address the topics, embedded assistance, messages, and graphics that will be needed to support the release.

**Outlines and topic models**

*Outlines* represent the structure of information, whether that information is a single task or an entire library. Outlines help you to identify all the pieces of information that you need and to put those pieces in the right order.

*Topic models* are a specific type of outline that shows the structure of topic-based information. To create a good topic model, start by identifying all of the goal-oriented user tasks and putting them in the most logical order. Then, determine the conceptual and reference information that is needed to support the tasks.

**Rough drafts**

Rough drafts are your starting point for writing the topics that support your product. During this information gathering phase, when you're identifying the pieces that you need to write, work at a very high level. Don't invest time in documenting details just yet. At this stage, your goal is simply to figure out which pieces of the puzzle you understand well enough to document and which pieces require more research before you can document them.

# Easy to use

Include notes in your outlines, plans, and rough drafts to identify information that you understand, and include questions to get more information about gray areas.

When you've gone as far as you can, you can do two things: you can start to document in detail the pieces that you do understand, and you can begin to solicit feedback from your technical experts on the pieces that you don't understand. The notes and questions that you added to your outlines, plans, and drafts will help technical experts to focus on the areas that need the most attention. Continue to fine-tune your rough drafts, plans, and outlines until you have a complete understanding of the subject.

The following outline describes the flow of a configuration process:

**Original**

1. Verify prerequisites.

2. Start the InfoViewer configuration wizard.

3. Select the multiserver environment option.

4. *and so on...*

The original outline doesn't include any notes, so reviewers might just skim the review and provide comments only about content that is blatantly inaccurate. Or they might spend time commenting on steps that the writer already understands.

## Revision

> **Note to reviewers:** Can step 1 be programmatically determined when the user starts the wizard, or must they do this step manually? If they must do this step manually, please provide the actions that they'll need to take.
>
> 1. Verify prerequisites *(The user has installed InfoViewer and has successfully run the installation verification program).*
>
> 2. Start the InfoViewer configuration wizard *(The user runs system/ivcfg.bat).*
>
>    **Note to reviewers:** For step 3, please provide details about the options that users will need to select for a multiserver configuration and any user interface elements (fields, parameters, etc.) that are unique to a multiserver configuration.
>
> 3. Select the multiserver environment option.
>
> 4. *and so on...*

The revision focuses reviewers' attention on steps 1 and 3. Presumably, the writer understands step 2 well enough to document that step without feedback from reviewers. The note prior to step 1 shows how writers can think beyond the documentation and suggest ways to improve the user experience.

## Interviews

Even if you're able to use the product extensively and participate in design and walkthrough activities, you will still likely need feedback from others to help ensure the accuracy of your information.

Technical experts are your primary source to validate how product functions work, and domain experts are your primary source for information about how or why users need those product functions. You need both perspectives to produce truly outstanding technical information. Involve both of these resources early in the writing cycle to eliminate disruptive false starts and wrong turns, and rely on them throughout the writing cycle to keep you on track. A few minutes interviewing a knowledgeable technical or domain expert is often more valuable than hours of research on your own.

Easy to use

# Verify the information that you write

A useful by-product of researching a subject before you write about it is that you get the raw information that you need, and you get it validated simultaneously. When a technical expert explains a concept to you, you can be reasonably sure that the explanation is accurate.

However, the explanation that you got from one technical expert during the initial design phase might conflict with feedback that you receive from other technical experts later in the cycle. Early feedback from customers might result in changes to the product. Product functions and features often change, get added, or get removed in response to schedule and resource changes. To ensure the accuracy of the information for the finished product, you must verify that information throughout the development cycle.

There are a number of ways to verify that information is accurate, including hands-on testing with nearly finished products, conducting formal technical reviews, and ensuring that your information is included in product testing.

## Hands-on testing with interfaces

Whenever possible, validate the embedded assistance that you write by using the actual product interface. There's no substitute for seeing and interacting with the UI content that you write just as your users will.

If you work in an agile development environment, you likely have opportunities to access the interface regularly. If you work in a traditional waterfall development environment, you might not be able to test the interface until late in the development cycle.

In traditional waterfall development environments, early iterations of user interfaces are often buggy or are not functionally complete. You need to be sure that all functions are present and are working as designed before you spend time validating the information that supports those functions. However, you still need to give yourself time to fix any problems that you uncover before the product deadline.

For example, after your team puts the finishing touches on the interface for your product, you devote some time to validating the embedded assistance that you added. One seemingly insignificant function that you haven't validated yet is the dialog for changing passwords. You open the dialog, display the hover help, and proceed to change your password to your favorite food, `macandcheese`, which meets the criteria indicated in the hover help.

Accuracy

Original

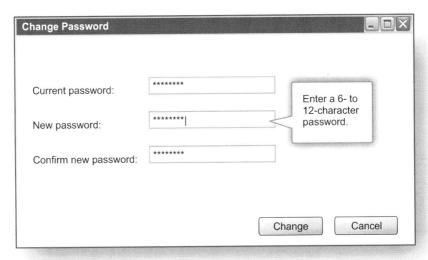

However, when you tab to the next field, you are presented with an error message that indicates that the password you entered is not valid. After further investigation, you find out that passwords must contain a number and a special character.

Revision

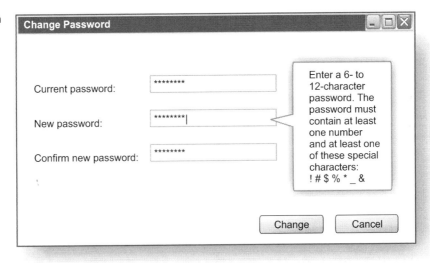

Fortunately, you still have time to change the hover help, as shown in the revision. You change your password to mac&ch33se, and all is well.

75

Easy to use

## Technical reviews and technical ownership

Almost all technical writing projects involve some sort of technical review. Reviews can take place near the end of the development cycle or iteratively throughout the development cycle. Remember that putting your content out for review doesn't necessarily mean that the outcome of that review will be completely accurate documentation. Technical reviews are effective only when reviewers do their job. Here are some tips to make your reviews more effective:

**Ensure that the right reviewers participate**

Typically, you know who the right people are: they are the technical experts whom you have developed a relationship with throughout the course of the project. However, don't overlook the feedback that you can obtain from other stakeholders.

For example, the people who work directly with customers in the field can often provide you with a more customer-centric perspective, which can help you make your information more usable. Your technical support team can be especially valuable reviewers of information about features or tasks that have a high incidence of support calls. They are also valuable reviewers of error messages and troubleshooting information because they deal with these types of content every day.

**Ensure that the reviewers focus on technical issues**

During reviews, technical experts sometimes get distracted by a missing comma or misspelled word, and they can't restrain themselves from commenting. Although their intentions are good, you need them to validate the technical accuracy of the content. The following tips can help prevent a technical review from turning into a copy edit:

- Make sure your review draft is as error-free as possible before you ask technical reviewers to review it. If your draft doesn't contain any distracting grammatical, spelling, or punctuation errors, reviewers are less likely to lose sight of their role in the review.

- When you send the draft to reviewers, provide guidelines for the review. For example, identify the specific content that you need reviewers to focus on. If the draft has not been edited, inform reviewers of this fact and tell them that the draft will be edited after the technical content is finalized.

- Use notes in the draft to help reviewers focus. Call their attention to individual sections or paragraphs that you need them to validate. Ask specific questions such as "In the following paragraph, are the restrictions for using the CHECK utility in a partitioned server environment complete and correct?"

**Assign technical ownership**

Assigning technical ownership of specific pieces of information to specific people helps ensure accuracy by instilling a sense of accountability in your technical experts. Because technical owners are responsible for vouching for the accuracy of the information that you write, they're likely to review the information thoroughly.

To assign technical ownership, you need to identify technical experts to be responsible for the accuracy of some area of the information, for example, the configuration topics, performance-related content, or all of the information that is related to a particular feature. Match potential owners with their area of expertise, ask them to commit to being a technical owner, and then notify them of what is expected:

- They will need to review the information that they own to make sure that it's accurate.
- If something isn't accurate, they will need to work with you until it is accurate.
- They will need to formally sign off that the information is accurate by sending you a note or by adding a comment in your project tracking tool.

Assign only one person per functional area so that there's no confusion about who is responsible for reviewing the content and formally vouching for its accuracy.

**Identify exit criteria for the review**

Well-defined exit criteria, which are the criteria that must be met before a review can be closed, are essential for determining how effective a review was and for taking action when a review is inadequate. By adopting exit criteria into your normal review processes, you remove any question about what constitutes a completed review. You can also use the exit criteria to establish corrective actions when necessary. The following list shows some typical criteria for exiting a review:

- All mandatory reviewers thoroughly reviewed the draft and sent a note to the writer who conducted the review.
- All critical issues were addressed.
- Concrete plans were established to address any remaining open items:
  - A technical owner was assigned to each open item.
  - The technical owner agreed to the plan for resolving the item.

- A due date was established for each item.
- A mitigation strategy was established if an item was not resolved by the due date.

When the criteria that you assign are met, you can be confident that the review was effective.

## Information in quality control tests

Testers are responsible for making sure that the product is working as designed and is free from bugs, errors, and design flaws that might cause problems for your users. The information that you write is an important part of the product, so it makes sense to include your information as part of formal product testing.

Ensure that your information is explicitly mentioned in test scripts and that testers use it to accomplish their testing. For example, make sure that test scripts include plans to validate the embedded assistance, including hover help, help text, error messages, and links to more information. Also, make sure that your quality control team uses your task topics when they install and configure the product.

## Maintain information currency

Information that's not current is not accurate. Information can become outdated from one release to the next or even within a development cycle, and some types of information are more prone to becoming outdated than other types of information. For example, reference information tends to be rather static, whereas installation procedures tend to need regular upkeep with each new release.

Technical information is susceptible to becoming out of date because technology is always changing and evolving. As a technical writer, you learn to accept, and even thrive on, this constant state of change. You anticipate that the interface, mobile app, or utility that you document today might look drastically different two years, or even two months, from now. Still, it can be difficult to make sure that everything you write stays as current as it needs to be.

For example, when you work on a new release, you're pretty good about adding new installation requirements and changing the description of how a utility works. But what about subtle changes, such as those to legal boilerplate information or trademarks? What about design changes that you aren't informed of? And what about changes that get overlooked because nobody recognizes the dependency between one piece of information and another? How can you make sure that *all* of your information stays current?

Maintaining information currency is a challenge, but the following guidelines can help:

- **Keep up with technical changes**
- **Avoid writing information that will become outdated**

### Keep up with technical changes

Making sure that the documentation stays in sync with product changes is one of the most important jobs for any technical writer. Users expect information to reflect the current state of the product.

Big changes tend not to get overlooked. You're not likely to neglect to document a major new scenario, task, or capability in the upcoming release. The changes that you need to watch out for are the ones that aren't always apparent.

#### Changes during a development cycle

Changes that occur between releases are usually obvious. After all, you're expecting things to change from one release to the next. Product planners identify new scenarios and functions, and you start thinking about the new documentation that will be required. But changes can also occur during the development cycle, after designs have been agreed on or after part of the product has presumably been finalized. These are the types of changes that tend to surprise you.

For example, in a traditional waterfall development environment, you typically use technical specifications to start writing. However, technical specifications can get out of sync with the product, even if the team tracks changes to the specifications. Changes creep in, but they often don't get communicated to all the team members who are affected by them.

In an agile development environment, where requirements and designs can change from one iteration to the next, you must be a regular participant in the design review process to stay current with changes as they occur.

For example, accuracy problems can result when you include graphics that show all or part of the user interface if the interface changes after the information is finalized. Assume that your writing team created a viewlet that shows users how to create steps in a process flow. The following window, which was designed based on the technical specification, plays a prominent role in the viewlet:

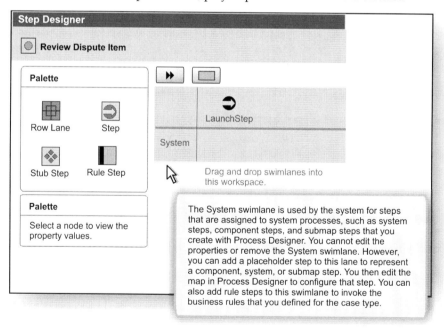

After the technical specification was written, however, the visual design team updated several of the icons in the palette, and the development team added a new icon to the palette. Unfortunately, word of these changes never made it to the writing team. The window that they included in the viewlet is now inaccurate. This problem is not limited to the viewlet. Task steps and embedded assistance that mention the contents of the palette are now also out of sync with the window. Ensure that information about the interface is current and that it matches what users will see and experience.

Attend design meetings, familiarize yourself with prototypes, or install iterations of the product as they become available throughout the development process so that you can compare the latest interface with information that you've written to support the interface.

**Changes to existing messages**

One type of information that gets overlooked in terms of currency is product messages. During information reviews, typically only new messages get reviewed, which means that older messages that become obsolete or inaccurate because of new changes often go undetected. For example, the following message accompanied version 5.1 of a product:

ABC0057E  The content element is not valid.

**Explanation:** The content element in the case deployment profile does not have the correct value. The solution cannot be deployed because the content transfer type document was not found. This document must exist in the object store directory.

**User response:** Open the case deployment profile in the InfoDBase configuration tool and ensure that the content element property has the value content transfer.

[ Close ]

However, the product function changed in version 5.2, as indicated in this review comment:

> **Comment – FDavis:** This message does not apply to v5.2.
> Deployment profiles were removed after v5.1. Also, users can no longer use the configuration tool to set the content element property.

Luckily, the writer included all of the product messages in the review and a diligent reviewer picked up on this change. Now the writer can determine whether this message is obsolete and should be removed or if it is still valid and needs to be updated to reflect the current behavior. The writer can also search all of the version 5.2 content for instances of *deployment profile* and *content element property* and address any other potential inaccuracies.

## Avoid writing information that will become outdated

You might think that the content that you write will provide value to your users indefinitely. Technology, however, has other ideas. The content that you write today might be outdated only several months later. Fortunately, some techniques can help you increase the lifespan of your content.

One technique is to avoid referring to features and functions as *new*. New is always a temporary state. Not only should you avoid explicitly using the term *new* to refer to a new feature, function, or component, you should avoid associating a feature, function, or component with a particular release. The only valid exception to this rule is when you write marketing information or *what's new* information whose sole purpose is to highlight the features of a new release of a product.

# Accuracy

The following table provides some examples for eliminating problematic constructions:

| Avoid this construction | Use this construction instead |
| --- | --- |
| 4. If you are using the new gateway feature, edit the configuration file and set the following configuration parameters: | 4. If you are using the gateway feature, edit the configuration file and set the following configuration parameters: |
| The REQTIME parameter has been added in this release to control the amount of time that the listening server waits for a request before timing out. | Use the REQTIME parameter to control the amount of time that the listening server waits for a request before timing out. |
| The latest release of InfoDBase includes enhancements to the UNLOAD utility that can result in improved performance when you unload large objects. | InfoDBase includes an UNLOAD utility that can result in improved performance when you unload large objects. |
| The newest models do not support extended trace capabilities. | Models 3500D, 4500D, and 4501D do not support extended trace capabilities. |

Another technique for prolonging the life of your information is to avoid referring to specific releases except when necessary. For example, the following web page provides general information about the InfoDBase product:

Original

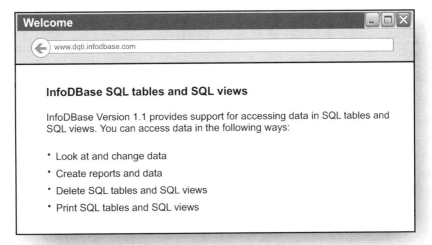

Specifying the release number for a product might be accurate for this release, but references to the release number will become inaccurate if they're not changed in the next release. If a statement about a product will be true in future releases, write the statement in a general way.

Easy to use

Revision

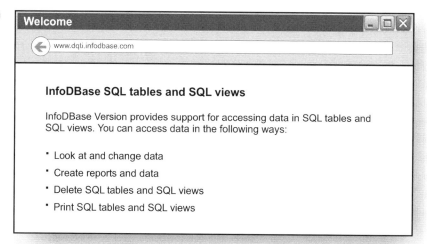

Some information, however, does require you to specify version and release details, so don't unilaterally stop referring to specific versions and releases. For example, you would likely find it impossible to document how to upgrade from one release to another without mentioning specific release information. Restrictions are another area where release-specific references might be needed, for example, "**Restriction:** The DBMON command is not supported by InfoDBase V4R1." Use your judgment to determine when it's appropriate to include references to specific product releases.

Also, be careful about including dates in graphics, examples, or code samples, especially in information that will likely not be updated frequently. Even if the information is still valid ten years later, users might suspect that the information is outdated based on a date, and they might waste time searching for a more current version of the information, which doesn't exist. For example, the following sample report, which is included in a reference topic, includes several dates:

Original

```
Date: 2014/06/15    Metadata Catalog Analysis Report Summary
Time: 13:50:30

Index Component Information
   Data set name:         JJR225.INDEX
   Version identifier:    410
   Date created:          2002/05/25 10:21:30
   Last updated:          2014/06/12 09:37:11
```

# Accuracy

Although these dates were current when the topic was written, reference information is often not updated regularly. Unless the dates are critical for users to understand the content, consider taking a more generic approach.

**Revision**

```
Date: yyyy/mm/dd    Metadata Catalog Analysis Report Summary
Time: hh:mm:ss

Index Component Information
   Data set name:        JJR225.INDEX
   Version identifier:   vrm
   Date created:         yyyy/mm/dd hh:mm:ss
   Last updated:         yyyy/mm/dd hh:mm:ss
```

In the revision, specific dates have been replaced with generic variables that won't ever go out of date, and the specific times have been replaced for the sake of consistency. The version identifier has also been replaced because the original value of 410 can become outdated as new versions and releases are introduced.

# Maintain consistency in all information about a subject

Many accuracy problems occur when information about a subject is not consistent in all of the places where that subject is mentioned. For example, users read a task scenario that describes how to set a group of parameters to achieve a performance goal; however, the reference topics for those parameters provide a different set of valid values than the values that are mentioned in the task scenario. Faced with inconsistent information, users don't know which instance of the information is correct.

Maintaining consistency across all of the information for a product or technology isn't easy, especially on large teams where different people are responsible for different parts of the product. Therefore, you need to be both proactive and diligent to ensure that your information is consistent. The following guidelines can help you achieve that goal:

- Reuse information when possible
- Avoid introducing inconsistencies

## Reuse information when possible

Content that is reused in multiple information deliverables should be written once and maintained in a common location or in a shared file that is reused whenever it is needed. Writing a single version of reused content reduces the potential for inaccuracies and inconsistencies across different deliverables and reduces the amount of maintenance that is required to keep the content current. If your information is translated, reusing information also reduces translation costs because reused information needs to be translated only once.

The simplest form of information reuse involves copying the information from one place and pasting it somewhere else. Choose this technique only if the reused information is very short, is reused in only two or three places, and is unlikely to change in the future.

For longer pieces of information that you do not want to copy verbatim, use a better approach for reuse: single sourcing. *Single sourcing* refers to the use of the same source, with few or no changes, for different outputs. Unless otherwise noted, the word *reuse*, when used in this book, refers to the use of a single source of information, not to the copying of information.

Methods for reusing content vary by authoring tool, but most tools offer some mechanism for reusing content easily. Depending on the authoring tool that you use, the method might be text insets, text entities, shared files, file entities, building blocks, or content references.

For example, you can use a single sentence in multiple places, such as in a topic in an HTML- or XML-based information set, in a printed document, and in embedded assistance. Rather than wasting your time synchronizing multiple copies of the same information, you can spend your time ensuring that the sentence is clear, accurate, and relevant for each deliverable that includes it.

The following figure shows how you might reuse a note in four separate pieces of information:

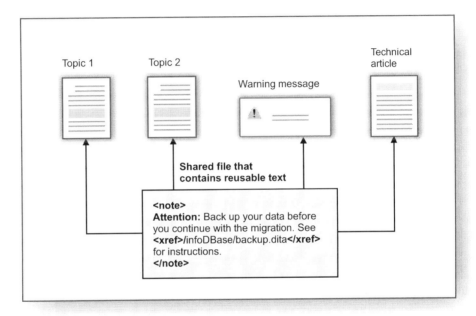

When updates are required, you can make the change in one place—the source text—rather than updating multiple copies of the text. The next time that you generate output, the changes will be implemented automatically.

Product and component names are a prime candidate for reuse because they tend to be ubiquitous in technical information and because they tend to change over time. When the name of a product or component changes, it's much easier to update that name once in a shared file and have that update automatically propagate across all of your files than it is to search and replace the name in every file that is part of an information set.

Other types of content that are good candidates for reuse include:

- Legal boilerplate text that must be the same in all of your organization's publications
- A task step that is common to several different tasks
- Lengthy commands, parameters, and environment variables that are duplicated in multiple topics
- Message text that is used in both the product interface and in a published message reference
- Version and release identifiers that need to be changed as new versions and releases become available

Not only does using a single source improve the accuracy of information, but it also eliminates the time that is needed to synchronize similar information. For more information about information reuse, see the style guideline "Commonly used expressions" on page 306 and *DITA Best Practices*, Chapter 10, "Content Reuse."

## Avoid introducing inconsistencies

Inconsistencies can occur when similar or related information is used in multiple places and the different instances of that information are not kept in sync. Inconsistencies ultimately lead to confusion.

When you follow a recipe that starts with "Preheat the oven to 325°" and ends with "Bake at 375° for 40 minutes," you don't know which temperature is correct, and you risk overcooking or undercooking your meal.

Inconsistencies that can indicate or cause accuracy problems include:

- Names or titles are incorrect, for example:
  - The content of a figure or table doesn't match its caption or surrounding text.
  - Topic titles don't accurately describe the content of the topic.
  - The column headings of a table are inconsistent with or don't accurately reflect the actual content of the column's cells.
- Field labels, hover help, tooltips, static text, and messages in the user interface do not match the names that are used in related help or task topics.
- The format of syntax elements in reference information does not match the format used elsewhere. For example, the date format of *yyyymmdd* is used in a reference topic, but the date format of *mmddyyyy* is used in related task topics.

- Information in marketing collateral, such as on a website or in a trade show brochure, isn't consistent with the related technical information.

The best way to avoid introducing inconsistencies is by reusing single-sourced information in multiple places. However, not all related information is a viable candidate for single-sourcing. Therefore, you must proactively synchronize related information that can't be single-sourced automatically.

Sometimes you need to include technical information in more than one place, often in more than one format. For example, perhaps you provide sample database tables with your product. One of those sample tables, Employee, contains data about employees:

Figure 4.1  Employee sample table

| EMPNO  | FIRSTNAME | LASTNAME | DEPT | JOB  | EDL |
|--------|-----------|----------|------|------|-----|
| 000010 | CHRISTINE | HAAS     | A00  | PRES | 18  |
| 000020 | MICHAEL   | THOMPSON | B01  | MGR  | 18  |
| 000030 | SALLY     | KWAN     | C01  | MGR  | 20  |
| 000040 | IRVING    | STERN    | D11  | MGR  | 16  |
| 000120 | SEAN      | CONNOR   | A00  | SLS  | 14  |
| 000140 | HEATHER   | NICHOLLS | C01  | SLS  | 18  |
| 000200 | DAVID     | BROWN    | D11  | DES  | 16  |
| 000204 | KAREN     | RUFFNER  | D11  | DES  | 18  |
| 000220 | JENNIFER  | LUTZ     | D11  | DES  | 18  |
| 000320 | RAMLAL    | MEHTA    | E21  | FLD  | 16  |
| 000330 | WING      | LEE      | E21  | FLD  | 14  |

The data in the sample Employee table is used in examples in the information set. The following example shows users how to list names of employees in a specific department from the sample Employee table:

**Original**

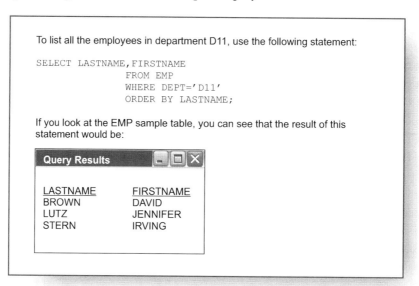

To list all the employees in department D11, use the following statement:

```
SELECT LASTNAME,FIRSTNAME
       FROM EMP
       WHERE DEPT='D11'
       ORDER BY LASTNAME;
```

If you look at the EMP sample table, you can see that the result of this statement would be:

**Query Results**

| LASTNAME | FIRSTNAME |
|----------|-----------|
| BROWN    | DAVID     |
| LUTZ     | JENNIFER  |
| STERN    | IRVING    |

In the original example, users who look at the sample Employee table to anticipate the results of the statement will be confused. They can see from the sample table that department D11 has four employees, but the result table in the original passage shows only three employees in department D11.

**Revision**

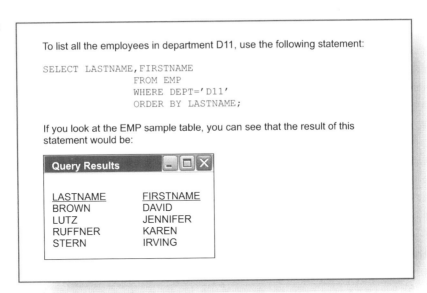

The revised passage accurately lists all four employees in the department; the example is consistent with the sample table and is therefore accurate.

The original example might have become inconsistent with the sample table for various reasons. Perhaps this is just an attention-to-detail error. Perhaps the sample table was changed after the examples were created. Or perhaps writing assignments changed, and the new writer was unaware of the dependency between the examples and the sample table. Regardless of the cause of this inconsistency, you can take steps to reduce the chance of inconsistencies finding their way into your documentation.

A good practice for keeping track of these types of dependencies is to insert comments into the source for your topics (if your authoring tool allows) or into a separate history file for your project. This practice is especially helpful on large projects with many writers and many different but related topics.

For example, consider the preceding situation in which the command output was inconsistent with the Employee sample table. In this case, the inconsistency might have been avoided if the source for the sample Employee table contained a note about the dependency. When writers change the sample Employee table, they know to also change the dependent examples in topics.

Another practice that can help you maintain the consistency of related information is to ask yourself each time that you update information: "Should I make other changes as a result of this one?" Follow up by searching your information set for potential related changes. This exercise is valuable because it causes you to consider other areas of the information that you might otherwise not think about.

For example, in the upcoming release of your product, you learn that valid values for the TIMEOUT parameter will be changed from 1-240 to 1-360. You update the reference topic for the TIMEOUT parameter accordingly, and then you search the product library for the string "TIMEOUT." Not surprisingly, you find other areas of the documentation that are affected by this change: a tip and several messages that instruct users to set the TIMEOUT parameter to 240. By identifying and making these changes, you've prevented your users from becoming confused, and you've saved your organization money by preventing support calls from users wondering what the correct settings are for the TIMEOUT parameter.

# Use tools that automate checking for accuracy

Spelling, grammar, punctuation, and typographical errors are typically thought of as style defects, and problems with links are most closely associated with retrievability defects. However, these types of problems can also affect the accuracy of the information.

Spelling, grammar, punctuation, and typographical errors can have a variety of effects. When users notice a minor error, their overall satisfaction with the information might decline slightly. If they notice many errors, they might question the overall accuracy of the technical information. If the errors are inaccuracies in code examples, procedures, statements of valid values, or syntax definitions—inaccuracies that result in significant problems—users are likely to lose confidence in the documentation and turn elsewhere for their information. They might even contact your support team, which ends up costing your company money. Similarly, broken links are a sign of information that hasn't been adequately maintained, and users are justified in questioning the overall accuracy and currency of that information.

Many tools are available to help writers improve the quality of their information by checking for spelling errors, grammar and punctuation problems, and other types of errors, such as invalid cross-references.

## Spell-checking tools

Probably the most common quality assessment tool is the spell-checking tool, which compares written content against a predefined list of terms and flags any terms that aren't in the list as being misspelled.

Spell-checking tools are useful, but they also have their drawbacks. For example, they can't spot misspellings that result in a legitimate word (*to* vs. *too* or *most* vs. *moist*), and in most cases they can't make up for lack of knowledge about commonly misused terms (*discreet* vs. *discrete*, *principle* vs. *principal*, or *loose* vs. *lose*).

Nevertheless, spell-checking tools can eliminate most embarrassing typos and misspellings, so they are worth running, especially if you don't have an editor or another colleague to check your work before it's published.

The following sentence demonstrates the strengths and weaknesses of spell-checking tools:

> The the plus sign is called a *binary* operater because of the use of two arguments, one argument one each side of the plus sign.

This sentence contains several problems:

- The word *the* is duplicated, and the word *operator* is misspelled. Spell-checking tools would catch these two errors.
- The second instance of the word *one* should be spelled as *on*. Most spell-checking tools would not catch this type of error, which reaffirms the need to have information edited by a human editor and not just by automated tools.

Most spell-checking tools also provide the ability to customize the terms that are checked, which means that you can add product and component names, obtuse technical terms, abbreviations, terms that your marketing department has decided to creatively capitalize, and any other terms that are specific to the subject matter that you document.

## Grammar-checking tools

Most authoring tools include some type of grammar-checking tool that can check for issues like passive constructions, negative constructions, gender-specific words, and even colloquialisms and jargon.

Some conventions that grammar-checking tools flag might conflict with the guidelines that your organization follows, so be careful about blindly accepting all of the advice that these tools offer. Many of these tools allow you to customize the tool to match your organization's standards. For example, the grammar-checking tool of one popular authoring program allows you to check for the use of the serial comma, which is a handy capability if your organization doesn't use the serial comma.

## Link-checking tools

A link checker verifies that the links in your content properly display a target page.

The use of link checkers is most valuable as part of normal maintenance, to identify links that no longer display a page. There's always a chance that the links that you add today might not work at all several months from now. Links become broken all the time, often simply because a company decides to reorganize or relocate their content. Therefore, you should check for broken links regularly so that you can find them before your users do.

# Accuracy checklist

Providing your users with accurate technical information is one of the most important parts of your job. Users expect all of the information that is part of a product to be accurate. If it's not, they lose confidence with not only the information, but with the entire product.

Use the guidelines in this chapter to ensure that your technical information is accurate. Refer to the examples in this chapter for practical applications of these guidelines.

You can use this checklist in two ways:

- As a reminder of what to look for, to ensure a thorough review
- As an evaluation tool, to determine the quality of the information

| Guidelines for accuracy | Items to look for | Quality rating<br>1=very dissatisfied,<br>5=very satisfied |
|---|---|---|
| Research before you write | • Information plans are established.<br>• Outlines and topic models are written.<br>• The appropriate technical experts have been identified.<br>• The product or product prototypes are available for evaluation purposes. | 1 2 3 4 5 |
| Verify the information that you write | • The information has been thoroughly reviewed by the appropriate technical experts.<br>• Technical owners have indicated that the information is accurate.<br>• Information has been validated as part of formal product testing. | 1 2 3 4 5 |
| Maintain information currency | • The final information includes late changes in the product.<br>• No information is labeled "new."<br>• Information from previous editions is current and valid.<br>• Interface descriptions are consistent with the actual interface.<br>• Product messages reflect current product behavior.<br>• Information does not include elements that can easily become out of date, such as dates and release identifiers. | 1 2 3 4 5 |

# Accuracy

| Guidelines for accuracy | Items to look for | Quality rating  1=very dissatisfied, 5=very satisfied |
|---|---|---|
| Maintain consistency in all information about a subject | • Duplicated information is consistent.  • No conflicts exist between separate but related topics. | 1 2 3 4 5 |
| Use tools that automate checking for accuracy | • Examples, syntax descriptions, procedures, and other types of information are free of spelling errors.  • Text is free of grammatical errors.  • All links display the intended content. | 1 2 3 4 5 |

To evaluate your information for all quality characteristics, you can add your findings from this checklist to "Appendix A, "Quality checklist," on page 545.

**CHAPTER 5**

# Completeness

*I try to leave out the parts that people skip.* —Elmore Leonard

From the user's point of view, completeness means that all of the required information is available. A subject is covered completely when:

- All of the relevant information is covered
- Each subject is covered in sufficient detail
- All promised information is included

Ensuring that your information is complete involves more than checking items off of a list—it starts with building the right list of what to include. You need to understand the purpose of the information that you are creating, for example:

- Who is your audience?
- What is your audience's level of expertise?
- What goals are they trying to achieve?
- What tasks do they complete to achieve those goals within your specific tool or product?
- What tasks do they complete to achieve those goals before or after they use your tool or product?

# Easy to use

Too often, technical writers create too much information about subjects that are not important and not enough information about what matters to users. If you write from a technical specification, you can easily write too much about how the product is constructed, which users don't need to know. If you follow old paradigms for how to deliver information, such as one help topic for each product window, you end up writing a lot of information that no one wants or needs to read. Instead, when you consider the goals of your audience and the tasks that they need to complete to reach those goals, you can discern which content matters to them and which content does not.

Delivering complete information is not a balancing act between adding too much information and not providing enough. Instead, through task analysis and by using the product, you learn exactly what information users need and when they need it. If you provide a help topic to explain each field in a window and no hover help for those fields, the information can seem incomplete because the information is not where the user needs it. Providing information in the wrong context makes it irrelevant. See the organization guideline "Separate contextual information into the appropriate type of embedded assistance" on page 332 to learn more about where information is the most effective.

As you learn what your users need, you add, subtract, and move information. Before you deliver your product, doing usability tests can help you identify where you have written too much information, where information is missing, and whether the information is where users expect it. After you deliver your product, you can monitor page hits and common searches to determine what users are looking for the most.

To make information complete, follow these guidelines:

- Make user interfaces self-documenting
- Apply a pattern for disclosing information
- Cover all subjects that support users' goals and only those subjects
- Cover each subject in only as much detail as users need
- Repeat information only when users will benefit from it

## Make user interfaces self-documenting

Before writing one word of topic documentation, make sure you are a part of the team that designs the user interface so that you can help ensure that it is as easy to use and intuitive as possible. Even if users don't read the topic documentation, they will always see and use the interface. One word or one change in an interface can save pages of documentation, which too often serves to "fix" deficiencies in the interface.

In an ideal situation, your user interface is based on the same scenarios as your documentation. When user interfaces are developed from scenarios, users are guided through tasks when appropriate so that they can stay on task and accomplish their goals. Scenario-driven design also ensures that interactions for common tasks are kept to a minimum.

For complicated scenarios that include tasks that are completed outside of the user interface, a wizard might provide instruction for those tasks and then assist users with completing the tasks that are inside of that particular user interface. For example, installation wizards often end with instructions for additional configuration steps or even open the next application or tool to complete that configuration.

# Easy to use

The following tool can benefit from more focus on design that supports the main usage scenario:

Original

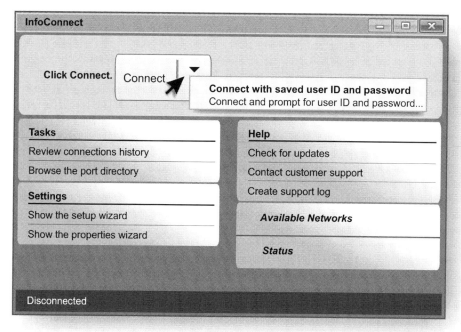

A common task in login interfaces is changing your password. If you look carefully at the original user interface, the only mention of password is in the drop-down list from the **Connect** button, a list which is itself hard to find.

The window includes a named section, **Settings**, which seems like it might offer some links for this task, but that task is not included in that section. Only by reading the help do you discover how to access the Change Password dialog—by selecting **Connect and prompt for user ID and password** from the **Connect** list.

Revision

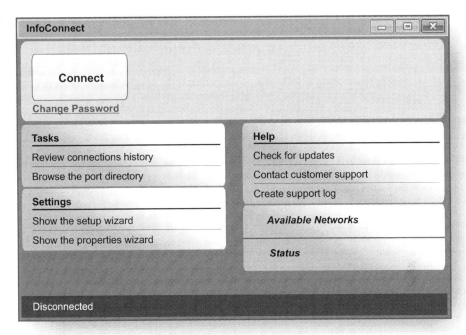

The redesigned user interface supports this common task, adding **Change Password** as a permanent link under the **Connect** button. Other ways to update this interface might include adding a **Change Password** link under **Tasks** or changing the **Connect** list choice to **Connect and Change Password**. Any of these changes could save confusion and extra documentation.

## Easy to use

After the user interactions are ideal, you can provide additional support for users without ever writing a word by incorporating programmatic assistance, as described in the following table:

**Table 5.1** Types of programmatic assistance

| Programmatic assistance | Explanation | Examples |
|---|---|---|
| Default values | Provide default values for fields or recommended selections for choices such as check boxes or radio buttons. These default values should support most of your users most of the time—if 25% or more of your users cannot use the default value, don't provide one. Instead, provide an example in the user interface. | This field provides a value for optimum performance rather than forcing users to read hover help or click a help link to find out what value to enter:<br><br>Crawl delay: 0.5 seconds between pages<br><br>Users can still read help to find additional information about this value, but they aren't required to read help to finish the configuration on this page.<br><br>As another example, the default distance measurement in cars is set based on the country where the car is sold. For example, US customers see miles by default instead of kilometers. |
| Detected values | Detect values or choices from the user's system. | Warning: InfoSheets is already installed on this computer.<br><br>Do you want to update it?<br><br>CAT123456     Close<br><br>Another example might be a new computer or device that detects the current time and time zone from a server that the computer or device is connected to. |
| Autocompletion | Autocomplete values that users are typing, thereby making meaningful suggestions for the value. Autocomplete is especially valuable in mobile applications because typing is more difficult on mobile devices. | ra\|    Cancel<br><br>Ramanand New Test<br>Ramanand Test1<br>Ramanand Text Search Brown |

**Table 5.1** Types of programmatic assistance *(continued)*

| Programmatic assistance | Explanation | Examples |
|---|---|---|
| Disable interface controls | Disable choices, fields, and buttons to prevent users from making mistakes. | In the following window, users create a folder in a content management repository. If folders aren't created in the correct location, they cannot be found or used later. The **Save in** field has a default value for the correct location, and it is grey so that users cannot change this value. Users are also unable to add the new folder until they supply a folder name, which ensures that a folder with a blank name is not created in error. If users try to add a folder without supplying the required folder name, they see a message that prompts them to add a folder name.<br><br>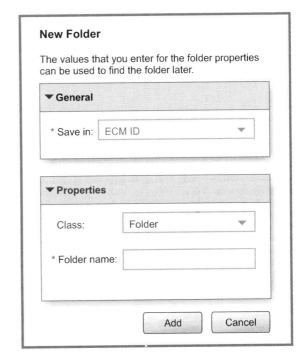<br><br>In a hardware context, many cars do not allow you to shift gears into reverse from fourth gear or move the car out of park before stepping on the brake. |

# Easy to use

Add programmatic assistance carefully and according to users' expectations.

**Original**

This field includes what looks like a default value, but it is actually an example. Likely, users won't realize it's an example and will attempt to accept the default and move to the next window. Then, they'll get an error message that states that the field value is not correct. Note also that the field is grey, and it's not clear how to edit the field so that you can change the portion that must be changed (`infodb_server:port`) to make the address functional.

**Revision**

| InfoDB server URL: | |
|---|---|
| | For example: http://infodb_server:port/wsi/IDBEWSMT |

In the revision, the example string is below the field and labeled as an example. Users can refer to it as they type in the field, which is no longer grey.

By starting with the user interface and bringing your skills as a writer to the design of the user interface, you help your team to improve the product experience. More importantly, you help users by reducing the amount of separate documentation they have to read and keeping them where they want to be—in the product, completing their tasks.

# Apply a pattern for disclosing information

To ensure that you cover all of the relevant information, apply a pattern for structuring your information.

"Embedded assistance" in Chapter 1 and the organization guideline, "Separate contextual information into the appropriate type of embedded assistance," introduce patterns and, in particular, a pattern for progressively disclosing information starting from the user interface. Following a *pattern* for progressive disclosure helps you determine where to put information so that it follows a natural progression and is exactly where users need it, when they need it.

Before you can apply the pattern, you need to know what assistance capability is available in your user interface so that you can define the pattern. Work with your team while the user interface is being designed to ensure that the necessary assistance capabilities are available to support embedded assistance. The following list provides several important capabilities to plan for:

**Static text**
> Ensure that you can insert static inline text when it's necessary. Determine the maximum length for static text and whether you can include highlighting and links.

**Tool or icon assistance**
> Determine how to display tool or icon assistance. In most web user interfaces, tool or icon assistance is provided as tooltips from the browser, which display in small yellow boxes. These tooltips usually have a maximum length of two to three words.
>
> Ensure that tooltips can be turned off if necessary. (Tooltips and hover help are defined in the Chapter 1 section "Embedded assistance" on page 4.)

**Control-level assistance**
> Determine how you can provide user interface control-level assistance. In web user interfaces, control-level assistance is provided most often as hover help. If this capability is not available, decide what mechanism you can use instead. Determine the maximum length for control-level assistance.
>
> Test whether you can provide hover help only for certain controls and not other controls. Because you don't need to provide hover help for all controls, you might want to use an icon, for example ❓, to indicate where hover help is included.
>
> Many teams want to enforce consistency by requiring hover help on all controls, which leads to a great deal of unnecessary help and a

# Easy to use

counterproductive customer experience. Most controls can and should be clear simply from their labels or any default or example values. Users do not want to see windows full of ❷ icons or with lots of hover help opening when they move their pointer around the interface.

Ensure that tooltips and hover help are not applied to the same user interface controls.

**Linking**

Verify that you can link from the user interface to separate contextual help topics, if necessary. You need a way to send users to help for additional information that will not fit within the user interface, as described in the retrievability guideline "Guide users through the information" on page 394.

**Messages**

Identify the various message display mechanisms, for example, dialogs, message status bars, and log windows. Determine how much text you can include and whether you can incorporate links. With your team, redesign message dialogs to accommodate all parts of the message, for example, the message text, explanation, and user response.

The following message dialog has a **Learn more** link, which opens a help topic:

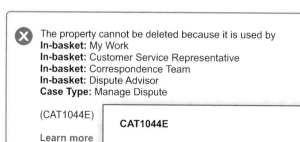

108  Developing Quality Technical Information

The only difference between the information in the dialog and the information in the help topic is the user response, which instructs users how to correct the problem. If the message dialog could display all of the message information, users would not have to click the link and be frustrated with this duplication. Instead, they could stay on task and fix their problem. Incorporating the full message into the message dialog not only solves the completeness problem of duplication, but also addresses the organization guideline "Put information where users expect it" and the retrievability guideline "Link appropriately."

**Help display**

Determine how and where linked help information will display. Not all assistance can fit within the user interface itself, so decide whether the assistance will display in an embedded help pane within the product or in a separate window or browser page.

After you have an idea of what assistance capabilities are available, you can define a pattern for progressively disclosing information such as the pattern in the following table:

Table 5.2  Sample pattern for progressively disclosing information in a user interface

| User interface element | Content |
| --- | --- |
| Labels (for fields, windows, buttons, group boxes, and so on) | Succinct nouns, using a short, well-managed list of product terms. |
| Static descriptive text at the top of windows | • Description of the overall action that users accomplish on the window if it's not obvious.<br>• Anything users must do before completing this window.<br>• Ramifications, if there are any, of the changes in this window.<br>• Link to additional information if needed. |
| Static descriptive text below fields | Examples for what to enter in a field. |
| Messages | Description of the situation. Provide an action for users to solve the problem or ask users to confirm their choices. Link to additional information if needed. |
| Hover help | • Syntax for what to enter in a field.<br>• Ramifications of the field change.<br>• Descriptive information for what to enter in the field.<br>• Link to additional information if needed. |
| Help topics | • Reference or concept topics that provide additional information that finishes a description begun in static descriptive text, messages, or hover help.<br>• Task, concept, or reference topics that provide contextual information from other user interface links or from other topics. |

By following a pattern, you can see where in the user interface you can provide certain information, should it be necessary. The word *necessary* is key because you should not clutter a user interface with information just because you have a map for where to put it. Instead, approach each window, command, or piece of hardware as if you can find a way to make it clear with no words at all. Often doing this work means working with usability engineers and product developers to adjust the design so that the user interactions will be more obvious and require less explanation, as described in the guideline "Make user interfaces self-documenting."

For example, does the user interface provide too many ways to navigate, such as tabs, breadcrumbs, a navigation tree, and a **Next** button? When the flow of tasks through the interface is streamlined, assistance can be briefer because you don't need to waste time explaining how to move around the interface for each task. After you are certain that the interactions are as clear as possible, add only the minimum number of words that are needed to make the interactions clear.

The following example shows an application that is not progressively disclosing messages:

Original

The original message includes multiple problems. It duplicates the message information in a message dialog and at the top of the window, a completeness issue, and provides too little specific information to solve the problem, a completeness and concreteness issue. Most importantly for this particular error, the message itself is missing some essential information: it does not indicate the name of the directory that does not exist. In addition, the user interface does not mark the field that caused the error, and so it does not provide an entry point as described in the retrievability guideline "Provide helpful entry points."

Easy to use

**Revision**

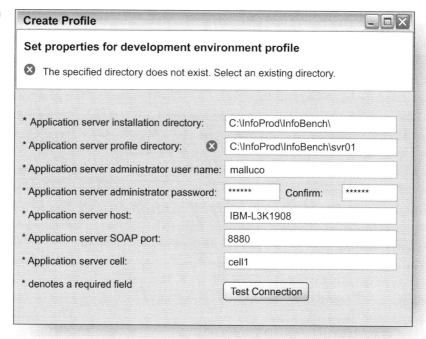

In the revised example, the writer worked with the usability engineer and product developers to eliminate the redundant error message display (the pop-up window) and to add an error icon next to the field that caused the error. Now the message text can remain as it is and work for either directory field, and the error icon on the field helps to pinpoint the problematic field.

When you follow a well-designed pattern for disclosing information in a user interface, users can more efficiently interact with the product. For example, the following user interface does not conform to its pattern for progressively disclosing information:

Original

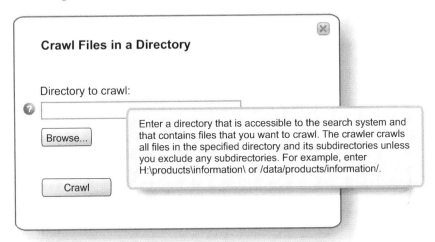

The hover help includes an example of syntax, but that example disappears when users move their pointers away from the hover help. Providing examples in hover help means that users must repeatedly go back and forth between the field and hover to get their syntax right.

# Easy to use

**Revision**

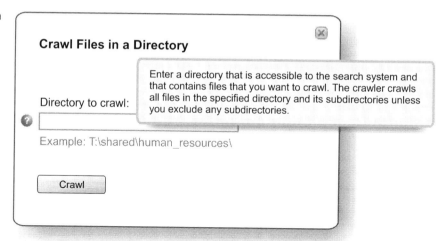

In the revised user interface, the example text has been removed from the hover help and added below the field in grey text so that users can see it while they enter values in the field.

As described in the organization guideline "Put information where users expect it," when information is in the wrong place, users either won't see it or won't be able to use it and stay on task. Let your pattern help you determine where information should go and look for places to add information or take information away.

# Cover all subjects that support users' goals and only those subjects

Before you decide what subjects to cover, you must identify your users, evaluate their goals, and determine the tasks that those users need to do with your product to meet their goals. By knowing your users and their goals, you know what subjects are relevant.

Both Chapter 1 and the task orientation guideline "Focus on users' goals" explain the difference between goals and tasks and the importance of understanding users' goals. Users complete tasks to achieve their goals, and if you focus only on tasks, you might omit important information. Placing too much emphasis on tasks is somewhat like trying to describe how to build a fence by writing information only about how to use a hammer. To write information that helps users achieve their goals:

- **Create an outline or topic model**
- **Include only information based on user goals**
- **Make sure concepts and reference topics support the goals**

## Create an outline or topic model

When planning what to write, you often either create an outline or build a topic model for the information about your product, whether your product is a software application, piece of hardware, or a training seminar. At a minimum, you must cover the major tasks, their subtasks, the reasons or conditions for doing them, and how to respond to expected errors.

Avoid setting the boundaries of your model at what tasks can be done with your product because users don't use products in isolation. Although you might be writing documentation for a television, you will frustrate users if

# Easy to use

you do not write about how to hook the television up to their cable box and recording and playback equipment, as shown in the following topic model:

**Original**

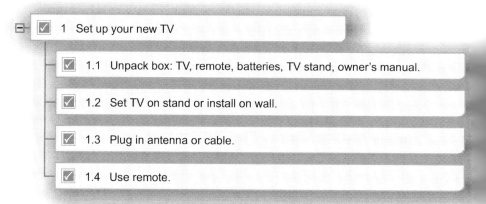

Most often, missing information is related to the gaps between products or outside of products. If you focus too closely on *what* you are documenting, you might create artificial boundaries around the information so that it describes only what happens within one product or even within a single window. Users can't see how it all fits together.

The easiest way to focus on user goals and avoid missing important tasks is to base your topic model or outline on user scenarios. Analyze your model to ensure that it includes the necessary content for users to achieve their goals regardless of how or with what tools they achieve those goals. Determine what tasks and subtasks your users will need to do. Some of this content might belong in a user interface, and some might end up as topic-based information. At this early stage, you are trying to get a high-level view of how users accomplish their goals and where your product fits in.

Start the model from the users' goals and the scenarios for achieving them, for example, how to watch movies on the television. Users often must complete steps in different products or tools. For example, to set up a television, users might need to assemble a TV stand and connect cables for other devices.

**Revision**

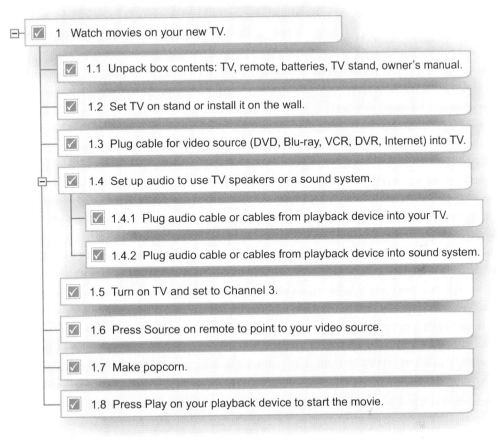

Ideally, enlist real users to validate both your understanding of their goals and of your topic model. Tell users all of the assumptions you made when you developed the scenario and model so that they can validate or correct your assumptions as well.

For example, you might be documenting for novices a product with a sophisticated text search feature and assume that the users won't understand search terminology such as *stemming*. During testing, you might learn that these users, though novices with your product, are not novices to searching or search terminology because they use LexisNexis products to do intensive searching as part of their jobs. With this information from real users, you can adjust your scenarios and models to provide the information that users actually need: how to take what they know about searching into the context of your product.

## Include only information based on user goals

The following welcome page in a product shows how easy it is to lose sight of users' goals and write too much about product function or about how to use the information, for example, how to click links to open a wizard:

Original

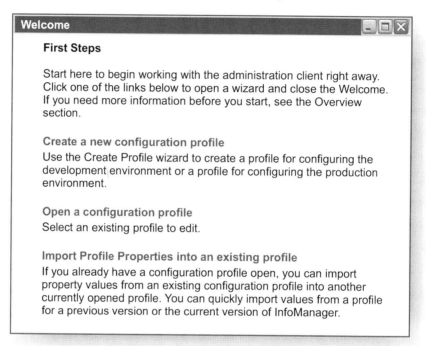

The welcome page tasks focus on the tool and building a profile instead of actual user goals. The profile saves settings for the tool so that users don't need to enter settings repeatedly. The profile is really a convenience mechanism but is being treated as if it was the entire goal of work in the tool. Users are getting a lot of information about the profile and no information about their real tasks.

Revision

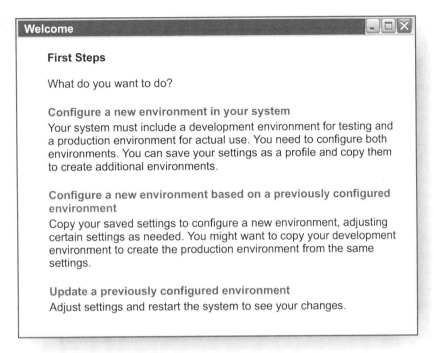

By contrast, the revision reorients the welcome page to the real user tasks: configuring an area of their system or updating a previously configured area. The users get the information they need about configuring their system, with the information about the profile reduced to its rightful place: as a convenience mechanism for saving settings.

Throughout the project cycle, refer to your scenarios and topic models to evaluate whether your information covers the necessary subjects, and continue to validate your work with users. Ask users:

- Are all the major tasks covered that users need to achieve their goals?
- Are all the subtasks described that users need to do to complete the major tasks?
- Are necessary steps outside of the tool referenced?
- Are all the reasons or conditions explained that are associated with optional tasks?
- Are problems anticipated and solutions described?
- Are any described tasks insignificant or needed by only a small subset of users?

The following task model shows a typical set of installation tasks:

**Original**

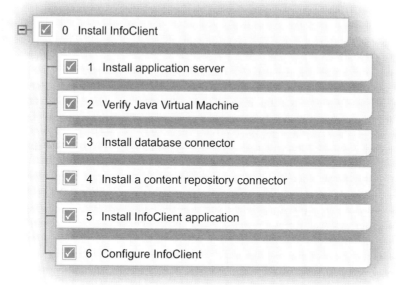

If users use the InfoClient by itself, the set of tasks seems complete. However, the scenarios for InfoClient indicate that this client will almost always be installed on systems that include many or most prerequisite products rather than installed on new systems where no other products are installed.

**Revision**

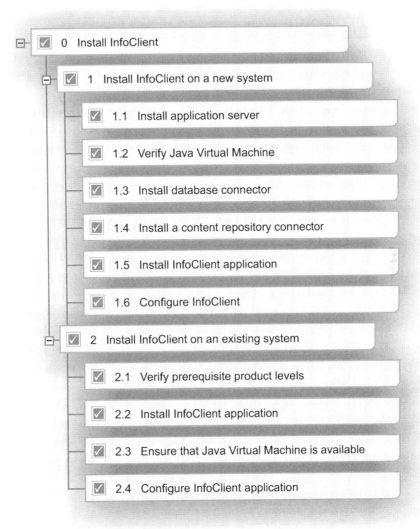

The revision adds the tasks for installing on an existing system, including previously missing tasks such as determining whether the existing system includes the correct prerequisites. This topic model shows how easy it is for writers who focus only on their own product and not on the users' goals or context to make assumptions that shortchange users. Always rely on thorough scenarios, grounded in users' goals and validated with users, rather than your own assumptions.

## Make sure concepts and reference topics support the goals

Sometimes you need to explain concepts as you introduce a subject, explain a process, or highlight benefits. Do not confuse concepts with information about how a product is built. Concepts help users to learn because they provide necessary vocabulary and context for what users are doing. For example, most communications protocols consist of multiple layers; however, users don't need to understand the detailed function of each layer to configure their router.

Similarly, reference information must support the tasks. For example, details about command parameters that provide context about using the command to complete a task are useful, but details about how the command parameters got their names are not. Unnecessary reference information is the most likely type to bloat your information set because it's easy to fall into the trap of documenting each field and variable, even those that are self-evident.

Before you write conceptual or reference information, ask yourself what users will do with the information. For example, will the information help users make a decision or do the task at hand? When users are configuring access control in a product, they probably need to understand the overall security model so that they know how each selection affects other choices. They don't need to know about preferences such as font size or full screen versus split screen windows. Information that isn't relevant to the users' task can frustrate and confuse users.

# Cover each subject in only as much detail as users need

After you know the subjects that you need to write about to address users' goals, you then need to decide how much detail to provide. You need to include enough detail for users to do their tasks but no more detail than they need. While you're analyzing tasks in the context of users' goals, you might not be able to anticipate all the details of a task. When you're writing, the details will become obvious.

Judgments about the completeness of detail are usually more subjective than judgments about which topics to include. For example, deciding whether to include a topic about "What's new in this release" is easy, but determining which features are worth highlighting is not.

To ensure that each subject includes only as much detail as users need:

- **Provide appropriate detail for your users and their experience level**
- **Include enough information**
- **Include only necessary information**

## Provide appropriate detail for your users and their experience level

Too much detail is frequently the result of writing to more than one type of user or to no particular user at all. Sometimes extra detail is included for the secondary audience, which can distract the primary audience. For example, many tasks can be completed by using more than one method, but documenting both methods in the same topic can be confusing for the users of each method.

The terms *primary audience* and *secondary audience* refer to the readers of a book or topic; these readers are a subset of all of your product users. If you are writing a tutorial, your primary audience consists of novice product users, and your secondary audience consists of more experienced product users who, for example, need to learn about a new function. Typically, you make these distinctions as you analyze your audience and design your information.

If your information has a secondary audience ask the following questions:

- How large is that audience?
- How (and how much) does the secondary audience differ from the primary audience?
- Is the document that you're writing the only source for this type of information for that secondary audience? Can the information for the secondary audience be omitted or moved elsewhere?

# Easy to use

The following passage shows installation information for a product that users can install by using a graphical interface or by using a command line. Presumably, most users will install the product by using the graphical interface and its embedded assistance. Command-line users, although a minority of product users, are the primary audience for the topic-based information in the following example:

**Original**

> **Configuring an InfoInstaller control point**
>
> To configure an InfoInstaller control point, you can use either the graphical program (InfinstGUI) or the command-driven program (InfinstCMD). If you use the graphical program, you must have installed InfoFinder.
>
> Before you configure an InfoInstaller control point, you need to:
>
> 1. Determine (or create) the InfoInstaller control database. If you are using InfinstGUI, right-click the CDB icon to see the name of the control database. If no CDB icon is on your desktop, you need to create the database. If you are using InfinstCMD, go to a command line and enter:
>
>    ```
>    create infinstCMD database
>    ```
>
>    If the database already exists, you will receive an error message.
>
> 2. *and so on...*

The primary audience of the topic-based installation information is individuals who use the command line. Each type of user will use only one installation method at a time, so providing details for both together burdens both audiences unnecessarily.

In the original passage, the first step addresses both the primary audience and the secondary audience, even though the actions that these users take differ.

**Revision**

> **Configuring an InfoInstaller control point**
>
> To configure an InfoInstaller control point, you can use either the graphical program (InfinstGUI) or the command line (InfinstCMD).
>
> | InfinstCMD (command line) | 1. | Create the InfoInstaller control database. At a command line enter:<br><br>`create infinstCMD database`<br><br>If the database already exists, you will receive an error message. |
> |---|---|---|
> |  | 2. | *and so on...* |
>
> | InfinstGUI (graphical user interface) | 1. | Install InfoFinder. |
> |---|---|---|
> |  | 2. | Double-click the InfinstGUI icon on your desktop, and follow the online instructions. |

The revised passage shows that two separate procedures address different users. The primary audience is not burdened with information that is directed at the secondary audience because users in the secondary audience can access the information that they need while they are using the graphical interface.

This pair of topics shows an important distinction between the primary audience for certain information and the primary users of certain product functions. When information has a specialized audience that is different from most users of the product, ignoring that majority or relegating them to a secondary position can be difficult. The revised topic acknowledges the majority of product users (the secondary audience for this information) and directs them elsewhere. Thereafter, the information targets its primary audience, who use the command line.

After you identify your audience or audiences, you must then investigate further. For example, you need to answer the following questions:

- Are your users experienced or inexperienced with your product or with similar products?
- Do you need to help novice users get up and running quickly but also help experienced users complete more complex tasks?
- Are the tasks something experienced users do infrequently, meaning that they might need more detail than for tasks that they do often?

Knowing the experience level of your audience helps you determine what information is relevant.

Be careful when determining how much focus you place on the needs of novices. Audiences and technology change over time, so do not emphasize the needs of novice users at the expense of experienced users. Novices do not remain novices for long—in fact, they might be novices only the first time they use the product that you are writing information for. As a result of this, you need to weigh how much focus you put on true novices, who are likely to need material directed at them only for a very short time.

Think about the technology changes you've personally experienced. If you're old enough, you remember when the mouse was new and users took a few moments to understand how to get the pointer to move across the screen if they ran out of table surface over which to move the mouse.

You might think about when web browsers or smartphones were new. How long did it take you to pick up the essential techniques you needed to use those devices to accomplish your goals? As you were learning to use these new technologies, so was everybody else. And because you were all going through this learning curve together, you were able to share what you learned with your colleagues, friends, and family. Latecomers to a technology benefit from the early adopters, who can show them how to get started. Technical documentation that focuses only on the needs of the true novice is of value for a limited time.

Writing about a product is easiest when your audience is homogeneous—either all experienced or all inexperienced. On one end of the spectrum, if your information is about a new product, you can assume that your audience is made up of novices, at least with this product. On the other end, if your information is about a well-established product that's been on the market for a long time and sells to relatively few new users, you can assume that most of your audience is experienced.

The challenge that writers frequently face is to successfully write to several different levels of experience. You don't want to make assumptions that cause major problems for novices. You don't want to bog down experienced users with lots of introductory information, as in the following topic:

**Original**

> **Building InfoTool files**
>
> Complete the following steps to build InfoTool files and automatically copy the necessary files onto the workstations in your network.
>
> 1. Log in to the control point host or managed host as user root. A root user has full authority to run commands in any directory.
>
>    ```
>    login root
>    ```
>
> 2. Change to the directory that contains the driver build script, for example:
>
>    ```
>    cd system_dir/infotool_dir/drivers/inft64/obj
>    ```
>
>    In this example, *system_dir/infotool_dir* represents the directory path where you installed InfoTool.
>
> 3. Run the driver build script:
>
>    ```
>    build_inft_driv
>    ```
>
> 4. Change to the directory with the watchdog build script.
>
> 5. Run the watchdog build script:
>
>    ```
>    build_inft_wdog
>    ```

The original passage assumes that users need a definition of root user and to know all of the commands for basic steps like logging in and changing directories. If your audience is experienced, these assumptions are probably not valid.

Easy to use

**Revision**

> **Building InfoTool files**
>
> To build InfoTool files and copy them into your network:
>
> 1. Log in to the control point host or managed host as root.
>
> 2. Run the script to build the InfoTool driver. In the path where you installed InfoTool (by default, `/opt/sys/InfoTool/`), change to the subdirectory `/drivers/inft64/obj` and run the script:
>
>    ```
>    build_inft_driv
>    ```
>
> 3. Run the script to copy the driver into the network. Change to the directory with the watchdog build script (by default, `/opt/sys/InfoTool_bak/drivers/inft64/obj`) and run it:
>
>    ```
>    build_inft_wdog
>    ```

The revised passage removes the unnecessary detail for users who know how to use the system but still need a reminder for the general flow of steps, locations, and command names.

The following figure provides a general guideline for the level of detail that you should provide for different types of tasks and audiences. Novices are unlikely to pay much attention to advanced tasks. However, when novices complete advanced tasks they need "do this, do that" instructions rather than in-depth conceptual framework and detail that they won't understand because they don't have context for it. They need more detail for basic tasks and more conceptual framework so that they can build their knowledge and experience. Experienced users are bogged down by basic tasks and extra detail for basics, but they prefer extra detail for advanced tasks to help build their understanding of how pieces fit together.

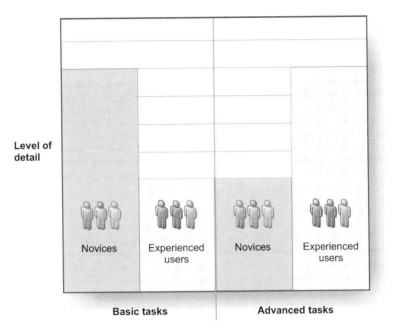

**Figure 5.1** Novice and experienced users need different detail levels for the same tasks

If your audience has a mix of experience levels, you need to decide how to satisfy varied information needs without causing problems for any group. These are some approaches that you might take for your topic-based information:

- Create separate topics for different experience levels. Link between the topics so that users can get what they need on request, regardless of their experience level.
- Put the information into two different formats, such as advanced tasks for experienced users and tutorials or video tours for novices.
- Within a single information set, segregate the basic topics from the advanced topics.
- Integrate the information, but graphically distinguish between information for different experience levels. For example, provide labeled tips or shortcuts for experienced users. Use this approach with caution; make sure that you don't end up with tasks where every step is labeled. A good rule of thumb is to completely separate information if more than a quarter of it requires a label for differentiation.

# Easy to use

- Programmatically customize information by audience type so that users see only the information that is relevant to them. For example, prompt users to answer questions and then return the information that meets their needs. Installation material can benefit greatly from customization because of the number of possible pathways through the information. For more information about using conditional coding in DITA-sourced documentation to customize content, see *DITA Best Practices*.

For embedded assistance, you do not have the luxury of providing alternative information. Instead, identify a sweet spot between the two experience levels—somewhere a bit beyond the novice users and a bit below the experienced ones—and write to that experience level. Provide links from embedded assistance to advanced topics and provide links from advanced topics to more basic information, as described in the retrievability guideline "Guide users through the information" on page 394. Novice users are more likely to seek out what they need.

## Include enough information

When you develop information or when you're editing information that you or others have written, look for logical holes: areas where some information might be missing. Follow these guidelines:

- ❑ **Provide information about what users *should* do**
- ❑ **Include complete instructions**

### Provide information about what users *should* do

You need to write the right information to help users succeed. You also need to provide the necessary information to help them when something goes wrong.

Rather than focusing on what users *can* do within a user interface, focus on what they *should* do. Chapter 1 introduced the concept of domain expertise, and it is described in more detail in the concreteness guideline "Consider the skill level and needs of users" on page 220. Domain expertise involves incorporating recommendations from experts in your documentation. One common example is field help that offers a range of possible values. While knowing the range is important, users still might not know what value is the best to choose for their situation without a recommendation from someone with experience. For example, how long should purchasers be allowed to hold tickets before purchasing them on an event website?

Suppose that you're documenting a product for performing legal analysis on large numbers of stored documents, including email. Because users of this product often need to assess thousands of documents, they need to be as efficient as possible. Tips for how to quickly move from document to document with as few clicks as possible are exactly what these users need.

Focus on the most likely questions users will have in their context. Users want information about how they can accomplish their goals in their environment. This type of information is the most difficult to discover, but it is often the most valuable for users. Your users don't have access to the expertise that you have access to. They don't have access to your development team, your support team, or those who help customers install and configure systems. When you write this type of material, you are like a reporter, asking questions of multiple experts and synthesizing the results to give users access to the expertise they're looking for.

That expertise can also help when you need to include information about what users should do when something goes wrong. The following topic provides some good information to show how users determine whether they've properly started the tool.

**Original**

> **Starting the InfoExporter on AIX:**
>
> To start the InfoExplorer on AIX:
>
> 1. On the InfoExporter server, log on as a user with write access to /bin and read, write, and execute access to the directory where you installed the InfoExporter.
>
> 2. Run the following command:
>
>    `install_directory/InfoExporter/FedExporter/bin/run_exporter.sh`
>
> The InfoExporter is running correctly if:
>
> - No exceptions or error messages appear in the command console.
> - The following message is displayed: `There is no schedule Please create a rule and schedule it using the administration client.`

The topic doesn't anticipate how to fix potential problems.

**Revision**

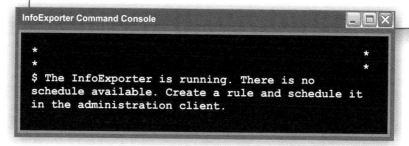

In the revision, the writer has worked with the development team to update the message in the user interface so that it now states clearly that the InfoExporter is running correctly. The topic no longer needs to repeat this information, but it does offer a tip to solve the most common problem and tells users where to look for additional information.

Because this information is in a command console, the message information cannot be easily linked from the user interface. Therefore, it's important to be clear in the messages and tell users how they can look up additional information (by message number). Make sure you document uncommon situations in the best location, as described in the organization guideline "Put information where users expect it." If message IFER1002E includes the following text in the command console, the writer doesn't need that detail in the revised topic either:

```
The InfoExporter is not started. Make sure you have write
access to the /bin directory and read, write, and execute
access to the installation directory.
```

Be careful, however, as you assess the need to document uncommon situations. Often those who develop and test a technology want to document every possible uncommon occurrence, but many of these cases are so rare that they shouldn't be part of the information. Remember that you are writing documentation for users to accomplish their goals, not a legal document to protect your company from users' mistakes.

## Include complete instructions

When you research instructions that you plan to write, follow every possible path in scenarios, including paths that lead outside a tool or application. Consider which of those paths, if any, require documentation that isn't provided by the related tool or product. Documentation often stops at the boundary of a tool or product with sentences like, "For more information about widgets, see InfoWidget information."

Seemingly practical concerns, for example, that the related product is developed by another company, do not need to stop you from testing the full path to ensure that any critical information that users need is available. If you're pointing to another set of information, users must be able to find what they need there.

When you focus on users' goals, you realize that accomplishing those goals often takes users to many different windows within a single application, to many different tools, and sometimes to steps that they complete outside of any tool or application. By focusing on the goals and scenarios and then testing procedural steps across the required tools, you quickly discover the gaps and assumptions that aren't explicit.

These gaps are even more obvious with hardware. For example, suppose your mailbox post is broken. You buy a new post for the existing mailbox. The instructions for the post explain how to install the new post, including the mounting board for a mailbox, but they do not explain how to mount the mailbox. A search on the company's website turns up instructions for your mailbox, but these explain only how to install the latch and flag onto the box, not how to install the mailbox on the post. Both the post and mailbox instructions are complete for their piece, but your goal is to have a functioning mailbox, not a new post with a mailbox on the ground next to it.

Keep the goals of your users in mind to avoid such situations. You can do several things to facilitate complete instructions:

- Consult with the team or writers who document the related parts or tools to decide how and where each of your information sets will talk about or link to the other.
- Read the information for the related products or tools as it relates to your product or tool. Make sure that, if possible, each set of information acknowledges the other and that transitions between the two sets make sense for users coming from each direction. If the related product is owned by another company, make sure your information is clear about what information users need from the related documentation. As described in the retrievability guideline "Use effective wording for links" on page 409, it is not enough to say, "See the documentation that came with your video camera" if supported video cameras do not provide any information about your television.
- Build an overall procedural map for achieving a given user goal. In the following example, upgrading a case management system called InfoManager is a user goal. You can use a procedural map for your own planning purposes, or you can publish it for users. Although the particulars of these maps aren't pertinent here, you can see procedures that start from two different starting points but arrive at the same finishing point: an upgraded system. The procedure at right includes two additional steps at the beginning for users who must upgrade their underlying platform system:

The procedures in these topic models are relatively straightforward, but many procedures can be much more complex, with many decision points. Tracking the instructions down each path will help you ensure that you provide complete instructions.

If you publish flow diagrams like these for users, you can also make them into image maps that link directly to the pertinent topics. As described in the visual effectiveness guideline "Illustrate significant tasks and concepts" on page 431, illustrations are a great way to orient users during complex procedures. Users can refer to the graphic at different times in the procedure, orienting themselves to where they are in the current complex flow.

## Include only necessary information

Too much information also creates problems of completeness:

- Unnecessary details slow down users, making information that users need more difficult to find and use.
- Unnecessary information can also annoy users, for example, when products confirm all decisions, even those that aren't destructive. Imagine that you select all checked out library books to renew them and then see a confirmation message asking, "Do you want to renew the selected items?" There's no disadvantage to renewing them all if you didn't mean to do it. However, potentially destructive actions, such as deleting, reformatting, or exiting without saving should always be confirmed.
- Additional information adds to your team's development and maintenance costs.

The idea that you can create information that is overly complete relates directly to the minimalist approach to technical writing. This approach was described by John Carroll in his 1990 book, *The Nurnberg Funnel: Designing Minimalist Instruction for Practical Computer Skill*:

> The key idea in the minimalist approach is to present the smallest possible obstacle to learners' efforts, to accommodate, even to exploit, the learning strategies that cause problems for learners using systematic instructional materials. The goal is to let the learner get more out of the training experience by providing less overt training structure.

Although Carroll originally focused on learners and tutorials, his techniques have been applied more broadly to other forms of information.

Often, too much information confuses users by making it difficult to decide what is important. The following topic maps out information that is obvious in each mobile environment.

Original

**Mobile browser simulator overview**

The mobile browser simulator is a web application that helps you test mobile web applications without having to install the device vendor native SDK. You can use the Mobile Browser Simulator to preview mobile applications on Android, iPhone, iPad, BlackBerry 6 and 7, Windows Phone 7, Windows Phone 8, and mobile web application environments.

**Tip:** When you preview a mobile application on an Android, iPhone, iPad, BlackBerry 6 and 7, Windows Phone 7, or Windows Phone 8 environment, only the devices for the selected environment are available. For example, if you preview a Worklight application on an Android environment, you can select only from the list of available Android devices. When you preview a Worklight application on a mobile web application environment, you can select from the list of available Android, iPhone, iPad, BlackBerry 6 and 7, Windows Phone 7, and Windows Phone 8 devices.

The original topic lists all mobile environments in the introductory paragraph, in the tip, and again in the examples in the tip. Not only is this a potential accuracy problem (as mobile environment release levels change), but it also makes the information longer than it needs to be. More importantly, it does not help users because the simulation choices that are available depend on the mobile environment.

Revision

**Mobile browser simulator overview**

The Mobile Browser Simulator is a web application that helps you test mobile web applications without having to install the device vendor native SDK. You can preview mobile applications on Android, iPhone, iPad, BlackBerry 6 and 7, Windows Phone 7, Windows Phone 8, and mobile application environments.

The revised passage excludes the tip, which in this case is extraneous, and it efficiently tells users what the simulator is and which environments they can use it in.

You might have noticed an additional change to the revised passage: the second sentence was shortened to remove the repetition of the component name.

# Easy to use

Many introductions to topics contain more detail than users need, as in the following topic that describes the accessibility features of a product:

**Original**

### Navigating windows by using the keyboard

 Find the Tab key in the upper left area of your keyboard.

Try pressing the key. It moves the cursor between two fields.

 Find the Backspace key in the upper right corner of the alphanumeric area.

Try pressing the key. It erases by going backward from where the cursor is.

You can also erase characters by pressing the long key, called the Spacebar, at the bottom of the keyboard. However, the Spacebar erases by going forward from where the cursor is.

The original passage includes too many details, both textual and graphical, about how to use a standard keyboard. Unless this is a product that teaches users how to type, these details are extraneous. The graphical elements break a related visual effectiveness guideline, "Use graphics that are meaningful and appropriate," by illustrating what is clearly obvious to keyboard users.

**Revision**

### Keyboard navigation

| Action | Keyboard key |
| --- | --- |
| Enter data into a field | Letter and number keys |
| Erase (backward) data entered into a field | Backspace |
| Erase (forward) data entered into a field | Spacebar |
| Move forward to the next field or control | Tab |
| Move backward to the previous field or control | Shift+Tab |

The revised topic tells users what keys to use to enter data in a window, but it excludes unnecessary details.

In user interfaces, too much information in the wrong place can detract from the usability of the product, as shown in the following web page:

Original

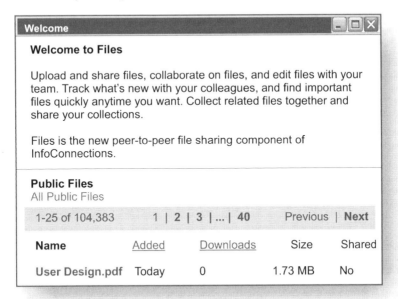

In this web page, the introductory information pushes down the main content of the page, which is the file list. To do the main task on the page, working with files, users must scroll, or worse yet, page forward. The behavior of the introductory box is problematic, but it also includes information that is redundant or unnecessary for the context, for example, "Files is the new peer-to-peer file sharing component of InfoConnections." This last sentence is marketing language that isn't appropriate for an audience that has already purchased the product, but this type of information creeps into user interfaces and documentation, especially when products are new.

Easy to use

**Revision**

In the revision, the introductory section includes a close button (the X in the upper-right corner) so that users can dismiss the introductory text after they're comfortable with the information. The static text has been shortened to remove redundancy, the marketing material, and the extra heading, none of which added value.

# Repeat information only when users will benefit from it

Repetition can be helpful in many situations, but it can easily become a problem in others.

**Helpful repetition**

Use repetition to:

- Orient users when they jump from another context, for example, from a link in a user interface to help content
- Appropriately emphasize and reinforce important points
- Eliminate the need for users to branch to another topic or page
- Provide information in different forms to support accessibility

Review the following user interface dialog and portion of an associated help topic:

**Original**

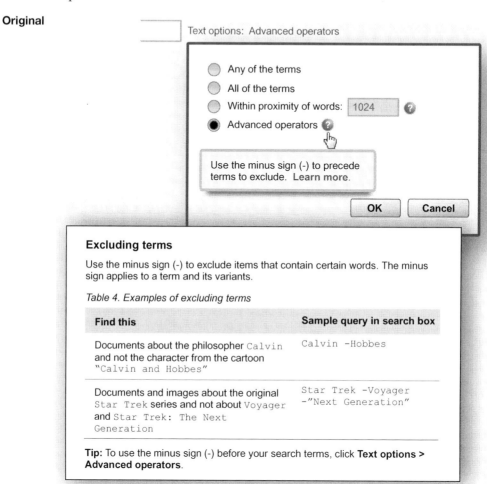

The original help topic contains no repetition, so the users of the help topic see examples of the minus sign but not how or where to enable it.

**Revision**

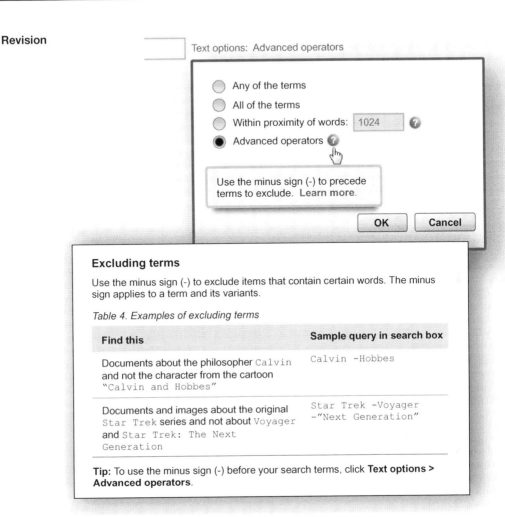

The revised passage includes the necessary information about the minus sign rather than forcing users to figure out from the examples what the minus sign does and how to access it in the user interface. You wouldn't want to copy everything from another information source. However, don't send users to a new context without providing some repetition to orient them.

As with a change of context from the user interface to linked help, you also need to consider the transitions between products or tools. All products that you document work with other products, and individual products often consist of multiple different tools or applications. Because of this, you might need to repeat details about another tool or product in the context of your own tool or product. Situations such as these call for a discerning use of repetition: explain

**143**

to users only what they need and hide what they don't need so that they can stay on task within their current context.

Use repetition to reinforce important information, such as when a product or technology doesn't behave as users might expect. In these cases, you document the exceptional behavior and then remind users of the exception in relevant locations in your information.

Another important usage for repetition is to save users from needing to follow a link unnecessarily, something that happens often between products, but also within a single product's documentation. On a mobile device, following links can cause you to needlessly lose your context, as shown in the following topic:

**Original**

> Choose **Save** to test the server connection. After the verification step is complete, the server will be saved and the My Rooms view will display, which shows all the rooms you currently own or manage on the new server.
>
> **Important:** You might receive the following warning:
>
> ```
> Trying to open a secure connection but there is an issue
> with the server's certificate.
> ```
>
> If you receive the warning, then the server you are trying to access uses an untrusted SSL certification. See Configuring meetings for Android devices for more information.

In the original topic, a link is provided to a separate topic for how to handle the warning about the security on the server. Users must follow the link, which is even more of a problem because they're on a mobile device.

**Revision**

> Choose **Save** to test the server connection. After the verification step is complete, the server will be saved and the My Rooms view will display, which shows all the rooms you currently own or manage on the new server.
>
> **Important:** You might receive the following warning:
>
> ```
> Trying to open a secure connection but there is an issue
> with the server's certificate.
> ```
>
> If you receive the warning, then the server you are trying to access uses an untrusted SSL certification. Turn on the **Allow Untrusted SSL** option in the **Settings** menu or have the administrator for the server secure it with a valid certificate.

In the revised topic, the troubleshooting information is included so that users can immediately go to the settings and make the change they need to get back on task.

A final case for repetition is related to accessibility: you might need to deliver information in more than one format or location for different audiences. For example, a video tour might be great for most of your audience, but visually impaired users won't benefit from it. They need access to that information, so you must provide the same information as a script that they can listen to by using a screen reader. The script should explain exactly what is happening in the video tour so that they can imagine the interactions that go with the scripted words.

## Unnecessary repetition

Repetition is bad when it does not provide necessary information; extraneous details waste reading time and might give a wrong impression about what is important.

Embedded assistance is particularly prone to the problem of too much redundant information, even though there's so little room available for assistance. One contributor to this problem is that the user interface sometimes requires or enforces a certain amount of assistance out of a mistaken notion of consistency.

For example, many user interfaces are coded to require hover help on all controls or no controls rather than allowing writers to appropriately pick the controls that require more explanation. When you can't negotiate a change to the technology, you might end up writing redundant assistance to fill the provided space, as shown in the following hover help:

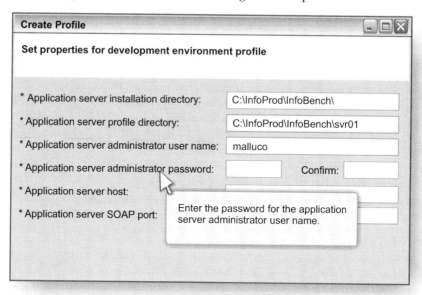

This practice has several unfortunate consequences:

- Writing teams waste time and are often confused about what to write, so they write redundant hover help.
- Testers waste time testing additional assistance.
- Translators waste time and money translating unnecessary text.
- Most importantly, users learn not to bother with hover help because there's nothing useful there.

When you repeat information unnecessarily, you not only interrupt and delay the users' task, but you might also confuse users and waste space—especially in mobile interfaces, where space is very limited.

The hover help for the following field shows how you can fill space with redundant information without answering users' questions:

**Original**

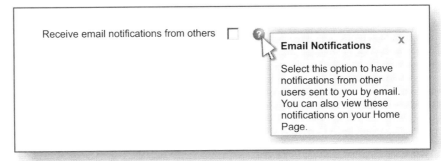

The field label is clear about the fact that notifications are received by email, but users are left wondering what notifications are. The hover help, rather than answering this essential question, repeats the information from the field label both in its title and first sentence. The second sentence is interesting new information, but it is obscured by the redundant information.

**Revision**

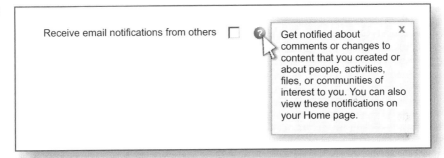

The revised hover help eliminates the space-consuming title and the information that is redundant with the label. Instead it answers the essential question about notifications and then provides the additional helpful detail from the original hover help.

# Completeness checklist

When you know your users and what they are trying to achieve, you can determine both the right information to provide and the right amount to provide.

From the user's point of view, completeness means that all of the required information is available, beginning with the information in any user interfaces. Well-designed user interfaces, which include helpful embedded assistance that follows a pattern of progressive disclosure, can reduce the amount of separate information that users need to achieve their goals. Information is complete when it covers all of the subjects that users need to understand to successfully achieve their goals. Complete information describes subjects at the right level of detail for the audience and the audience's experience level.

You can use the following checklist in two ways:

- As a reminder of what to look for, to ensure a thorough review
- As an evaluation tool, to determine the quality of the information

| Guidelines for completeness | Items to look for | Quality rating<br>1=very dissatisfied,<br>5=very satisfied |
|---|---|---|
| Make user interfaces self-documenting | • Common tasks are easy to find and accomplish in the user interface without documentation.<br>• The user interface uses effective and consistent programmatic assistance. | 1 2 3 4 5 |
| Apply a pattern for disclosing information | • Expected embedded assistance mechanisms are available.<br>• Explanations of parallel information provide a similar level of detail.<br>• Embedded assistance follows the pattern established for the product. | 1 2 3 4 5 |
| Cover all subjects that support users' goals and only those subjects | • Information contains no extraneous topics such as product internals.<br>• Embedded assistance and topics cover the information that users need to achieve their goals.<br>• The amount of conceptual information is appropriate for the audience.<br>• Information that doesn't directly support a user task includes an explanation for why that information is necessary. | 1 2 3 4 5 |

| Guidelines for completeness | Items to look for | Quality rating<br>*1=very dissatisfied,*<br>*5=very satisfied* |
|---|---|---|
| Cover each subject in only as much detail as users need | • Explanations contain the right amount of detail.<br>• Procedures or tasks are complete; no steps are missing.<br>• Only the most common uses are described in detail.<br>• New concepts are sufficiently developed.<br>• Introductory information is sufficiently detailed for the audience. | 1 2 3 4 5 |
| Repeat information only when users will benefit from it | • Information is repeated when it is important for the user to stay on task or establish context.<br>• Information does not include unnecessary repetition.<br>• User interfaces do not require or include unnecessary assistance. | 1 2 3 4 5 |

To evaluate your information for all quality characteristics, add your findings from this checklist to Appendix A, "Quality checklist," on page 545.

# PART 3
# Easy to understand

Users appreciate technical information that can be understood with as little effort as possible. After all, people don't read technical information for pleasure; they read it to accomplish a task quickly. Information that is easy to understand is clear, unambiguous, concrete, and consistent where it counts.

| | |
|---|---:|
| **Chapter 6. Clarity** | **153** |
| Focus on the meaning | 155 |
| Eliminate wordiness | 161 |
| Write coherently | 174 |
| Avoid ambiguity | 180 |
|     Use words as only one part of speech | 180 |
|     Avoid empty words | 183 |
|     Use words with a clear meaning | 187 |
|     Write positively | 189 |
|     Make the syntax of sentences clear | 194 |
|     Use pronouns correctly | 199 |
|     Place modifiers appropriately | 201 |
| Use technical terms consistently and appropriately | 205 |

|  |  |
|---|---|
| Decide whether to use a term | 205 |
| Use terms consistently | 207 |
| Define each term that is new to the intended audience | 210 |
| Clarity checklist | 212 |

## Chapter 7. Concreteness — 215

|  |  |
|---|---|
| Consider the skill level and needs of users | 220 |
| Use concreteness elements that are appropriate for the information type | 223 |
| Use focused, realistic, and up-to-date concreteness elements | 240 |
| Use scenarios to illustrate tasks and to provide overviews | 243 |
| Make code examples and samples easy to use | 247 |
| Set the context for examples and scenarios | 251 |
| Use similes and analogies to relate unfamiliar information to familiar information | 253 |
| Use specific language | 256 |
| Concreteness checklist | 259 |

## Chapter 8. Style — 261

|  |  |
|---|---|
| Use active and passive voice appropriately | 263 |
| Convey the right tone | 267 |
| Avoid gender and cultural bias | 273 |
| Spell terms consistently and correctly | 276 |
| Use proper capitalization | 280 |
| Use consistent and correct punctuation | 284 |
| Apply consistent highlighting | 296 |
| Make elements parallel | 302 |
| Apply templates and reuse commonly used expressions | 305 |
| Use consistent markup tagging | 311 |
| Style checklist | 314 |

CHAPTER 6

# Clarity

*One should not aim at being possible to understand, but at being impossible to misunderstand.*   —Quintillian

Clear information is information that users can understand the first time. They don't need to reread it to untangle grammatical connections, sort out excess words, decipher ambiguities, discover relationships, or interpret the meaning. Clarity in technical information is like a clean window through which people can clearly see the subject.

Clarity is important for native speakers of English and nonnative speakers alike. If a sentence, list, table, button label, or menu is confusing to native speakers, that element will likely be even more difficult for a nonnative speaker or translator to comprehend.

The first time that most writers try to write about something, words don't flow easily. Often, to break through a block, writers write as if they were speaking, but this approach introduces problems such as unnecessary words, vague referents, and long sentences. People can get visual and auditory cues when you're speaking, but written words must convey a message on their own.

Clear information is mainly the result of rewriting, which requires replacing, adding, and deleting parts to achieve clarity. Clear information requires close attention to words, phrases, sentences, lists, and tables—whether that information is in documentation, in user interfaces, or on hardware—to make sure that each word contributes appropriately to the message.

Clarity provides the rationale for many decisions about style. For example, you might choose bold for certain uses of words, and you might choose italics for certain other uses. Consistently applying these style decisions enhances clarity by making information easier for users to understand.

To make information clear, follow these guidelines:

- **Focus on the meaning**
- **Eliminate wordiness**
- **Write coherently**
- **Avoid ambiguity**
- **Use technical terms consistently and appropriately**

# Focus on the meaning

The most important goal of information to get right is the meaning. To make information meaningful, you must remove the wordiness and grammar problems, but you also need to pay attention to what the information is supposed to convey. Even a sentence that has no grammatical, punctuation, or ambiguity problems can still confuse users if that sentence is long, complex, or convoluted.

Adding too much content into one sentence adds an extra cognitive load on users and forces them to slow down as they read. In technical information, sentences can sometimes grow because you need to be descriptive and precise. However, you must balance the need for precise descriptions with how much users can reasonably be expected to understand in one sentence or one paragraph.

## Proximity of verbs to their subjects

When too many words are added between a subject and its verb, the sentence can be difficult to understand. Users must remember the subject, comprehend all the subsequent words, connect the subject to its verb, and then decipher the relationships between all the words. You can help users focus on the meaning of your information by keeping your subjects close to their verbs.

Can you find the subject and verb of the main clause in this sentence?

**Original**

> The procedure to enable searching for text in documents that are archived in InfoDBase Manager after you install the text search support component includes steps with predefined settings.

The original sentence has 21 words between the subject (*procedure*) and its verb (*includes*).

**Revision**

> The procedure to enable text search includes steps with predefined settings. If you enable the text search component, you can search for text in documents that are archived in InfoDBase Manager.

In the revision, the first sentence has only four words between the subject and verb. The ideas are broken into two sentences, which makes the passage considerably easier to understand. Although there's nothing wrong with using prepositional phrases and dependent clauses, take care to keep subjects reasonably close to their verbs by moving intervening phrases elsewhere.

The following sentence contains a long *which* clause between the subject and verb:

**Original**

> An electronic document, which might be a single file, a digital photo, or a set of related files that can be treated as one object (for example, an email message and attachments), can be stored in object stores and other repositories.

The *which* clause adds a whopping 29 words between the subject *document* and its verb *can be stored*.

**Revision**

> An electronic document is a single file, a digital photo, or a set of related files that can be treated as one object. For example, a set of files might be an email message and its attachments. You can store electronic documents in object stores and other repositories.

In the revision, several shorter sentences are used so that the subjects stay closer to their verbs.

## Long gerund phrases

Another way to make a sentence more difficult to read is to use a very long gerund phrase as the subject of the sentence. Gerunds, which are verb forms that end in *-ing* and are used as nouns, can be especially difficult to translate. Many languages don't have quite the same grammatical structure as the English gerund. For example, the English gerund must often be changed into an infinitive to be translated into Italian.

The following passage contains a long gerund phrase as the subject of the sentence:

**Original**

> Importing data by using a CSV file from the database repository is required if you do not use the InfoImport tool.

The full subject of the sentence is *Importing data by using a CSV file from the database repository*, which is 11 words. Even though the sentence is not that long, the long subject makes the user wait to be able to understand what the sentence means.

**Revision**

> If you do not use the InfoImport tool, you must import data by using a CSV file from the database repository.

The revision replaces the long gerund phrase as the subject with a straightforward subject and verb (*you must import*), which makes the sentence easier to understand and translate.

## Long coordinated phrases and clauses

One way to obscure the meaning of a sentence is to string together lots of words, phrases, or clauses joined by *and* or *or*, which are called coordinate conjunctions. To keep users engaged and to focus on the meaning, reduce the number of coordinated phrases by using lists or shorter sentences.

**Original**

> You must configure categories that affect functions, including calculation of facility and system condition values, configuration of results data, creation of tasks and projects from submitted planned work requests, descriptions of building system condition or risk, and identification of opportunities or deficiencies, before you use the Building Manager and Blueprint Reviewer tools.

The original sentence, which has 52 words, is mostly one long list of items connected by *and*.

# Easy to understand

**Revision**

> Before you use the Building Manager and the Blueprint Reviewer tools, you must configure categories. Those categories can affect how you do the following tasks:
> - Calculating facility and system condition values
> - Configuring results data
> - Creating tasks and projects from planned work requests
> - Describing building system condition or risk
> - Identifying opportunities or deficiencies

In the revision, the numerous coordinated phrases are replaced by an unordered list and an introduction to that list, which makes the information much easier to scan.

Meaning is also obscured if a sentence contains too many ideas joined by the word *or*.

**Original**

> To normalize data from month to month and year to year to see whether the carbon footprint is affected by global warming or daily temperatures or if the data reflects an inefficiency in the HVAC equipment or any of the assets that are located in a facility, you create climate logs for each facility.

The sentence is introduced by two long phrases joined by *or*:

> *To normalize data from month to month and year to year to see whether the carbon footprint is affected by global warming or daily temperatures*

> *if the data reflects an inefficiency in the HVAC equipment or any of the assets that are located in a facility*

And each of those phrases contains more phrases joined by *or* and *and*.

**Revision**

> To determine whether the carbon footprint is affected by global warming or daily temperatures, you can normalize data month to month and year to year. If you normalize the data, you can also determine whether the HVAC equipment or other assets in a facility are inefficient.

To make the sentence easier to read, the revised passage uses two shorter sentences, which also moves subjects closer to their verbs and reduces the cognitive load.

## Too many modifying phrases and clauses

Technical information must be precise. However, if a sentence contains too many modifying phrases or clauses, the information is difficult to comprehend. Shorter is better, but creating short sentences in technical information can be challenging.

Be careful about stuffing sentences with too many relative clauses, which are clauses that start with *who*, *which*, or *that*.

**Original**

> Use the **crawlSecure** command, which you can run from the index server, to start crawlers that will index websites that are protected by a firewall.

The original sentence starts simply enough, but it then uses three relative clauses.

**Revision**

> Use the **crawlSecure** command to start one or more crawlers to index websites that are protected by a firewall. Run the command from the index server.

The revised passage turns two of the relative clauses (*that* clauses) into a simple clause and a short phrase:

| Original | Revision |
|---|---|
| *which you can run from the index server* | *Run the command from the index server* |
| *that will index websites* | *to index websites* |

Avoid truncating relative clauses to participles because relative clauses are easier to translate. For example, don't change *that are protected by a firewall* to *protected by a firewall* even though the second phrase is shorter.

# Easy to understand

Prepositional phrases can also add too much modification in a sentence. Prepositions are necessary for connecting and modifying words and phrases, but too many in one sentence can make the sentence difficult to comprehend.

**Original**

> To address long response times during periods of heavy traffic on the Support site, the team studied the effects of keyword placement in the company's web pages.

The original sentence has five prepositional phrases:

*during periods*
*of heavy traffic*
*on the Support site*
*of keyword placement*
*in the company's web pages*

**Revision**

> To address long response times when Support site traffic was heavy, the team studied the effects of keyword placement in web pages.

In the revised sentence, two of the prepositional phrases (*during periods* and *of heavy traffic*) were converted to a clause, which has a subject and verb: *when Support site traffic was heavy*. Also, *on the Support site* was reworded to avoid another prepositional phrase. The revised sentence contains only two prepositional phrases.

# Eliminate wordiness

Wordiness buries the meaning, adds to translation costs, reduces customer satisfaction with your product, and wastes users' time. Wordiness is even more detrimental in user interfaces, where you have less room to explain concepts or provide instructions. To help users understand your information and complete their tasks, make every word count. Eliminate words that do not contribute to the meaning.

## Roundabout expressions

Roundabout expressions, such as *due to the fact that*, use several words where one or two or none will do. In speaking, people often use extra words to gain time to think about the next thing to say. In writing, however, these extra words hamper the meaning.

How many roundabout expressions can you find in the following sentence?

**Original**

> Given the fact that you have created an object, it can operate in a manner independent of the other objects based on the same class.

The original sentence uses several roundabout expressions:

*given the fact that*
*in a manner*
*based on*

The sentence also uses a long verb phrase (*have created*), which is unnecessary and adds to the wordiness.

**Revision**

> After you create an object, it can operate independently of other objects of the same class.

The revision cuts away the excess words and more succinctly gets to the point.

Easy to understand

The following roundabout expressions are wordy and can often be reduced to one or two words:

| Roundabout expression | Concise term |
|---|---|
| a great many | many |
| a variety of | various, different, many, several |
| as long as | if |
| as well as | and |
| at present, at the present time | now |
| at this (that) point in time | now, then |
| despite the fact that | although, even though |
| due to the fact that | because |
| during the course of | during |
| for the most part | usually |
| given the condition that | if |
| in an efficient manner | efficiently |
| in case of a | in a |
| in conjunction with | with |
| in many cases | often |
| in order for | for |
| in order to | to |
| in spite of | despite |
| in the event that | if, when |
| is possible that | maybe, perhaps |
| of an unusual nature | unusual |
| on account of the fact that | because |
| once in a while | occasionally, sometimes |
| on the other hand | however, conversely, alternatively |
| quite a few | several |
| whether or not | whether |
| with regard to | about |

The expletive construction is another type of roundabout expression. An *expletive* is a grammatical construction that places the subject of the sentence after the verb, for example, "There are expletives in technical information." English has three expletives: *there is/are*, *here is/are*, and *it is*. Although expletives are grammatically correct and useful in some circumstances, they are wordy, contain weak verbs, and make your ideas less direct.

Pairing an expletive with a relative clause is a common problem that is easy to fix. For example, *there are* is an expletive in the following sentence, and that expletive is followed by a relative clause (*that are available for late-model engines*):

**Original**

> There are several types of turbo chargers that are available for late-model engines.

**Revision**

> Several types of turbo chargers are available for late-model engines.

The sentence is much more succinct when the expletive and the relative pronoun (*that*) are deleted.

You might not be able to remove every expletive, but you should weed out as many as possible, especially the ones that occur with the relative pronouns *who*, *which*, and *that*.

## Redundancies

Redundancies occur when you say the same thing twice, such as *group together* or *step-by-step procedure*. To keep your writing succinct, eliminate redundant text.

**Original**

> The DXD specification describes the minimum records management requirements that must be met based on current regulations.

Easy to understand

In the original passage, the idea of a requirement is repeated: a requirement is the same thing as something that must be met.

**Revision**

> The DXD specification describes the minimum records management requirements based on current regulations.

You can correct the redundancy by removing one of the repeated items. In the revised sentence, the relative clause *that must be met* is removed.

The following words are often used together, especially in speaking, but their meaning is so close that one word is enough.

| Redundancy | Succinct expression |
| --- | --- |
| adequate enough | sufficient, enough, adequate |
| advance planning, warning, or preview | planning, warning, preview |
| by means of | by |
| cancel out | cancel |
| combine together, connect together | combine, connect |
| create a new | create a |
| end result | end, result |
| entirely complete | complete |
| every single | every, single |
| exactly the same | the same |
| grow or increase in size | grow, increase |
| group together | group |
| integral part | part |
| involved and complex | complex |
| is currently | is |
| new innovation | innovation |
| one and only one | one |

# Clarity

| Redundancy | Succinct expression |
|---|---|
| period of time | period, time |
| plan in advance | plan |
| reason...is because | reason is that..., ...because... |
| red in color | red |
| refer, reply back | refer, reply |
| repeat again | repeat |
| round in shape | round |
| share in common | share |
| sequential steps | steps |
| subject matter | subject |
| summary conclusion | summary, conclusion |

User interfaces can also suffer from redundancies. For example, for input fields, menus, tables, or other areas where users enter or select data, you don't typically need to add a verb to tell users that they should do some action.

**Original**

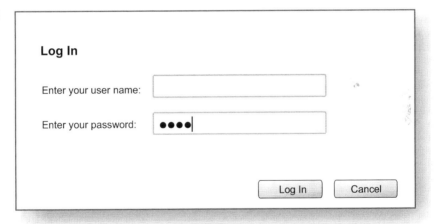

# Easy to understand

**Revision**

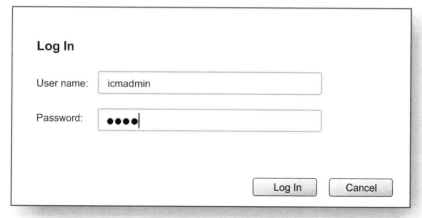

Users of software products rarely need a description of the action that they must take to enter or select something. A verb such as *enter* for a field label is often redundant. Additionally, removing the extra verb saves space, which is crucial for mobile device interfaces.

Logically circular statements are another form of redundancy. *Circularity* means that a statement starts with a premise but then ends by repeating that premise.

**Original**

> 3. If you want to specify an attribute restriction, specify the attribute to restrict.

The original sentence seems to chase its own tail by stating the same notion twice: that if you want a restriction, you specify a restriction. But that makes no sense.

**Revision**

> 3. Optional: Specify attributes to restrict.

By replacing the conditional statement (*If you want to specify an attribute restriction*), which is really an option, with *Optional*, the revised sentence removes the circularity.

## Unnecessary modifiers

Unnecessary modifiers, such as *very* or *actual*, sometimes appear in technical information because of the way people speak. However, these words take up valuable space and make users wade through unnecessary words.

Some unnecessary modifiers are used to intensify or add drama to writing, but users of technical products don't need drama: they need facts. If you write marketing information, some of the intensifying modifiers might be worth keeping, such as *highly*. But be careful not to overrun your information with these modifiers because they can quickly lose their impact.

The following modifiers are often unnecessary in technical information:

| | | |
|---|---|---|
| absolutely | definitely | particularly |
| actual | easy | perfectly |
| actually | easily | quite |
| all | existing | really |
| any | fairly | significant |
| basic | just | significantly |
| basically | highly | simply |
| certainly | modern | specific |
| completely | of course | specifically |
| current | overall | totally |
| definite | particular | very |

Such words are meant to intensify a word or phrase, but your information is clearer without them. When you consider using a modifying word, try the sentence in your mind with and without it; think about the effect on the meaning of the sentence and surrounding sentences. If the word adds no value, remove it.

**Original**

> Enter the actual values for your environment.

# Easy to understand

**Revision**

> Enter the values for your environment.

The word *actual* is often used for emphasis in speech, but it adds no meaning to the sentence and so should be removed.

Unnecessary modifiers should also not be used in other areas, such as titles, headings, user interfaces, lists, or tables.

**Original**

In the title of the original dialog (*Change Current Password*), the word *Current* is unnecessary. What other password would a user change?

Revision

The revised dialog title omits the modifying word *current*, which doesn't contribute any useful information.

## Imprecise verbs

A concise, powerful verb more clearly conveys your intended meaning and promotes a more active style. English speakers prefer reading forceful verbs because those verbs are typically engaging and unambiguous.

Weak verbs, on the other hand, are sometimes wordy and less direct. Weak verbs come in several forms: some are hidden in passive sentences; others are buried beneath weighty nouns; and others are simply longer variations of other shorter verbs.

Some weak verb phrases are formed by a passive verb followed by an infinitive, such as *is configured to do something*. Although a passive verb can sometimes be appropriate, you should generally use more powerful, active verbs to improve clarity and reduce wordiness.

# Easy to understand

Be wary of passive sentences that might be hiding precise verbs, for example:

| Imprecise passive verb | Precise verb |
|---|---|
| The system *is configured to run* on Windows. | • The system *runs* on Windows.<br>• The system *can run* on Windows. |
| The software *was created to process* record policies. | The software *processes* record policies. |
| InfoDBase *is designed to reduce* the need for archiving resources. | InfoDBase *reduces* the need for archiving resources. |
| The project manager *is designated to schedule* the meetings. | • The project manager *schedules* the meetings.<br>• The project manager *must schedule* the meetings.<br>• The project manager *will schedule* the meetings. |

The revised sentences are shorter, but another benefit of using the stronger, more concise verb is that the sentence stays focused on what's important. For example, the point is not that something *is designed to do* some task, but rather that it *does* some task.

Turning a strong verb into a noun plus a weak verb is called *nominalization*. For example, *making a decision* pairs the weak verb *make* with the noun *decision*. The more concise and direct way is to use the strong verb *decide*. Nominalizations are wordy, can obscure your meaning, and make the information less engaging.

**Original**

> After the inspectors perform an examination of the testing equipment, they must do a verification of the new safety guidelines.

The original sentence uses the following nominalizations:

*perform an examination*
*do a verification*

Clarity

**Revision**

> After the inspectors examine the testing equipment, they must verify the new safety guidelines.

The original sentence hides two powerful verbs: *examine* and *verify*. Removing the nominalizations not only reduces the number of words and makes the writing more vibrant, but it improves the tone. Too many nominalizations can create a pretentious or overly formal tone.

Watch for nominalizations that use weak verbs such as *be*, *have*, *perform*, *make*, and *give*. Replace the nominalization with a more precise verb as shown in the following table:

| Imprecise verb | Precise verb |
| --- | --- |
| be in agreement | agree |
| be capable of | can |
| carry out an inspection of | inspect |
| conduct an investigation of | investigate |
| do a verification of | verify |
| draw a conclusion | infer, conclude |
| give an answer | answer |
| give rise to | cause |
| have a requirement | require |
| have knowledge of | know |
| have plans | plan |
| have the ability to | be able to, can |
| have the capability to | can |
| hold a meeting | meet |
| keep track of | track |
| make a decision | decide |
| make a distinction | distinguish |

**171**

# Easy to understand

| Imprecise verb | Precise verb |
|---|---|
| make a proposal | propose |
| make a suggestion | suggest |
| make changes to, make a change | change |
| make contact with | meet, contact |
| perform the installation | install |
| provide assistance | help |
| render inoperative | break |
| serve to define | define |
| show improvement | improve |

For less complex, single-word verbs, the English language often has two or more verbs that have the same meaning. Often the longer verb derives from Latin, such as *assist*, and the shorter word derives from Anglo-Saxon, such as *help*. When you have a choice, use the shorter verb, which is usually more direct:

| Word derived from Latin | Word derived from Anglo-Saxon |
|---|---|
| accomplish | do |
| construct, fabricate | make |
| discover | find, realize |
| enumerate | list |
| initiate, commence | begin, start |
| locate | find |
| perform | do |
| possess | have |
| present | give, show, tell |
| require | need |
| retain | keep |
| terminate | end |
| transmit | send |
| utilize, employ | use |

You might be tempted to replace some Latin-derived words with phrasal verbs, which consist of a main verb plus a preposition, for example:

| Word derived from Latin | Phrasal verb |
| --- | --- |
| accelerate | speed up |
| align | line up |
| complete | fill in, fill out |
| configure | set up |
| defer | put off |
| disconnect | cut off |
| exclude | leave out |
| review | look at |

Phrasal verbs pose a problem for nonnative speakers of English because they can be difficult to learn and because they might have multiple definitions. Also, you can keep the parts of a phrasal verb together, or you can insert an object between the main verb and its preposition. For example, you can express the following idea in two ways:

- You can shut down the server.
- You can shut the server down.

This flexibility can make the sentence more difficult to translate. If you're writing information that will be translated or that will be used by nonnative speakers of English, avoid using phrasal verbs that might be confusing. However, some phrasal verbs, such as *check in/out*, *shut down*, and *turn on/off*, are so ingrained in technical information that you should continue to use them.

# Write coherently

Coherence broadly describes the quality with which ideas are connected and comprehensible to users. That comprehension is driven partly by how sentences and phrases are linked grammatically and by whether the ideas in a paragraph, illustration, or even a user interface window are logical, ordered, connected, and relevant.

Achieving coherence can be challenging in technical information because such information often states facts about objects, concepts, or tasks. However, those facts must be connected in such a way that users can see how each statement or paragraph relates to the one before it, the one after it, and the overall idea that those statements and paragraphs convey.

Sentences, paragraphs, lists, tables, illustrations, user interface text, and examples that are clear on their own don't necessarily result in clear information overall.

These elements must work together to emphasize the most important ideas that you want to convey. The ideas that are expressed in each element shouldn't have equal weight. Instead, some ideas should reinforce other ideas, and thus contribute to the overall clarity of the information.

You can make sentences and paragraphs coherent by:

- Connecting sentences by using transition words such as *next, because, however, so, therefore, also,* and *finally*
- Ensuring that each paragraph and topic focuses on one idea
- Subordinating or coordinating ideas appropriately

In text, the lowest level of coherence is the paragraph. Ensure that the flow and connections between sentences in a paragraph make sense and help the user understand the overall point.

**Original**

> The adjuster opens the **Properties** tab that is embedded in the editor. The adjuster enters information into a form that was set up by the company administrator. The form is a Word document, which was created from a template. Administrators can configure templates from the configuration manager, which is also where they set the permissions for who has access to the template. The adjuster completes the form. The adjuster adds the document to the company repository where all open claims are stored. Other employees can access the document.

The original paragraph is a series of statements that don't seem to be connected, and the paragraph digresses by describing how administrators create a template. The paragraph also doesn't effectively subordinate some ideas to others and therefore makes every statement of equal value.

**Revision**

> In the **Properties** tab, which is embedded in the editor, the adjuster enters information into a Word document that was set up by the company administrator. When the adjuster completes the form, the adjuster then adds the document to the company repository. Now, other employees also can access the document.

The revision improves the coherence of the paragraph in several ways:

- Each sentence is logically connected to another with words such as:

  *then*

  *when*

  *now*

- Subordinate and relative clauses are used to show that some ideas are subordinate to others:

  *which is embedded in the editor*

  *that was set up by the company administrator*

  *when the adjuster completes the form*

- The paragraph stays on point by omitting the information about where administrators configure templates.

Just as you should connect one sentence to another, you should also connect one paragraph to another. Sentences and paragraphs must also logically integrate with and transition to lists, tables, examples, and illustrations.

For example, to connect sentences in a paragraph, you use transitional words or phrases at the start of sentences, such as *for example*, *therefore*, *however*, and *as a result*. To integrate lists, tables, illustrations, or examples, you can add a sentence to introduce them, and you can provide meaningful captions for illustrations and tables.

# Easy to understand

**Original**

> ### Placing objects by using a grid
>
> A grid is a set of horizontal and vertical lines that appear on InfoProduct forms. A grid is active only when objects are placed or moved on forms.
>
> You can display a grid on forms to help you place components. You can control the size and visibility of the grid lines. The grid is useful when you draw or place components as they are created or when you move and resize them.
>
> If you want the grid lines to be visible, set the Visible property to True. To change the space between grid lines, enter new values in the Height and Width fields. These values measure the distance in twips (twentieth of an inch point) between grid lines.
>
> *Table 1. Properties of a grid*
>
> | Property | Possible values | Default | Description |
> | --- | --- | --- | --- |
> | Active | True | True | Makes an object align with the grid lines |
> |  | False |  | Lets you place an object anywhere on the form |
> | Visible | True | True | Makes the grid visible |
> |  | False |  | Makes the grid invisible |
> | Height | 45 to 1485 twips | 5 pixels | Sets the vertical space between lines |
> | Width | 45 to 1485 twips | 5 pixels | Sets the horizontal space between lines |

Taken individually, each sentence in the original topic is clear. However, the paragraphs are not clear, and the table is not integrated. Judging by the first sentence in each paragraph, users might surmise that the first paragraph describes a grid, the second paragraph explains what users can do with a grid, and the third paragraph tells them how to use the properties. However, a closer inspection shows the flow of ideas:

**Paragraph 1**
- Definition of grid
- Restriction about when the grid is available

**Paragraph 2**
- Usefulness of a grid
- Properties of a grid that you can customize (or control)
- Usefulness of a grid

**Paragraph 3**
- How to make grid lines visible
- How to change the space between grid lines
- Measurement of the space between grid lines

**Table**
- How to make grid lines active
- How to make grid lines visible
- How to set the vertical space
- How to set the horizontal space

As you can see, some ideas are unnecessarily repeated, some ideas aren't logically connected, and some ideas aren't subordinate to other ideas.

Easy to understand

The following revision integrates the information more coherently while avoiding repetition:

**Revision**

> **Placing objects by using a grid**
>
> You can display a grid, which is a set of horizontal and vertical lines, on forms to help you place objects.
>
> You can control the following attributes of the grid:
> - Whether objects automatically align with the grid
> - Whether the grid is visible
> - The size of the vertical and horizontal space between grid lines
>
> You will probably want the grid to be active and visible so that you can use it as you create objects or move them. However, objects will also align with the grid when it is active and invisible.
>
> The following table provides the possible values for the properties of the grid.
>
> *Table 1. Properties of a grid*
>
> | Property | Possible values | Default | Description |
> |---|---|---|---|
> | Active | True | True | Makes an object align with the grid lines |
> |  | False |  | Lets you place an object anywhere on the form |
> | Visible | True | True | Makes the grid visible |
> |  | False |  | Makes the grid invisible |
> | Height | 45 to 1485 twips* | 5 pixels | Sets the vertical space between lines |
> | Width | 45 to 1485 twips | 5 pixels | Sets the horizontal space between lines |
>
> * A twip is a unit of measure that represents one-twentieth of a point. There are 72 points in an inch (or 72 points in a centimeter), 1440 twips in an inch, and 567 twips in a centimeter. Use twips when you need a measurement that is independent of the screen resolution.

The flow of ideas in the revision is:

**Paragraph 1**

- Usefulness of a grid (including definition)

**Paragraph 2 and list**
- What attributes of a grid you can control

**Paragraph 3**
- Relationship of active and visible
- Relationship of active and invisible

**Paragraph 4**
- Introduction to the table

**Table**
- How to make grid lines active
- How to make grid lines visible
- How to set the vertical space
- How to set the horizontal space

The revision includes necessary information without being redundant. It includes the definition of a grid but emphasizes the value of a grid to the user. The revision also includes the definition of *twip* as a note to the table rather than as a sentence in one of the paragraphs before the table; the definition is important only in the context of the information in the table.

The revision relates the text to the table by using a list to emphasize the meaning of the properties. The list describes all of the properties, and then a paragraph provides more detail about the relationship between the Active and Visible properties. This detail is warranted because the relationship is not obvious.

You'll probably need to use fewer transitions in reference information, which users tend to scan rather than read linearly, or in user interfaces, where space is at a premium. In these situations, use effective headings and labels to change a subject or to express the purpose of a chunk of information.

Easy to understand

## Avoid ambiguity

This guideline could be called "Have mercy on translators and nonnative speakers." These people are more likely than native speakers to have difficulty when information is ambiguous. As of this writing, only 25% of Internet users are English speakers, but over half of all website content is in English.

Ensure that your content is free from ambiguity by following these guidelines:

- ❑ Use words as only one part of speech
- ❑ Avoid empty words
- ❑ Use words with a clear meaning
- ❑ Write positively
- ❑ Make the syntax of sentences clear
- ❑ Use pronouns correctly
- ❑ Place modifiers appropriately

### Use words as only one part of speech

When a word is used as more than one part of speech, users and translators can get confused, especially when the word is used as more than one part of speech in the same sentence or paragraph. Similarly, technical terms such as command names or user interface controls that are used as if they were plain English can be confusing. To make your information clear and easy to translate, use words as only one part of speech.

#### Words that can be nouns, verbs, or adjectives

Each of the following words can be used as a verb, noun, or adjective:

| branch | fix | log | search |
| catalog | function | move | set |
| change | index | name | team |
| check | input | output | trigger |
| copy | issue | permit | type |
| design | key | port | upgrade |
| display | link | process | use |
| document | list | record | version |
| file | load | register | work |

Developing Quality Technical Information

Although the following sentence is short, you might need to read it several times to understand it:

**Original**

> Record the date of the record.

In the original sentence, the first instance of *record* is a verb, and the second instance is a noun. But the repeated word makes the sentence awkward.

**Revision**

> Enter the date of the record.

The revision keeps the noun (*the record*) but changes the verb (*enter*). In general, you can more easily find a synonym or similar word for a verb than a noun. For example, the verb *record* could mean *register*, *write*, *enter*, or *remember*. However, finding a synonym for the noun is more difficult. Finding a synonym won't make the information inaccurate, but using a synonym in place of the original word might.

In the following sentence, the nouns are hard to distinguish from the verb:

**Original**

> The team documents design changes.

**Revision**

> The members of the team document the changes in the design.

Easy to understand

The revision inserts articles before the nouns (*changes* and *design*) and expands the noun string (*design changes*), making the verb (*document*) more obvious. The change in the subject to *members of the team* also helps reduce the possible interpretation that *team* modifies *document*.

**Keywords as plain text**

Another common problem in technical information is to use technical keywords, such as commands, API names, parameters, file names, or user interface controls, as if they were normal English words. Technical keywords should generally be treated as proper nouns or adjectives.

You can cause unexpected clarity problems, especially for translation, when you use these terms as other parts of speech, such as using them as verbs. For example, you wouldn't take the proper noun *Michelle* and turn it into a verb as in *Michelled* or *Michelling*.

The following table shows some unclear use of commands, interface controls, and other technical terms:

| Keyword type | Incorrect use | Correct use |
|---|---|---|
| Button name | You can specify that your scanner is **Active** if you want to scan all files. | • You can specify that the scanner is active if you want to scan all files.<br>• To scan all files, click **Active** on the Scanners page. |
| Command name | You can reinit the server after you upgrade the database. | • You can restart the server after you upgrade the database.<br>• You can restart the server after you upgrade the database by using the `reinit` command. |
| Command name (Uppercase) | After the database was upgraded, the personnel table was DROPped. | • After the database was upgraded, the personnel table was dropped.<br>• You can use the `DROP` command to drop the personnel table after you upgrade the database. |
| Technical term | To investigate a possible torn meniscus, the surgeon will arthroscope the knee. | • To investigate a possible torn meniscus, the surgeon will perform an arthroscopy on the knee.<br>• To investigate a possible torn meniscus, the surgeon will use an arthroscope on the knee. |

When you work with and develop technical products, watch out for technical terms that are being used in confusing ways. You and your team might understand the term, but that doesn't mean that users outside your organization will.

Be especially careful about trademarked terms. Changing a trademark in any way violates the trademark and can lead to legal problems. For example, don't alter words such as *Google* or *Xerox* to turn them into verbs or adjectives.

For more information about using technical terms, see "Use technical terms consistently and appropriately" on page 205.

## Avoid empty words

An *empty word* is a word that has no meaning until it's further qualified by other words or phrases. An empty word could mean almost anything, which means it conveys nothing at the same time. For example, the word *aspect* simply means a particular feature or part of some thing, but that definition doesn't convey anything concrete or familiar.

Empty words often make the information unclear, and they contribute to wordiness because they require more words to make the point. An empty word is sometimes convenient to use when you're writing a draft and can't yet find a more precise term. However, when you're ready to finalize your content, look for empty words and replace or eliminate them.

The following paragraph contains empty words that add no value:

**Original**
> One factor that can affect the degree of sharing is minimizing system unavailability. This aspect of the data-sharing design is influenced by factors such as requiring users to connect to servers that are in a specific geography.

The words *factor*, *factors*, and *aspect* in the original paragraph unnecessarily inflate the word count and possibly leave users guessing about what the *factor* and *aspect* refer to.

**First revision**
> Minimizing the time that the system is unavailable can affect the degree of data sharing. Data sharing is also affected by requiring users to connect to servers that are in a specific geography.

# Easy to understand

The first revision rids the paragraph of two of the empty words, but the paragraph still uses the empty word *degree*, which doesn't provide useful information.

**Second revision**

> You can improve the performance of data sharing by doing these tasks:
> - Minimizing the time that the system is unavailable
> - Requiring users to log in to servers that are in a specific geography

The final revision is shorter, unambiguous, and easier to read.

The words *basic* and *advanced* are commonly misused in technical information. These are empty words because *basic* and *advanced* are subjective qualities. To avoid overcrowding windows or dialogs, user interface designers use these terms to separate controls or functions. These vague terms are also sometimes used to make the dialog or section of a page purposely vague so that more functions can be added later without having to find a better location for the controls.

However, because users typically have no idea what constitutes a function that is basic or advanced, these words are effectively empty words.

**Original**

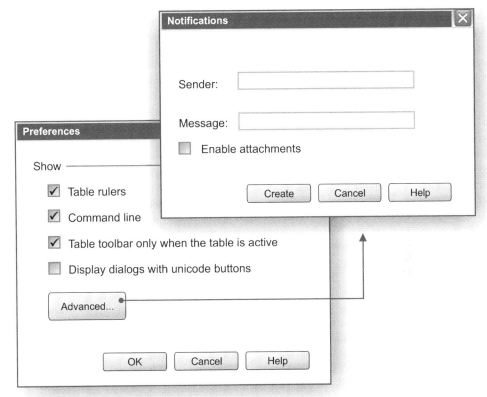

In the Preferences dialog, how would users know what happens when they click **Advanced**? When they do click that button, they see a dialog to create notifications. But how are notifications advanced? Also, the **Advanced** button label might lead novice users to ignore this function when it really does apply to the task that they need to complete.

# Easy to understand

**Revision**

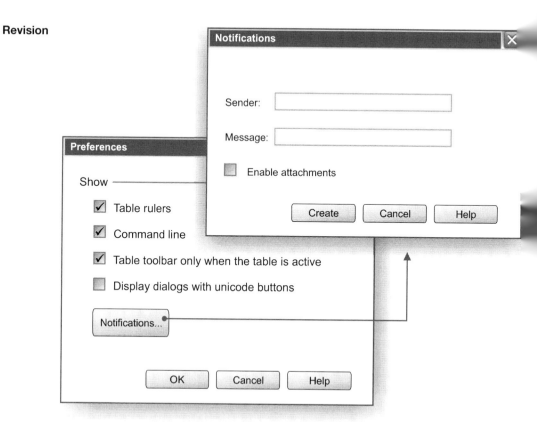

A button labeled **Notifications** is much clearer than the empty term **Advanced**.

The following table contains examples of common empty words that can be ambiguous or simply extraneous:

| Empty word | Original | Revision |
|---|---|---|
| area | Most problem areas are found when the database is indexed. | Most problems are found when the database is indexed. |
| by or in nature | The processor is fast by nature. | The processor is fast. |
| condition | The error condition occurs when too many users log in at the same time. | The error occurs when too many users log in at the same time. |
| factor | The resources factor can affect the schedule. | Lack of resources can affect the schedule. |

| Empty word | Original | Revision |
|---|---|---|
| feature | Configure the email archiving feature after you install the client software. | Configure email archiving after you install the client software. |
| function | You can use the text search function to find documents. | You can use text search to find documents. |
| process | The process of folder deletion might take one hour. | Deleting folders might take one hour. |

## Use words with a clear meaning

Because of the richness of the English language, the words that you choose might have more than one meaning. Some potentially ambiguous words are needed to convey information about relationships between clauses. Problems arise when words such as *since, once, as,* and *while* could indicate either time or something else, such as cause or contrast.

Words that have more than one meaning can be confusing when more than one meaning fits the context. You can reduce the confusion over which meaning a word has in a particular context by using the word consistently with only one of its meanings. For example, use *replace the part* to mean only *put the part back* and not also to mean *substitute a new part for the existing part*.

The use of *since* in the following sentence is ambiguous:

**Original**

> Since you created the table, you have authority to change the table.

Does the word *since* refer to when the table was created or to why the table was created?

**Revision**

> Because you created the table, you have authority to change the table.

A native speaker of English would probably recognize that *since* in the original sentence refers to cause rather than time. However, using the unambiguous word (as in the revision) is safer than relying on the context to help users make the correct choice. For example, with machine translation, when you translate the original sentence from English to French and German, the first part of the sentence might be translated as follows:

| English text | French translation | German translation |
|---|---|---|
| Since you created the table | Depuis la création de la table | Als Sie die Tabelle erstellt haben |
| Because you created the table | Parce que vous avez créé la table | Da Sie die Tabelle erstellt haben |

The first French and German translations indicate time, and if you translated those clauses back to English from French or German, you might get *Since the creation of the table*, which is the wrong meaning. If you translate the clause with *because*, the clause is translated more accurately.

If your content is translated by skilled translators, they might notice the subtle difference and translate the content appropriately. However, if you use only software to translate your content, you're likely to get inaccurate translations when you use such ambiguous words.

*May* and *can* can be confusing when they are used interchangeably. In technical information, *may* rarely has the meaning of permission, but it's sometimes used ambiguously to indicate possibility or capability, as in the following sentence:

**Original**

> Administrators may restrict project area users from logging in to the web application server.

Does this sentence mean that administrators have the ability, the permission, or the potential to restrict users? Users won't know unless you use *may* and *can* conscientiously in the surrounding information and thus prepare the user for the intended meaning, which should be about the ability of administrators to do the action.

**Revision**

> Administrators can restrict project area users from logging in to the web application server.

Reserve *may* for the permission to do something, *can* for the capability to do it, and *might* for the possibility of doing it. In most cases, you should use *can* or *might*. The word *may* is rarely needed in technical information and tends to be used more often in legal text.

The following list shows some words that can be ambiguous and their clearer alternatives:

| Ambiguous word | Clear alternative |
| --- | --- |
| as | because |
| as long as | if |
| in spite of | regardless of, despite |
| may | can, might |
| once | after, when |
| since | because |
| through | finished |
| while | although, whereas |

## Write positively

A sentence with more than one negative word in it can be hard to understand. If you put two negative words (such as *not unlike*) in a sentence, you make the sentence positive, as is true in mathematics. However, most people have more trouble understanding a sentence that has double negatives than an equivalent positive sentence.

**Original**

> A parent folder cannot be deleted if it is not empty.

# Easy to understand

**Revision**

> You can delete a folder only if it is empty.

To understand the original sentence, you have to do some mental jumping jacks to decide what the sentence is trying to convey. The positive revised sentence is more quickly understood.

Long negative sentences require even more effort to comprehend.

**Original**

> Do not install InfoConnect until you check that your computer does not have these conflicting programs.

The idea in the original sentence is difficult to grasp because of the two sets of negative expressions.

**First revision**

> Before you install InfoConnect, check that your computer does not have these conflicting programs.

The first revision changes the negative *not...until* in the first part of the sentence to a positive form, so the negative in the second clause is easier to understand. Telling users what to do is much stronger and clearer than only implying it by saying what not to do.

Some negatives mask a need to supply more information. The first revision might be fine for users who don't find any conflicting programs, but what about those who do?

# Clarity

**Second revision**

> Before you install InfoConnect, check your computer for these conflicting programs.

This second revision gets rid of *does not have* and sets the stage for information about what to do if you find the conflicting programs on your computer.

Negative expressions can take any of several forms. The more obvious ones are *no*, *not*, *none*, *never*, and *nothing*.

| Negative expression | Positive expression |
|---|---|
| does not accept | rejects |
| does not allow | prevents |
| does not have | lacks |
| not able | unable, cannot |
| not different | similar |
| not exclude | include |
| not include | exclude |
| not many | few |
| not possible | impossible, cannot |
| not the same | different |
| not...unless | only if |
| not unlike | like |
| not...until | only when |

# Easy to understand

Another common way for negatives to get into a sentence is through a negative prefix, which is attached to the beginning of a word.

| Negative prefix | Examples of words that use the prefix |
|---|---|
| de-, dis- | detach, deactivate, deemphasize, disentangle, disable, disagree |
| ex- | exclude, extinct, extinguish |
| in-, ir- | ineffective, inefficient, ineligible, irregular, irresponsible |
| non- | nonexistent, nonlinear, nonsurgical |
| re- | reduce, refuse, reject, reverse |
| un- | unavailable, undo, unlike, unpredictable, unrelated |

Some other negative words that are less obvious have negative meanings, although their form looks positive: *avoid, limit, wrong, fail, doubt*.

**Original**

> Transitions cannot occur between states that are inherited from different classes.

**Revision**

> Transitions can occur only between states that are inherited from the same class.

The revision provides the same information but avoids the negative construction (*cannot*).

The positive approach is also important in error messages, which tend to be written negatively.

**Original**

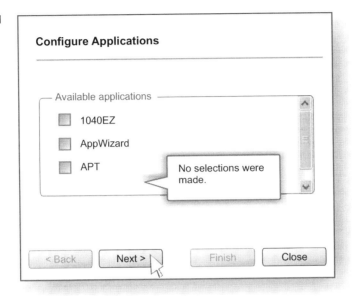

**Revision**

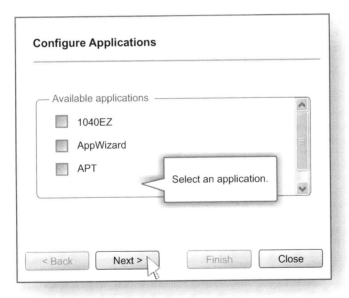

Rather than using *No selections were made*, the revised message can be more direct and clearer by using the positive, more active sentence *Select an application*.

However, some negative expressions are necessary and useful. For example, the most straightforward way to state a restriction or warning is by saying *Do not do...* or that some thing *cannot contain...* as shown in the following dialog:

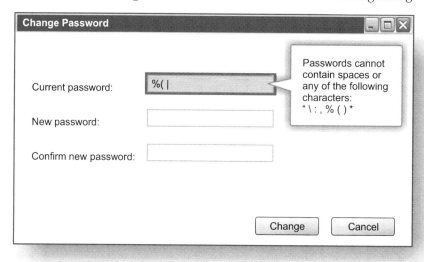

## Make the syntax of sentences clear

In casual or informal writing, words such as *that* or *the* are often dropped to save space. Sometimes sentences are further truncated because a verb or noun is dropped. This shortening of sentences is called an *elliptical style*. An elliptical style can impair readability for many users, especially for translators and nonnative speakers of English. Avoid the elliptical style in most technical information except in areas where brevity is required, such as tables or mobile apps, where you have less space to add text.

Some words, such as conjunctions, articles, prepositions, and relative pronouns, carry little meaning in themselves, but they glue together the more significant parts of a sentence. These *syntactic cues* help users, particularly translators and nonnative speakers, understand how the words in a sentence relate to each other. For example, an article (*the, a, an*) can help users recognize a noun, which might be a subject or an object in the sentence. These cues can help translators create more accurate translations.

Most machine-translation systems also depend on syntactic cues to analyze information. For example, it's common in English to drop the word *that* after verbs such as *ensure* or *make sure*. However, inserting or dropping *that* can create different translations:

| English text | German translation | German to English translation |
|---|---|---|
| Ensure you install the components. | Stellen Sie sicher, die Komponenten zu installieren. | Make sure to install the components. |
| Ensure that you install the components. | Stellen Sie sicher, dass Sie die Komponenten installieren. | Ensure [or make sure] that you install the components. |

For German, dropping the word *that* in the English sentence can create a nonsensical German sentence. When that German sentence is translated back into English, that English sentence is almost as nonsensical.

However, in some cases, you have little choice but to drop some words because you have limited space. These little words are sometimes left out of instructions, lists, glossary definitions, tables, and user interfaces.

Original

In the original user interface, the labels for the buttons **Open as a PDF file** and **Send as a Link** are clear in English, but they are too long for a button label. And when they're translated to other languages, the labels can double in length.

**Revision**

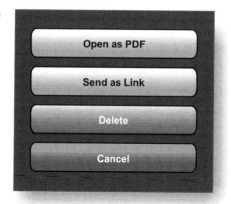

When space is limited, you can omit syntactic cues, but be sure to weigh the advantages and disadvantages. The shorter labels in the revision are clear in English, but be sure that the translated labels make sense in other languages.

The following verbs require the syntactic cue *that* when these verbs take a noun clause as the object:

| | | | |
|---|---|---|---|
| announce | explain | require | understand |
| assume | indicate | specify | validate |
| be sure | mean | suggest | verify |
| ensure | recommend | suppose | write |

**Original**

Verify different users can access the object store content.

In the original sentence, you might, at first glance, read *verify different users* as if these users needed to be checked for identification.

**Revision**

> Verify that different users can access the object store content.

With the addition of *that*, you can see that the action is to make sure that users can access the object store content, not that users need to be verified.

Perhaps the most troublesome deletion, even for native speakers of English, is *that* when it introduces a relative clause. Many writers have learned to leave it out, as in the following sentence:

**Original**

> Each illustration accurately depicts the function it was designed to illustrate.

**Revision**

> Each illustration accurately depicts the function that it was designed to illustrate.

The addition of *that* in the revised sentence helps clarify how the parts relate to each other without sounding stilted.

Consider what happens when the sentence becomes more complex:

**Original**

> Ensure that each illustration accurately depicts the object, concept, or function it was designed to illustrate and does so as simply as possible.

# Easy to understand

**Revision**

> Ensure that each illustration accurately depicts the object, concept, or function that it was designed to illustrate, and that it does so as simply as possible.

The revision inserts *that* just after the word *function* and after the word *and*. These simple additions make the sentence easier to read and translate.

Another way to make the syntax of sentences clear is to avoid sentence fragments. A sentence fragment is a group of words that is missing a subject, a complete verb, or both. For example, the following phrases are fragments:

*Invalid parameter*

*No file found*

*Search completed*

*No items to display*

Fragments are generally acceptable in tables, lists, tooltips, and table and graphic captions, where space is limited. However, you should avoid them in paragraph text and some types of embedded assistance, such as messages and static text, because fragments can be difficult to understand and translate.

Using fragments in error, warning, and informational messages can be especially problematic. If you don't make sentences complete, the message can be difficult to parse. The following warning message leaves out so many words that it's difficult to comprehend the real problem:

**Original**

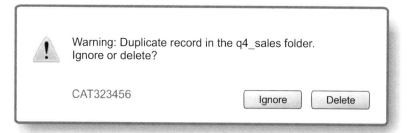

In the original message, several words were dropped, and there's no full verb phrase, which creates a sentence fragment. Dropping these words can slow reading speed because users must still mentally insert words to make the sentence readable.

Revision

Users are much more likely to read quickly if the words that glue a sentence together are present, as in the revised message.

## Use pronouns correctly

Incorrect use of pronouns can cause clarity problems for all users, especially translators. Ensure that pronouns clearly refer to the nouns that they replace and that you use *that* and *which* appropriately.

### Vague referents

A pronoun can be ambiguous if the noun that it replaces isn't immediately apparent. If this reference, or connection, isn't clear, the error is called a *vague referent*.

Original

> InfoDBase is a virtual database that you can create from the target file system. It consists of a set of statements that provide input to the projection utility.

The original passage is confusing: does the pronoun *it* refer to *InfoDBase* or to *the target file system*?

Revision

> InfoDBase is a virtual database that you can create from the target file system. This database consists of a set of statements that provide input to the projection utility.

The revision clarifies that the referent is the database.

# Easy to understand

Consider the referents for the pronouns *each* and *one* in the following sentence:

**Original**

> You can have multiple catalogs for a single source; however, each can access only one.

In the original sentence, the referents for *each* and *one* aren't clear. Some users might guess correctly what these words refer to, aided by the order of the nouns in the first clause. However, other users waste time deciding whether the clause might mean that *each source can access only one catalog* or *each catalog can access only one source*.

**Revision**

> You can have multiple catalogs for a single source; however, each catalog can access only one source.

By changing each pronoun to a modifier and repeating the nouns (*catalog* and *source*), you can remove any possible ambiguity.

A pronoun such as *it*, *this*, or *that* at the start of a sentence can cause confusion, as in the second sentence of the following paragraph:

**Original**

> Copy the file so that you can edit and save your version locally. This is especially important when InfoConnect is installed on a network and you want to run a customized version of the product.

The original paragraph refers to a noun that is only implied in the preceding sentence. There is nothing for the pronoun *this* to replace.

**Revision**

> Copy the file so that you can edit and save your version locally. A local copy is especially important when InfoConnect is installed on a network and you want to run a customized version of the product.

The revised passage replaces the pronoun *this* with the noun phrase *A local copy*, which eliminates the vague referent.

### *That* and *which*

Relative pronouns introduce relative clauses, which can be restrictive or nonrestrictive. *Nonrestrictive clauses*, or *nonessential clauses*, can be dropped from a sentence without changing the meaning. However, *restrictive clauses*, or *essential clauses*, can't be dropped from the sentence because they add information that's necessary for users to understand the intended meaning.

Use *that* with restrictive clauses and *which* (plus commas) with nonrestrictive clauses. For example, each of the following sentences means something different, and both can be correct:

| Sentence | Explanation |
| --- | --- |
| The indexing servers, which use InfoDBase JDBC drivers, have shut down unexpectedly. | This sentence means that all of the indexing servers have shut down, and all use InfoDBase JDBC drivers. |
| The indexing servers that use InfoDBase JDBC drivers have shut down unexpectedly. | This sentence implies that some of the indexing servers use InfoDBase JDBC drivers, and some do not. Only the servers with InfoDBase JDBC drivers are shutting down unexpectedly. |

Careful users and translators recognize the distinction between *that* and *which*. If you don't use these pronouns correctly, your users and translators might misinterpret your information.

## Place modifiers appropriately

The English language is governed by syntax, or word order. But English is also flexible, which means that you can place modifiers in different spots in a sentence and that you can turn nouns into adjectives. However, because of this flexibility, you must be careful about how you use and where you place modifiers.

### Dangling modifiers

A *dangling modifier* is a word, phrase, or clause that seems to modify a word that's not clearly stated in the sentence. Dangling modifiers are often created when an introductory participial phrase (*-ing* phrase) is combined with a passive sentence.

You can correct dangling modifiers in several ways. For example, use full clauses, which have a subject and verb. Also, be sure that the modifying phrase is next to the word that it's supposed to modify.

## Easy to understand

| Dangling modifier | Revision |
|---|---|
| *Assembling the processor*, the connections were damaged. | • Assembling the processor, we damaged the connections.<br>• When we assembled the processor, we damaged the connections.<br>• The connections were damaged when the processor was assembled. |
| An error occurred *creating or opening the directory*. | • An error occurred when the directory was created or opened.<br>• *More direct*: The directory cannot be created or opened. |
| *When installing InfoDBase Version 2.2.1*, your existing EmailTemplate.xsl file will be overwritten. | • When you install InfoDBase Version 2.2.1, your existing EmailTemplate.xsl file will be overwritten.<br>• When InfoDBase Version 2.2.1 is installed, your existing EmailTemplate.xsl file will be overwritten. |

### Misplaced modifiers

*Misplaced modifiers* occur when a modifying phrase is next to the wrong word. Unlike dangling modifiers, misplaced modifiers can generally be fixed simply by moving the modifying phrase to another place in the sentence.

| Misplaced modifier | Revision |
|---|---|
| Select the configuration server to which you want to link the list manager servers *in the Configure Server Name field*. | In the **Configure server name** field, select the name of the configuration server that you want the list manager servers to link to. |
| By changing some settings *that are based on your hardware environment*, you can improve performance. | To improve system performance, you can change various settings. The changes that you make depend on what hardware you are using. |
| You can create an optional task for your process *that can be manually started*. | For your process, you can create an optional task that can be manually started. |

The word *only* is perhaps the most common source of ambiguity in technical information. The word *only*, like many other adverbs, can modify almost any other word and can be placed in almost any position in a sentence. However, the position of *only* can radically affect the meaning of a sentence, as in the following sentences:

# Clarity

> *Only* an administrator can open the XML files in the dashboard.
>
> An administrator can *only* open the XML files in the dashboard.
>
> An administrator can open *only* the XML files in the dashboard.
>
> An administrator can open the XML files *only* in the dashboard.

Each sentence means something different, so be sure to put *only* next to the word or phrase that it's supposed to modify.

## Squinting modifiers

Another type of a misplaced modifier is called a squinting modifier. A *squinting modifier* occurs when you place a word (usually an adverb) between two other words, phrases, or clauses. Squinting modifiers seem to modify two items at the same time, which creates ambiguity.

| Squinting modifier | Revision |
| --- | --- |
| Programs that produce errors *consistently* require regular maintenance. | • Programs that consistently produce errors require regular maintenance.<br>• Programs that produce consistent errors require regular maintenance.<br>• If a program produces errors, it almost always requires regular maintenance. |
| Using the content cache *sometimes* improves the speed at which clients gain access to content. | • Using the content cache can improve the speed at which clients gain access to content.<br>• If you use the content cache occasionally, you improve the speed at which clients gain access to content. |
| Administrators can notify the support team *after several hours* that new users must reenter their credentials. | • After several hours, administrators can notify the support team that new users must reenter their credentials.<br>• Administrators can notify the support team that new users must wait several hours before they can reenter their credentials. |

## Easy to understand

### Ambiguous coordination

When you connect two or more items with a conjunction, such as *and* or *or*, be sure that the coordinating items are unambiguously modified. In some cases, you might need to repeat the modifier so that your meaning is clear.

Each of the following ambiguous sentences can be rewritten in different ways to have different meanings:

| Ambiguous sentence | Clear sentence |
| --- | --- |
| To run the ABS sensor and solenoid, use a 3.0 Amp fuse. | • To run the solenoid and the ABS sensor, use a 3.0 Amp fuse.<br>• To run the ABS sensor and ABS solenoid, use a 3.0 Amp fuse. |
| Use the InfoDB tool to index the database and save your settings. | • Use the InfoDB tool to index the database and then save your settings.<br>• Use the InfoDB tool to index the database and to save your settings. |

### Ambiguous noun phrases

In technical information, the names of products, components, utilities, features, tools, or technologies often use nouns to modify other nouns, such as *program request handler*, *attribute properties window*, and *network terminal option*. These noun combinations can cause confusion about what modifies what. For example, in the noun phrase *program request handler*, is the handler for program requests, or is the request handler for programs? Does the middle noun go with the first noun or the last noun, or are the three equal? When possible, rewrite phrases that have three or more nouns.

If the first pair of nouns in the phrase serves as an adjective for the third noun, consider using a hyphen, as in *character-set identifier*. If, however, the pair of nouns is established as having no hyphen, you need to observe that convention.

| Ambiguous sentence | Clear sentence |
| --- | --- |
| The following list describes the *control room reactor operator functions*. | • The following list describes the operator functions for the control room reactor.<br>• The following list describes the reactor operator functions for the control room. |
| The administrator must coordinate the *new web application server upgrade project*. | • The administrator must coordinate the project to upgrade the new web application servers.<br>• The administrator must coordinate the new project to upgrade the web application servers.<br>• The administrator must coordinate the project for the new upgrade of the web application servers. |

# Use technical terms consistently and appropriately

If you document a large number of products, services, or technologies, it's important that you manage terminology across your company or organization. Otherwise, it can be challenging for teams on different projects to keep terminology in sync. For example, one team who works on search technology might use the term *spider* to refer to a type of software application that crawls websites. Another team might use *crawler* to mean the same thing.

To ensure that terms are consistent, you should implement a terminology management system, which means creating a repository of terms and having a process to review, add, delete, or change terms as required by the product. Such a system could be a simple web or wiki page where you store terms, or it could be a more sophisticated database system. You can also invest in terminology management tools to scan your information for problematic terms.

Managing terminology in technical information helps you and your users in several ways:

- Reduces confusion or misinterpretations, which improves customer satisfaction
- Reduces inaccuracies in translations, which can reduce cost
- Makes writing teams more efficient, which means that each writer isn't inventing new terms or spending more time researching terms

To use technical terms consistently and appropriately, follow these guidelines:

- ❑ **Decide whether to use a term**
- ❑ **Use terms consistently**
- ❑ **Define each term that is new to the intended audience**

## Decide whether to use a term

Whenever you can reasonably avoid introducing a specialized term, do so. Unless a new term really helps users, it is just one more thing for them to learn.

When you must decide whether a term is necessary and appropriate, first ask these questions:

- Does a suitable term for the thing that you want to name exist? If a widely used and accepted term does exist, use it. You don't need to coin a new term, even if you're tempted to replace a technical term with a nontechnical term.
- Do you intend to use a term only once or twice? If so, you might be able to use a descriptive phrase instead.

# Easy to understand

- Are your users highly skilled administrators, or are they novices with little technical knowledge? Changing even the most unruly jargon can be confusing for your users if those users are already familiar with the jargon.

Are all the terms necessary in the following paragraph?

**Original**

> Each keyword describes one type of program failure. A set of keywords, which is called a *keyword string*, describes a specific problem in detail. Because you use a keyword string to search a database, a keyword string is also called a *search argument*.

The original paragraph introduces a term, *search argument*, that the user can probably get along without. *Search argument* is used in many other contexts, such as in web searches, programming languages, and database queries, so its meaning is already diluted.

**Revision**

> Each keyword describes one type of program failure. You use a set of keywords, which is called a *keyword string*, to describe a problem in detail and search a database for a similar problem.

The revised example uses generic words and focuses on the uses of the keywords rather than on adding to the user's vocabulary. Depending on the experience of the audience, you might also drop *keyword string* as unnecessary.

But sometimes your audience expects a more technical term. You might think that you must always use the simplest term; however, some users are so accustomed to a specific vocabulary that changing that vocabulary doesn't make sense.

For example, administrators who work with the z/OS operating system understand jargon such as *IPL* (to start the system), *abend* (an abnormal end or crash), and *terminate* (stop). Similarly, users who do frequent, complex database searches, such as paralegals or research scientists, are likely to understand terms such as *Boolean*, *stop words*, and *stemming*.

Writing information that excludes these terms makes your information more difficult to understand for these users. Therefore, ensure that you understand your users' vocabulary before you make terminology decisions.

## Use terms consistently

If you decide to use a term that your audience is not familiar with, carefully choose it, introduce it, and use it consistently. If your field or industry uses two or more terms for the same thing, pick one and use it consistently. In some types of writing, you are encouraged to use different words for the same thing for variety. In technical information, however, using more than one term for the same thing causes confusion and can lead to inaccurate translations. And using two terms can suggest that the two terms refer to different things.

Ensure that terms used in a user interface are consistent across all the components of the product and consistent with the documentation.

The following user interfaces are from a web interface and its mobile app counterpart:

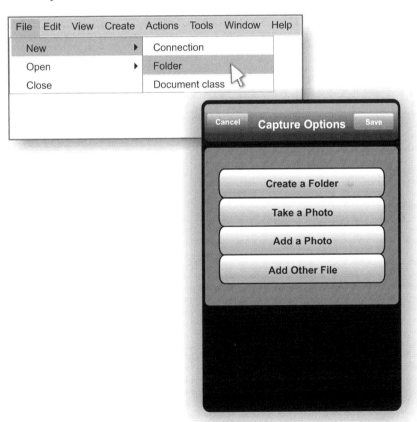

In the original interfaces, the web interface of this product uses the terms *New* and *Folder*, as shown in the cascading menu, to refer to the action of adding a new folder to a repository. However, the mobile app for the same product uses the button label *Create a Folder* to refer to the same action.

**Revision**

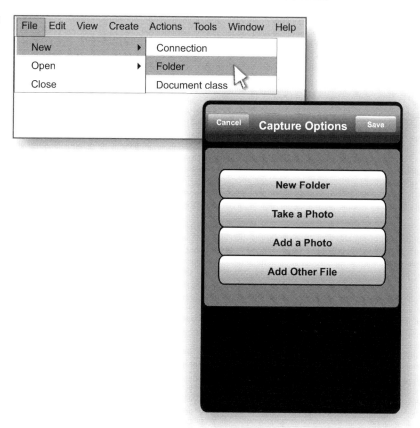

The revised interfaces use the same term (*New Folder*) consistently so that users who use both the mobile app and the web interface won't be confused by the different terms.

For complex or large products, such as enterprise database software, ensure that terminology is consistent across all topics, user guides, and websites.

**Original**

> The secure data model provides security features and properties for your repository. The secure installation is the most common data model for InfoDBase because it uses the minimal number of properties but provides a minimal level of security for database objects.

Two different terms are used even though they mean the same thing: *secure data model* and *secure installation*.

**Revision**

> The secure data model provides security features and properties for your repository. The secure data model is the most common data model for InfoDBase because it uses the minimal number of properties but provides a minimal level of security for database objects.

The revised passage uses the same term, *secure data model*, consistently.

Carefully choose terms that indicate distinctions among things. For example, in database technology, a decision was made many years ago to use the terms *physical records* and *logical records*. However, a better contrasting term for *physical records* might have been *nonphysical records* rather than *logical records*, which suggests illogical records as its opposite. Physical records and logical records are common terms today and should not be changed. However, new users of these terms might understand them more easily if the distinction between them was more apparent.

Also, be aware of any terms that are jargon. *Jargon* is used by people in a particular area of expertise. Such a term might pervade your project or your workplace and might be inadvertently added to the user interface or documentation. Such terms can result from:

- Playing off another word, such as *automagically* plays off *automatically*.
- Having a different meaning from what's generally expected. For example, *hit* and *check* have a specialized meaning when used to refer to using a user interface.
- Truncating or abbreviating a standard term, such as *stats* from *statistics*, *typo* from *typographical*, or *admin* from *administration* or *administrator*.
- Using shorthand in internal communication about a product or component. For example, terms such as *indexer* for an indexing server, *jClient* for a Java client application, or *app server* for a web application server might end up in the information unless you carefully monitor your terminology.

# Easy to understand

Avoid using jargon unless it's appropriate for your audience and no other word fits the situation. For example, poorly written error messages sometimes use terms that seem to have little to do with the problem.

**Original**

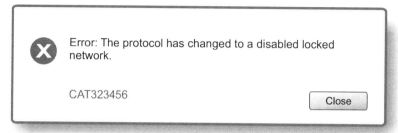

The original error message uses jargon to refer to something more straightforward. The terms *protocol* and *disabled locked network* might be terms that software developers use, but those terms aren't something that users need to see.

**Revision**

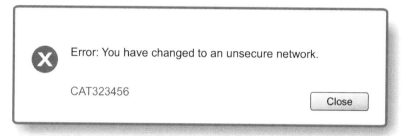

The revised message replaces the jargon with clear, straightforward language that's more appropriate for the intended audience.

## Define each term that is new to the intended audience

In all writing, the successful communication between the writer and the user of the information depends on the vocabulary that they have in common. When users first encounter a new subject, it might seem like a foreign language, with its own vocabulary and rules for combining the concepts that the terms express. For example, native speakers of English who have never developed a mobile app might feel as lost as if they were in a foreign country the first time that they try to use an application development tool.

Some terms might be familiar, but their definitions might differ from what users expect. For example, *object* and *class* are familiar words, but they have special meanings in object-oriented programming. Or terms such as *repository* and *server* are commonly used interchangeably, but each might mean something different depending on the product.

Define terms that are new to users or are used differently from what users expect. Such terms include acronyms and abbreviations. In the text where you first use a new term, explain the term in a meaningful way and highlight it. Try not to use the term before you explain it.

Include a definition where you first use a new term. You can also add definitions to glossaries and indexes. When you write definitions, try to use terms that are familiar to the user.

Ensure that the definitions aren't circular. Circular definitions, as in the following passage, are frustrating for users:

**Original**

> **index entry**
> A key and a pointer paired together.
>
> **key**
> One half of the pair that makes up an index entry.
>
> **pointer**
> One half of the pair that makes up an index entry.

**Revision**

> **index entry**
> A key and a pointer paired together.
>
> **key**
> One or more characters in a data record; identifies the record and establishes its sequence.
>
> **pointer**
> An address or other indication of location.

The original definitions of *key* and *pointer* merely paraphrase the definition of *index entry*. The revised definitions are informative and are independent of each other.

Make sure that new terms are compatible with each other and with existing terms. If you define two terms separately, define the combination only if it has a special significance. For example, you don't need to define *class attribute* if you've defined *class* and *attribute*. However you'd probably define *data type* rather than the individual terms *data* and *type*. *Data type* is a widely used term with its own meaning, which is separate from the meanings of *data* and *type*.

Easy to understand

# Clarity checklist

Clarity in technical information covers a broad range of issues, most of which you can resolve at the level of words and sentences. You need to be aware of possible meanings that someone might read into what you write, other than what you intended. This awareness is particularly important when translators and nonnative speakers might use your information.

You can use the following checklist on in two ways:

- As a reminder of what to look for, to ensure a thorough review
- As an evaluation tool, to determine the quality of the information

| Guidelines for clarity | Items to look for | Quality rating<br>1 = very dissatisfied,<br>5 = very satisfied |
|---|---|---|
| Focus on the meaning | • Subjects are close to their verbs.<br>• Sentences are short and don't contain complex grammatical structures.<br>• Modifying phrases and clauses are kept to a minimum. | 1 2 3 4 5 |
| Eliminate wordiness | • Sentences are free from wordiness problems, such as roundabout expressions, redundancies, and unnecessary modifiers.<br>• Verbs are powerful, active, and concise. | 1 2 3 4 5 |
| Write coherently | • Sentences and paragraphs use effective transition words that tie ideas together.<br>• Sentences and paragraphs effectively transition to tables, lists, and illustrations.<br>• Paragraphs focus on one idea.<br>• Some ideas are subordinate to others. | 1 2 3 4 5 |
| Avoid ambiguity | • Most words are used as only one part of speech.<br>• Technical keywords aren't used as common nouns or verbs.<br>• The text is free from ambiguous words, such as *may* and *as*.<br>• Ideas are stated positively.<br>• Sentence syntax is clear because words such as *that* and *the* are not left out.<br>• Pronouns are used correctly.<br>• Modifiers clearly modify the right words or phrases. | 1 2 3 4 5 |

# Clarity

| Guidelines for clarity | Items to look for | Quality rating<br>1 = *very dissatisfied*,<br>5 = *very satisfied* |
|---|---|---|
| Use technical terms consistently and appropriately | - Use of terms is consistent.<br>- Terms are appropriate for the audience.<br>- Terms that are used are necessary.<br>- Jargon isn't used unnecessarily.<br>- Terms are clearly defined.<br>- Acronyms and abbreviations are defined.<br>- Definitions are easy to understand. | 1 2 3 4 5 |

To evaluate your information for all quality characteristics, you can add your findings from this checklist to Appendix A, "Quality checklist," on page 545.

CHAPTER 7

# Concreteness

*We learn by example and by direct experience because there are real limits to the adequacy of verbal instruction.* —Malcolm Gladwell

When you document a complicated procedure, a new concept, or a default value for an input field, you can make that information more concrete by relating it to an experience, skill, or knowledge that users already possess. By using examples, scenarios, or even just precise terminology, you can provide specific and practical details that help your users truly understand the unfamiliar.

What makes concreteness so important in technical writing is that many technical subjects, such as relational database access, inheritance among Java classes, and the effects of a chemical compound on the human nervous system, are abstract. People cannot experience abstract subjects directly, as they can with physical objects such as a building, a bicycle, or a cheeseburger. However, people can successfully master abstract concepts, terms, and ideas much more easily if they can relate those concepts, terms, and ideas to something real in their experience. Good writers look for ways to make abstract information more concrete.

Writing concrete information, however, can be a challenge for a variety of reasons:

**Lack of subject knowledge**

You need experience with a subject to know what examples or scenarios to use and the creativity to recognize the possibilities for making

# Easy to understand

information more concrete. Writers often do not have time to master a subject, or the subject is complicated.

**Lack of knowledge about the audience**

You must also understand your audience well enough to choose concreteness elements that the audience can relate to. For example, an analogy that involves blueprints will likely resonate more effectively with an audience of civil engineers than it will with an audience of financial planners.

**Lack of resource**

Adding the elements that make information more concrete, such as examples, samples, or graphics, is often regarded as extra work, something to do after you write the basic information. You might choose not to include concreteness elements in an information plan or outline, and concreteness elements that are missing might not be so obvious during a review as a missing step or task. However, a lack of concreteness elements will definitely slow users down when they use the information.

**Added maintenance**

Writing concretely can be risky because something that is concrete, such as a detailed code example, is more prone to becoming inaccurate or dated than something abstract. Keeping concreteness elements up to date and relevant often requires regular maintenance. You must be willing to take this risk and employ strategies to reduce it.

These challenges are worth overcoming because concrete information is so much more useful than abstract information. When elements of concreteness are used appropriately, they can improve all types of information, including traditional book-based information that consists primarily of task, concept, and reference topics; help topics; videos; and content that appears in a user interface, such as window titles, field labels, and error messages.

Think of concreteness elements as tools in a writer's toolbox. Just as a skilled carpenter knows which tool to use in certain situations, you must know when and where to apply the concreteness element that is appropriate for your audience and the information that you are writing. The concreteness elements that are available to you include:

## Scenarios

Scenarios are stories that are based on real-life situations that your users might encounter. Scenarios provide users with a big-picture perspective of complex task and overview information.

A good scenario is based on users' actual goals, skill level, and any other relevant factors. For example, a scenario might describe how an experienced systems architect can use several functionally related products together to achieve a business goal, such as strengthening a security environment or improving performance. Or a scenario might describe how a new user of a software application can get up and running quickly with the three or four most common functions of that application.

*Scenario-based information* is not a concreteness element, but rather the output of a process in which information is designed and written to address specific scenarios. Before a single word of documentation is written or a single user interface window is coded, writers, technical experts, and usability engineers research how an audience will use a product or a technology in the real world. Based on the results of that research, scenarios are developed that will guide writers as they determine which types of information are required to meet users' needs.

For example, one scenario for a workplace safety program might reveal the three most common tasks that must be performed during a safety audit, as well as the typical experience level of the auditor. Based on that information, the writer can accurately determine which tasks need to be documented and can determine the appropriate level of detail for those tasks.

## Programmatic assistance

Programmatic assistance eliminates the need for written information, which from your users' perspective is a good thing. Programmatic assistance automatically performs tasks for your users. For example, when users select an option in a user interface entry field, other options that are related to the option that they selected can be automatically populated with default values. Users can then accept those default values, or they can provide values of their own. Instead of writing documentation about the values that your users need, you provide that information in the interface.

## Examples

Examples are perhaps the most common type of concreteness element. An example shows your audience one way to do something or describes a concept from a particular perspective. Your audience can then apply the example that you provided to help them complete their tasks. Examples can take many forms, such as a code snippet that users can modify or simply a sentence that begins with the phrase *For example*.

## Samples

Samples that you provide with a product or technology enable your users to become productive as quickly as possible. A sample is a tangible object that users can interact with to complete a task. For example, many software products provide sample commands or applications that users can run unchanged to achieve a basic understanding of how things work. Users can then take those samples and modify them to complete their specific tasks.

## Similes and analogies

Similes and analogies relate the unfamiliar to the familiar by means of comparison. A simile is a figure of speech that uses the words *like* or *as* to make a direct comparison, for example, "the central processing unit is like the human brain."

Analogies are usually used to explain more complex subjects or relationships. Analogies demonstrate how something or some set of things is similar to something that the user is familiar with, for example, "The environment consists of a processing unit (the brain), two types of storage (long-term memory and short-term memory), an input device (senses that are used to take in information, such as sight and hearing), and an output device (the sense of speech, the ability to write, nonverbal cues)."

## Visual elements

Visual elements such as icons, illustrations, and the use of color, shading, and highlighting help users to quickly understand a piece of information that would otherwise require them to read text. For example, an illustration that depicts the many components of a software environment provides users with a visual understanding of that environment and saves users from having to read several pages of written information. Similarly, the use of red, yellow, and green status indicators in a user interface provides users with information about the health of their system without requiring them to do any unnecessary reading. The visual effectiveness guidelines explain how to effectively use visual elements to make information more concrete. See Chapter 11, "Visual effectiveness," on page 421 for more information.

**Precise language**

Terminology can run the gamut from the abstract to the concrete, as shown in the following illustration:

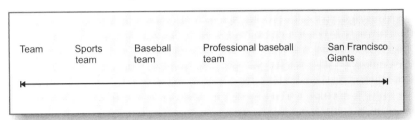

Using precise language requires a writer to be diligent about determining the most appropriate words and phrases for the intended audience. Using a word such as *resources* when you really mean *storage* or a word such as *power supply* when you really mean *120 volt AC adapter* can make information more abstract than it needs to be and even hide the meaning of that information. The lack of precise language can also impact clarity.

In addition to making a product or technology more usable, concreteness elements such as examples, scenarios, and similes can help eliminate the dullness that many people consider typical of technical information.

Because of the importance of scenarios, examples, and samples to users, this chapter emphasizes these means of achieving concreteness. This chapter also focuses on the quality characteristics (particularly task orientation, accuracy, clarity, and retrievability) that make scenarios, examples, and samples effective.

To make information concrete, follow these guidelines:

- **Consider the skill level and needs of users**
- **Use concreteness elements that are appropriate for the information type**
- **Use focused, realistic, and up-to-date concreteness elements**
- **Use scenarios to illustrate tasks and to provide overviews**
- **Make code examples and samples easy to use**
- **Set the context for examples and scenarios**
- **Use similes and analogies to relate unfamiliar information to familiar information**
- **Use specific language**

# Consider the skill level and needs of users

One challenge of providing useful concreteness elements is being able to satisfy an audience that consists of widely different skill levels (novice to expert), that has different needs (a casual user versus a power user), and that has different backgrounds (for example, cultural differences). As with all the guidelines in this book, the key principle is to *know your users*.

## Domain expertise

The most effective way to understand the needs of your users and to develop useful examples is to work with user data that you obtain from experienced users or that you obtain from colleagues who work directly with experienced users. The knowledge that these subject matter experts provide is called *domain expertise*, which is built up over years of experience with a product, technology, industry, or discipline. There is no substitute for domain expertise: it can't be developed overnight, it can't be gleaned from a book, and without it, you might have trouble developing examples that will help your users complete their tasks as quickly and efficiently as possible.

For example, if you are writing a tutorial to help nurses learn how to use a new medical charting application, you would rely on somebody with expertise in the domain of healthcare. Most likely, that person will be a nurse, a medical records specialist, or a healthcare manager. These domain experts could provide you with realistic information about a patient's temperature, blood pressure, heart rate, and other vital signs that are required to make the tutorial as concrete as possible. Without input from a domain expert, you might be tempted to make your best guess about what a realistic blood pressure should be for a patient on a certain combination of medications. Or you might take the lazy approach and document the interface with unhelpful instructions such as "Enter the blood pressure in the **Blood pressure** field." A domain expert could tell you precisely what a realistic blood pressure reading should be in that situation.

Domain experts can also help you to write effective and useful examples, samples, and scenarios by helping you understand what users want and need to do with the technology that you are writing about. How do users prefer to interact with the technology? What are their goals? What criteria constitute a positive user experience? If you can obtain solid answers to these questions, you are well on your way to helping your users to be successful with your product.

Notice the level of detail in the **Example** section of the following parameter description:

Original

> **RESOURCE TIMEOUT**
>
> The RESOURCE TIMEOUT parameter controls the number of seconds that lapse before a resource is canceled.
>
> **Range:** 1 - 360
> **Default value:** 30
> **Example:** `RESOURCE TIMEOUT=60`

The original example merely illustrates the proper syntax, which is fairly simple and likely obvious to most users. The user's real need is knowing how to set this parameter appropriately for their situation, and only somebody with practical knowledge of this technology knows what the appropriate settings are for different situations.

Revision

> **RESOURCE TIMEOUT**
>
> The RESOURCE TIMEOUT parameter controls the number of seconds that lapse before a resource is canceled.
>
> **Range:** 1 - 360
> **Default value:** 30
> **Examples:**
>
> > Use the following examples as guidelines when you initially establish timeout values. You likely will need to fine-tune these settings over time.
> >
> > **RESOURCE TIMEOUT=30**
> > Accept the default value of 30 for noncritical applications and applications that run multiple times during the day. Cancelations occur most often during peak usage periods, so applications that run multiple times during the day will likely be rerun during an off-peak usage period when system resources are more readily available and a timeout is less likely to occur.
> >
> > **RESOURCE TIMEOUT=180**
> > Use a value of 180 for applications that are run on an ad hoc basis, applications that are not mission critical, and applications that do not consume a high level of system resources. For example, 180 is appropriate for an application that generates reports.
> >
> > **RESOURCE TIMEOUT=360**
> > Use a value of 360 for mission-critical applications.

# Easy to understand

The revised parameter description contains several useful examples that get users a lot closer to completing a task.

## Skill levels of the audience

Depending on your audience, you might need to provide different examples and scenarios for different skill levels and about various subjects. An example that was written to address the needs of a novice data-entry clerk will almost certainly not be as useful to an experienced software administrator.

For example, you might need to provide novice users with a scenario that helps get them up and running quickly with the most basic functions of a product. You also might need to provide experienced users with one or more scenarios to help them understand some of the more complex features or procedures of that same product.

Users of new products and technologies always need examples and scenarios because all users will be unfamiliar at first. However, remember that novice users do not remain novice for long, so be cautious about devoting too much effort to documenting basic tasks and functionality. See the completeness guideline "Provide appropriate detail for your users and their experience level" on page 123 for more information about how to design information that meets the skill levels of your audience.

## Cultural backgrounds of the audience

Technical information in the twenty-first century can be accessed virtually anywhere by anybody who has a laptop or mobile device. Therefore, you should always assume that your information will be used by an international audience. When you write scenarios and examples, avoid using terms that are specific to your particular culture. For example, the terms *inch*, *dollar*, *social security number*, and *zip code* might not be relevant or familiar to a non-U.S. audience. Instead, use more culturally generic terms such as *centimeter*, *price*, and *identification number*.

For more information about writing for a culturally diverse audience, see the style guideline "Avoid gender and cultural bias" on page 273.

# Use concreteness elements that are appropriate for the information type

The examples, samples, scenarios, and other concreteness elements that you provide must be appropriate for the type of information that you are writing.

Users have different needs and expectations depending on the type of information that they are using. For example, a detailed code sample will be less useful in a high-level conceptual topic than it will be in a task or reference topic because users read conceptual information to gain broad understanding of a subject. They read task and reference topics to put that understanding into practice. Similarly, user interface elements have unique concreteness requirements.

## Embedded assistance

Embedded assistance is arguably the most important type of written content for those products that have user interfaces, which includes both software and hardware products and technologies. Because embedded assistance is so valuable to users' productivity, you should strive to make it as concrete as possible.

Embedded assistance elements are defined in "Embedded assistance" on page 4. The following sections describe those embedded assistance elements from a concreteness perspective.

### Programmatic assistance

As described in the completeness guideline "Make user interfaces self-documenting" on page 101, providing high-quality programmatic assistance is one of the biggest favors that you can do for your users. Programmatic assistance is more than simply providing a hint in an input field. You can increase the likelihood that your users will be successful with your product and will have a positive experience by including default values, automatically entering information or values for users so they don't need to manually enter that information, and automatically completing input fields.

Whenever you can choose between using programmatic assistance and using another type of embedded assistance, always use programmatic assistance. Programmatic assistance is the most concrete form of embedded assistance.

## Text labels

Make text labels for user interface elements, such as input fields, radio buttons, and drop-down lists, as descriptive as possible so that users know exactly what they need to enter:

**Original**

```
XYZ Utility ------------- RENAME OBJECT -------------
Command ===>

Name: _____
New name: _____
Check for dependencies: _ Specify Y for Yes or N for No.
```

The field labels in the original example are vague. Users who need more information about these fields might be expected to click the help button to find what they need. But why send users to the product help when you don't need to?

**Revision**

```
XYZ Utility ------------- RENAME OBJECT -------------
Command ===>

Object to rename: _____
New name for the object: _____
Generate a list of objects that are dependent on this
object? Y (Y or N)
```

A better solution is to provide more descriptive labels by using specific language, as shown in the revision. Notice that the purpose of the last input field has been broadened to address not only checking for dependencies, but also listing any dependent objects that are found.

Unless you are restricted by space constraints that limit the number of characters that you can include in the interface, always use as much space as is needed to describe each user interface element. Notice also that the last field in the panel includes a default value. Most users will want to know about dependencies between these objects, and checking for these dependencies will do no harm, whereas not checking for dependencies can cause problems.

The purpose of some user interface elements, however, is so complex that a concrete text label on its own is not enough to help users specify the right value or make the right choice. In these situations, consider including an example that shows a correct value or choice. In the following dialog, an example is provided for the **Subject or content** field because users must enter a value that is difficult to explain through the text label alone and that must also follow a specific syntax:

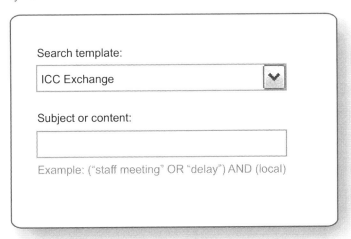

See "Use specific language" on page 256 for additional examples of how the careful use of text elements can improve concreteness in user interfaces.

## Messages

Messages that are displayed in the interface are useful for communicating things like the status of a task, the progress of an operation, or a value that was entered incorrectly or not at all. Because users typically care about messages only when something has gone wrong, it's important to give them as many details about the error situation as possible.

**Original**

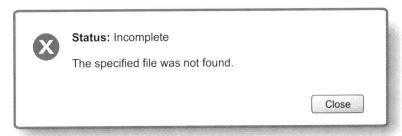

The original message informs users that something is not right, but it doesn't give them enough detail to correct the situation.

**Revision**

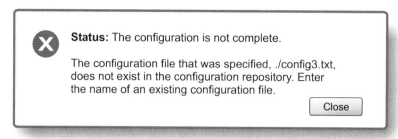

The revised message describes the error from the users' perspective and informs them specifically what has gone wrong. More importantly, it provides a specific action for them to take to correct the error on their own.

## Control-level assistance

Control-level assistance is help that is associated with a particular user interface control. Control-level assistance typically takes the form of tooltips or hover help. The following illustration shows two examples of control-level assistance: a tooltip for an icon and hover help for a radio button:

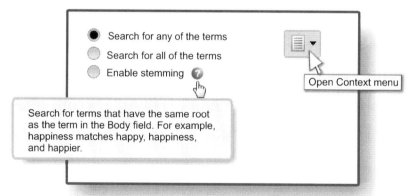

The language in these elements must be concise and specific. Examples are extremely useful to help users understand the purpose of a user interface control and the effects of using that control, as demonstrated in the hover help for the **Enable stemming** radio button.

# Task information

In task information, examples are important both within steps and for the overall task. Within a step, users appreciate examples of the things they must do and examples of the results of their actions.

**Original**

> 1. Identify the version and release of the ProgLang product that you use. To find the version and release, compile a program by using the LIST option. The version and release are shown as the last item at the top of the listing output.
>
> 2. Enter keywords and the identifier for any error messages that you received while the application was being compiled.

The original version makes the steps seem more difficult than they are because it gives no examples.

**Revision**

> 1. Identify the version and release of the ProgLang product that you use. To find the version and release, compile a program by using the LIST option. The version and release are shown as the last item at the top of the listing output. For example, you might see the following line:
>
>    ```
>    CC 5648-A25 PROGLANG 2.1.2
>    ```
>
> 2. Enter keywords and the identifier for any error messages that you received while the application was being compiled. For example, you might enter:
>
>    ```
>    ABEND0C4 XYAZDS0084-S
>    ```

In the revised version, the example of output in the first step and the example of input in the last step help users know what to expect.

You can't provide an example for every step, and you shouldn't need to. If users can see the actual interface while doing a step, they probably don't need a screen capture of the interface.

Given time constraints, providing an example for every task is probably impossible and perhaps not desirable from the users' point of view. How do you decide what tasks most need examples? You can consider certain characteristics of the tasks, as shown in the following figure:

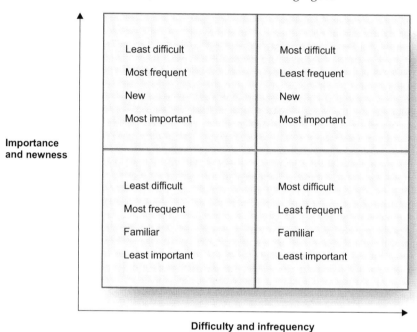

Figure 7.1  Criteria to consider when you select examples

Focus particularly on the following kinds of tasks (as shown in the upper-right quadrant of the figure):

- Tasks that are complex or difficult
- Tasks that are important but performed infrequently
- Tasks that are critical to using the product
- Tasks that are new or that have changed

How useful do you think the task example in the following step is?

> 2. Click **Color > *Color name*** to change the color of any of the windows. For example, click **Color > Red** to make the window red.

The action in this step falls in the lower-left quadrant of the previous figure: it is extremely simple and probably not critical to using the product. Most users can likely get along just fine without it.

The following step also involves defining a color scheme, but in a much more important context:

> 2. Define the characteristics of the different alert levels by adding the relevant parameters to the alert definition file. For example, you might define the following characteristics for the HIGH alert level like this:
>
> ```
> ALERT = HIGH (COLOR="RED", FREQ="30", SEND="ALL");
> ```
>
> Based on this definition, high alerts will be displayed in red text on the console; they will be reissued every 30 seconds until the operator acknowledges the alert; and notifications will be sent to all personnel who are defined in the alert notification list.

Defining alerts that affect the health of a system is certainly considered a critical action. Although the syntax of the ALERT statement might not seem too complex, defining an alert incorrectly could result in serious problems. Also, users are not likely to define or change alerts frequently, which means that they're unlikely to recall the details for completing this task when they do need to work with alerts.

## Conceptual information

In conceptual information, you can use examples to help define concepts and ideas. For example, when you define a new idea, you describe it in specific terms. Examples, however, help to make the description even more specific.

Original

> A shock absorber *dampens* the amount and severity of rebounding that occurs when a vehicle travels over rough or bumpy surfaces.

This definition of a shock absorber introduces the concept of *dampening* in relation to the function of a shock absorber. Because this application of the term *dampen* is specific to automobile suspensions, some users might not be familiar with it.

# Easy to understand

**Revision**

> A shock absorber *dampens* the amount and severity of rebounding that occurs when a vehicle travels over rough or bumpy surfaces.
>
> For example, when a car passes over a bump, the weight of the car causes the springs to compress. Shock absorbers limit, or dampen, the kinetic energy that is released when the springs decompress, therefore reducing the amount that the car bounces.

The revision adds an example that illustrates the concept of dampening in action. Providing an example of how the concept works in a practical setting helps users to more easily understand the meaning of the concept.

Similes are another effective tool for describing abstract concepts in terms that users understand.

**Original**

> A relational database is a set of tables in which each table is a set of relations.

This description relies on users knowing what is meant by the terms *table* and *relations* within the context of relational databases.

**Revision**

> A relational database is like a file drawer in which each folder holds related papers. A database table is like one folder in the file drawer.

The revision relies on similes to relate the abstract concepts of database tables and relations to things that users are familiar with: file drawers and file folders.

# Concreteness

You can also use graphics to make the meaning of a concept more concrete.

**Original**

> The Utility Console environment consists of the following components:
>
> - The Utility Console client, which runs on any supported web browser
> - The Utility Console server, which runs on any supported application server
> - The communication plug-in, which runs on the mainframe and provides distributed clients access to third-party plug-ins through a centrally controlled standard TCP/IP socket
> - Policy definitions, which define the general characteristics of each third-party plug-in, as well as security and performance settings
> - One or more third-party plug-ins
>
> The Utility Console client communicates with the Utility Console server via a secure HTTPS connection, and the Utility Console server connects to the communication plug-in through an SSL TCP/IP connection.

The original passage relies on words alone to define the various components of a product, to indicate their role in the environment, and to describe how the components interact with each other.

# Easy to understand

**Revision**

The Utility Console environment consists of the following components:

- The Utility Console client, which runs on any supported web browser
- The Utility Console server, which runs on any supported application server
- The communication plug-in, which runs on the mainframe and provides distributed clients access to third-party plug-ins through a centrally controlled standard TCP/IP socket
- Policy definitions, which define the general characteristics of each third-party plug-in, as well as security and performance settings
- One or more third-party plug-ins

The Utility Console client communicates with the Utility Console server via a secure HTTPS connection, and the Utility Console server connects to the communication plug-in through an SSL TCP/IP connection.

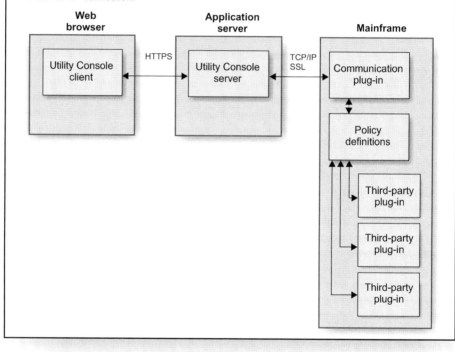

The revision includes an illustration that depicts the physical location of each component, which makes the relationships much easier to understand.

Scenarios are another effective tool for demonstrating broad concepts, such as the benefits that a product or technology can provide to an organization or how a group of functionally related products can be used to provide a comprehensive solution. See "Use scenarios to illustrate tasks and to provide overviews" on page 243 for more information about how to use scenarios effectively.

## Reference information

In reference information, samples and realistic examples are essential to the usefulness of the information. Reference information that contains high quality samples and examples is a valuable tool that can speed users toward completing their tasks. Examples of programming statements such as methods and commands give users valuable information about syntax and usage.

**Original**

> In MathML, use the <apply> element to indicate the operator to apply to the arguments in a mathematical expression. The <apply> element takes the place of parentheses.

The original passage requires users to read the explanation carefully and does not provide an example for users to follow.

**Revision**

> In MathML, use the <apply> element to indicate the operator to apply to the arguments in a mathematical expression. The <apply> element takes the place of parentheses, as in the following example:
>
> ```
> <apply><times/> <ci> a </ci> <ci> b </ci> </apply>
> ```
>
> The tagging resolves to (*a x b*).

The revised passage shows users what they need to type if they specify the elements. The passage provides both an example of the syntax and the equivalent result, in case users are unfamiliar with the elements in the example.

# Easy to understand

You might also provide a sample (actual code or data for using a product) such as:

- A sample database so that users can practice using your product to populate a database
- A sample application so that users can familiarize themselves with key product features and the product interface

Samples can take the place of descriptive information and give users a chance to learn by doing, as in the following pair of examples:

**Original**

> **Guidelines for heading styles**
>
> - Define headings by using a CSS or the style tag. Do not set the style on individual heading tags.
> - Use only heading levels 1, 2, and 3.
> - Use a sans-serif font.
> - Increase the size of the heading by approximately one third from one heading level to the next.

The original passage certainly does explain the guidelines for HTML heading styles, but it could do a much better job of illustrating what users really need to do to define heading styles.

**Revision**

> **Guidelines for heading styles**
>
> - Define headings by using a CSS or the style tag. Do not set the style on individual heading tags.
> - Use only heading levels 1, 2, and 3.
> - Use a sans-serif font.
> - Increase the size of the heading by approximately one third from one heading level to the next.
>
> The following sample conforms to the guidelines for heading styles. You can use this sample as a starting point to define the headings for your library.
>
> ```
> <!DOCTYPE html>
> <html>
> <head>
> <style>
> h1 {font:215% arial,sans-serif;}
> h2 {font:165% arial,sans-serif;}
> h3 {font:125% arial,sans-serif;}
> </style>
> </head>
> <body>
> <h1>Sample heading level 1</h1>
> <h2>Sample heading level 2</h2>
> <h3>Sample heading level 3</h3>
> </body>
> ```

The revised passage provides an opportunity for users to experiment with real code. By doing so, they are much more likely to be successful in applying the guidelines.

## Troubleshooting information

Because users access troubleshooting information only when something is not working properly, it is critical that this type of information is as concrete as possible. Failure to provide concrete troubleshooting information increases the chances that users will become frustrated with your product and will contact your technical support department, which results in a financial cost to your organization.

The two most common forms of troubleshooting information are troubleshooting procedures and error messages.

Easy to understand

## Troubleshooting procedures

Troubleshooting procedures are task-oriented topics that describe the symptoms and possible causes of a problem and include actions that users can take to resolve the problem.

Provide your users with troubleshooting procedures for the most common and the most disruptive error situations, and make sure that the procedures that you provide are detailed enough for your users to resolve the problem.

**Original**

> **Recovering from disk failure**
>
> If you experience a disk failure, complete the following steps:
>
> 1. Ensure that no incomplete I/O requests exist for the failing device.
> 2. Check the disk status.
> 3. *and so on...*

The original troubleshooting procedure doesn't do a very good job of providing the details that users need. For starters, this procedure is missing explanations for the symptoms and causes of the error condition. How can users be certain that this procedure pertains to the problem that they are experiencing? Also missing is a necessary level of detail in the steps for resolving the problem. Remember that users are stuck and want to get unstuck as quickly as possible. They don't want to comb through reference information looking for the commands that they need to solve the problem.

Revision

> **Recovering from disk failure**
>
> **Symptoms**
> No I/O activity occurs for the disk address. Databases and tables that reside on the unit are unavailable. Error message ABC2002E is issued, which indicates the unit address of the device.
>
> **Causes**
> Typical causes of disk failures include hardware problems, network communication problems, and the introduction of incompatible hardware or software into your environment.
>
> **Resolving the problem**
>
> 1. Ensure that no incomplete I/O requests exist for the failing device by issuing the following command, where xxx is the unit address that was provided in message ABC2002E:
>
>    ```
>    VARY xxx,OFFLINE,FORCE
>    ```
>
>    The following console message is displayed, which indicates that no incomplete I/O requests exist.
>
>    ```
>    UNIT TYPE STATUS VOLSER VOLSTATE
>    xxx  3390 O-BOX  XTRA02 PRIV/RSDNT
>    ```
>
> 2. Check the disk status by issuing the following command:
>
>    ```
>    D U,DASD,ONLINE
>    ```
>
> 3. and so on...

The revision explains the symptoms and potential causes of this error condition, and the steps for resolving the error situation include the actual commands that users must enter to complete the procedure.

## Error messages

Messages for simple error conditions, such as entering an incorrect password, often need only a single sentence to get users back on track. However, messages that address complicated error conditions should include the following components:

- The message text, which succinctly yet completely describes the problem
- Explanatory text, which provides a detailed explanation of the problem
- If applicable, information about the state of the system or product environment as a result of the problem
- An action or actions that users can take to remedy the problem

All of these components should be as concrete as possible.

Original

The message text, Required Value, is vague. In fact, the way it is phrased doesn't even indicate that there's a problem. And the explanation doesn't tell the user specifically which value is missing or what to enter.

Revision

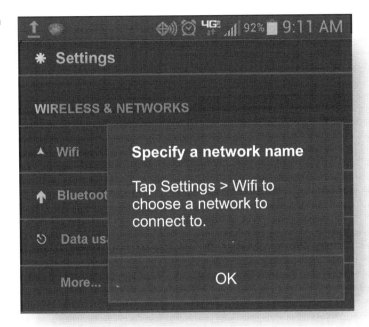

The revised message includes concrete information that helps users understand why the error occurred and how they can resolve the error:

- The vague message text has been rewritten to provide a more concrete description of the problem and includes the specific name of the affected function.
- The explanation has been replaced with a concrete action that users can take to remedy the problem on their own. Providing a detailed explanation of the cause of the problem helps users avoid making the same mistake again.

For complicated error conditions, the knowledge that is required to make these types of improvements is sometimes outside the bounds of a writer's expertise. Therefore, work with your technical experts to identify the concrete steps that help users to help themselves.

Easy to understand

# Use focused, realistic, and up-to-date concreteness elements

The effectiveness of the examples, samples, scenarios, and other concreteness elements that you use depends in large part on these elements being focused, realistic, and current.

## Focused elements

One of the main purposes of examples, samples, and scenarios is to help users make progress as quickly as possible. A short, well-focused code example can often replace multiple paragraphs that explain the syntax of a piece of code, which relieves users from doing unnecessary reading. Therefore, when you include concreteness elements, try to pare them down to include just the information that is essential to the point that you want to make.

**Original**

```
2. Edit the DBXPARSE sample job and specify the three required REPORT commands:

   //DBX    EXEC  PGM=DBXA90,PARM='OBJECT,NODECK,NOBATCH,LIST'
   //SYSLIB       DD DISP=SHR,DSN=SYS1.MACLIB
   //SYSPRINT     DD SYSOUT=*
   //SYSUT1       DD UNIT=SYSDA,
   //             DISP=(,DELETE),SPACE=(CYL,(1,1))
   //SYSIN DD *
   //* REPORT01 DD SYSOUT=A,DCB=BLKSIZE=&PRTBLK,OUTLIM=0
   //* REPORT02 DD SYSOUT=A,DCB=BLKSIZE=&PRTBLK,OUTLIM=0
   //* REPORT03 DD SYSOUT=A,DCB=BLKSIZE=&PRTBLK,OUTLIM=0
   //SUMMARY DD SYSOUT=A,DCB=BLKSIZE=&PRTBLK,OUTLIM=0
   //SYSUDUMP DD SYSOUT=A
```

The sample code in the original example includes the three REPORT commands that are mentioned in the step, but it also includes a lot of statements that aren't related to this step. This example requires users to scan a lot more information than they need to.

**Revision**

```
2. Edit the DBXPARSE sample job and specify the three required REPORT commands:

   //SYSIN DD *
   //* REPORT01 DD SYSOUT=A,DCB=BLKSIZE=&PRTBLK,OUTLIM=0
   //* REPORT02 DD SYSOUT=A,DCB=BLKSIZE=&PRTBLK,OUTLIM=0
   //* REPORT03 DD SYSOUT=A,DCB=BLKSIZE=&PRTBLK,OUTLIM=0
   //SUMMARY DD SYSOUT=A,DCB=BLKSIZE=&PRTBLK,OUTLIM=0
   //SYSUDUMP DD SYSOUT=A
```

The revision is much more focused. It includes only the three REPORT commands that are required and the one statement that precedes these commands so that users know where to add these commands in the file.

## Realistic elements

Users appreciate examples, samples, and scenarios that they can relate to and that reflect their goals. Consider the following guidelines when you create concreteness elements that try to meet your users' needs:

- Use concreteness elements that pertain to the particular environment or circumstances in which the product or technology will be used. Also, consider common exceptions to the standard usage of a product or technology, for example:
    - If a software product will be used on thousands of similar workstations, make sure that any samples that you create are based on a large-scale implementation.
    - If an industrial chemical is affected by or performs differently in extreme heat or cold, make sure that your examples take these temperature differences into account. Don't provide examples only for the median temperature.
- Use *realistic*, rather than *real* data in examples, samples, and scenarios. Using real samples and examples can make your organization potentially liable if a sample or example causes problems with a user's system. For example, performance data can vary widely depending on environmental considerations. If you use real performance data in an example and your users don't experience the same performance benefits, your users might feel that they were misled.
- Avoid the use of fictitious names of organizations in examples, samples, and scenarios because sometimes these fictitious names turn out to be real, and might even be the trademark of an organization. Instead of using a fictitious name, such as *XYZ Savings and Loan*, use a generic descriptive name instead, such as *a medium-sized regional bank*.
- Similarly, don't use fictitious web addresses. Even if a web address seems safe to use (that is, you checked it, and it doesn't display a page), you can't be sure that the web address won't be used by somebody in the future. Instead, use one of the web addresses that have been designated by the Internet Assigned Numbers Authority (IANA) exclusively for use in examples. These web addresses are www.example.com, www.example.net, www.example.org, and www.example.edu.

# Easy to understand

## Current elements

Concreteness elements can become outdated and inaccurate if they contain functions that are no longer supported. Code examples and samples can become outdated if they include items from popular culture or numbers (such as prices and dates). If you must include dates, try to make them far enough in the future that they won't soon seem outdated. Consider how long the information must last.

To reduce the work that is needed to maintain examples and samples, remove items that are likely to show their age quickly. Consider what item is likely to become dated in the following example and how you might change it:

**Original**

> This statement creates an object, an instance of the class car. Unless a later statement provides a different name, the product will use the name myBronco for the object and assign the name of the variable *mycar*.
>
> ```
> car.new('myBronco', mycar)
> ```

**Revision**

> This statement creates an object, an instance of the class car. Unless a later statement provides a different name, the product will use the name myFord for the object and assign the name of the variable *mycar*.
>
> ```
> car.new('myFord', mycar)
> ```

The original example uses a model name for a car, which is more likely to change or become obsolete than the manufacturer name, which the revised example uses.

Whatever examples and samples that you use, be sure to make the explanatory text match the example or sample.

# Use scenarios to illustrate tasks and to provide overviews

A *scenario* depicts a series of events over time, usually in the context of a fictitious but realistic set of circumstances. It shows one path through one or more tasks or products. A scenario can range from a high-level overview of the key benefits of using a product to a detailed explanation of the steps that are required to set up a complex hardware or software environment.

You use scenarios primarily to teach (as in a tutorial) or to describe and motivate (as in marketing and technical overview information).

Scenarios that are used to teach are referred to as *task scenarios*. Task scenarios provide users with a hands-on perspective of interacting with a product or technology and show how a product or technology can contribute to solving a larger business goal. Unlike the lists of steps in a procedure, task scenarios follow a story that depicts the actions in a real-life setting.

Scenarios that are used to describe and motivate are referred to as *business scenarios*. Business scenarios focus on the application of one or more products or technologies to achieve a large or complex business goal, such as improving data security across a dozen different data centers in a dozen different countries. A business scenario can't possibly address all of the steps that are required to achieve a complex goal, but it can illustrate different methods and approaches for reaching the goal and the benefits for pursuing one method or approach over another.

To create an effective scenario, you need to create a scene with characters who have similar tasks or similar problems to those of your users. When you write scenarios, focus on goals that users are likely to want to achieve. Scenarios that are based on interesting and relevant goals are more helpful than those that focus on rarely used or esoteric features. You might be able to get ideas for scenarios from usability engineers who use scenarios for testing or from colleagues who demonstrate your product to customers.

## Task scenarios

Task scenarios are an effective way to illustrate how users can achieve a specific goal that requires them to complete multiple tasks or that requires them to choose between multiple options. Task scenarios truly make a task concrete by presenting it in a real-world context. The audience for task scenarios is typically the users of the product or technology.

# Easy to understand

The following passage introduces users to a tutorial lesson:

**Original**

> In this lesson, you learn how to write an application to process data for more than one application server.
>
> You can manage the transfer of data in one of the following ways:
> - Use virtual memory as a buffer for holding a few rows of data.
> - Use disk space as a buffer for holding many rows of data.
> - Use multiprocessing to run each update in a different thread.

The original passage presents only the tasks without a context in which to understand them.

**Revision**

> In this lesson, you learn how to write an application to process data for more than one application server.
>
> A major retail company periodically sends a copy of the employee table that is stored at headquarters to the regional branches. The application that you are writing needs to update the local copies. However, because a database manager can be connected to only one application server at a time, the application must manage the transfer of data.
>
> An application can manage the transfer of data in one of the following ways:
> - Use virtual memory as a buffer for holding a few rows of data.
> - Use disk space as a buffer for holding many rows of data.
> - Use multiprocessing to run each update in a different thread.

The revised passage presents a task scenario that puts the user in a realistic situation with a need to use the task information.

If you simply add a thin story layer to what is still hard-to-understand information, as in the following topic, you haven't created a real task scenario:

Original

> **Connection ID usage**
>
> In this scenario, a two-party call exists between party A and party B. The Make_Call from party A to party B results in connection IDs being generated to represent their participation in the call. Party A then issues an Extend_Call program call to extend the call to party C. The Extend_Call places party B on hold and results in two new connection IDs being created: one for party A's connection in the new call (party ID 1) and one for party C (party ID 2). To join all three parties in a conference call, Party A issues a call to InfoConference and requests party ID 1 and party ID 2.

The original scenario might represent a real situation that users encounter with the telephone software, but it doesn't represent how most people interact with the software. Users might need to read this narrative several times just to understand whether the parties in it are people and what they are doing. The manner in which this scenario is presented—a solid block of text that contains impersonal references such as party A and party B—also makes this scenario difficult to parse and understand.

Revision

> **Conference call and connection IDs**
>
> Alice needs to set up a conference call with Subir and Calvin so that all three of them can have a discussion. Alice calls Subir, puts him on hold while she calls Calvin, and then connects Calvin in to the call.
>
> The implementation of this conference call in terms of connection IDs looks like this:
>
> | Event | Connection IDs |
> |---|---|
> | 1. Alice calls Subir | a1, b1 |
> | 2. Alice extends the call to Calvin | a2, c1 |
> | 3. Alice calls InfoConference | a1, b1, c1 |

The revised scenario clarifies the situation and the relationship between what the people do and what the software does in response. Scenarios can emphasize actions and actors without the imperatives that are usual in technical information. Writing scenarios is probably as close to narrative writing as writers of technical information get.

## Business scenarios

Unlike task scenarios, business scenarios describe each action at a very high level and do not provide the specific steps that are required to complete each action. Business scenarios are typically more conceptual than task scenarios.

# Easy to understand

Note that although business scenarios are commonly used to describe events in a business context, they can apply to other contexts as well, such as education and science.

An effective use of business scenarios is in overview content that presents a broad set of information in a realistic setting. For example, business scenarios are ideal for demonstrating how multiple products can be used together to achieve a solution, which might require a database system, application server middleware, a network environment, and a mobile client device. Individual task and concept topics can be written to provide details for implementing what is described in the scenario. The audience for business scenarios is typically people who are involved with evaluating a product or solution to determine if that product or solution is appropriate for their organization's business goals.

The following passage comes from information for evaluating a product:

**Original**

> With a distributed relational database, you can benefit from accessing data as if it were located wherever you are.

**Revision**

> With a distributed relational database, you can benefit from accessing data as if it were located wherever you are.
>
> Consider a large multinational insurance company with headquarters in one city and sales offices in many other cities around the world. Each sales office has a relational database for its own transactions, and each month the office must send data to headquarters for the sales summary report. By using a distributed relational database that merges the data from all of the sales office databases, headquarters can access the sales data directly.

The original passage leaves to the user's imagination what the benefits might be. The revised passage presents a business scenario that illustrates a specific benefit that users can apply to similar situations.

The scenarios that are used in this book as examples are necessarily brief. However, scenarios in technical information can be quite long, depending on the complexity of the goal that you are attempting to document. Scenarios that deal with the use of more than one product are particularly likely to be longer.

## Make code examples and samples easy to use

Code examples and samples help users do their own work because users can copy the code or other content and adapt it to their own needs.

Examples and samples in software information often contain code and commands. For example, a C++ routine does not describe C++ code; it *is* C++ code for a particular purpose. When you describe the C++ language, use examples of C++ code.

Remember that you cannot provide an example or sample for everything, so when you prioritize which information would benefit from a related example or sample, focus on the most difficult, frequent, important, and new information.

If your examples or samples are large sections of code, provide them online or as files with the product media rather than providing them only in the product documentation. Users must transcribe large code examples that are not online or included with the product before they can copy or adapt the code. Also, an advantage of including samples and examples online is that you can update them more easily; if they are included in printed information, they are more likely to become outdated due to the amount of time that it takes to publish a new version of printed information.

When you provide examples and samples, include an appropriate level of instruction in comment lines. The following excerpt is from a sample installation response file, which is a tool that is used to install an identical instance of a program on many computers:

**Original**

```
<agent-input clean="true" temporary="true">
<server>
  <repository location='http://www.ibm.com/software/repmanager' />
  <repository location='https://www.ibm.com/software/prodname/repmanager' />
<!-- <repository location='https://www.mycompany.com/repositories/' />
<!-- <repository location='/home/user/repositories/websphere/' />
<!-- <repository location='loc_rep'/>.
  .
  .
  .
```

The original version includes no instructional text whatsoever. Although experienced users might be able to figure out how to modify and run this sample, less-experienced users would likely be forced to consult additional reference materials before being able to run this sample.

**Revision**

```
##### Instructions for using this sample ###################################
#
# Use this sample response file to install the client on multiple machines.
#
# Follow the instructions in the file to modify the code for your needs.
# When you are finished modifying the code, save and run the file to install
# the client.

##### Attributes for agent-input ###########################################
#
# The clean and temporary attributes specify the repositories and other
# preferences and whether those settings will persist after the silent
# installation finishes.
#
# Valid values for clean:
#    true = Use only the repositories and other preferences that are
#           specified in the response file.
#    false = Use the repositories and other preferences that are specified
#            in the response file and the installation utility.
#
# Valid values for temporary:
#    true = Repositories and other preferences specified in the response
#           file do not persist in the installation utility.
#    false = Repositories and other preferences specified in the response
#            file persist in the installation utility.
############################################################################
<agent-input clean="true" temporary="true">
.
.
.

##### Local Repositories ###################################################
# Uncomment and update the following line to use a repository
# that is located on your own machine. Replace <local_repository>
# with the full directory path of the repository. Enclose the path
# in single quotation marks.
############################################################################
<!-- <repository location='<local_repository>'/>
```

The revision provides the following improvements over the original:

- A thorough description of the purpose of the sample
- Detailed instructions for modifying each section of the sample
- Descriptions of the attributes that users must specify

Often, when you provide these three things, you can eliminate some or all of the task information that is related to a sample.

Another important difference between the original and revision is how variables are presented. A *variable* is a placeholder that users must replace with their own value. The original version includes the variable `loc_rep`, which is not presented in any way that indicates that it is a variable. Also, it is not named intuitively. How would users even know that this variable needs to be changed?

In the revision, the variable has a more intuitive name of `local_repository` and is enclosed in less than and greater than symbols (< and >), which is a common convention for indicating variables. Alternative methods for indicating variables are the use of brackets ([ and ]) and braces ({ and }), which display correctly in all text editors.

When you add comments to code examples and samples, consider where those comments will be most helpful—probably not far from what they're explaining.

**Original**

> The following example creates a user-defined function (UDF) named `map_scale` that calculates the scale of a map. Notice that the UDF identifies map as the data type to which it can be applied. The code that implements the function is written in C and is identified in the EXTERNAL NAME clause.
>
> ```
> /* */
> CREATE FUNCTION map_scale (map)
>     RETURNS SMALLINT 's'
>     EXTERNAL NAME 'scalemap'
>     LANGUAGE C
>     PARAMETER STYLE DB2SQL
>     NO SQL
>     NOT VARIANT
>     NO EXTERNAL ACTION
> ```

The original passage uses sentences outside the code to give information that is more clearly presented in the revised passage as comments in the code. Users can easily miss the information when it is separate from the code. Also, users can see the comments if they access the code online.

# Easy to understand

**Revision**

> The following example creates a user-defined function (UDF) named `map_scale` that calculates the scale of a map. Notice that the UDF identifies map as the data type to which it can be applied. The code that implements the function is written in C and is identified in the EXTERNAL NAME clause.
>
> ```
> /* map is the data type to which map_scale applies */
> CREATE FUNCTION map_scale (map)
>    RETURNS SMALLINT 's'
> /* scalemap is the name of the C function that implements the UDF */
>    EXTERNAL NAME 'scalemap'
>    LANGUAGE C
>    PARAMETER STYLE DB2SQL
>    NO SQL
>    NOT VARIANT
>    NO EXTERNAL ACTION
> ```

When you add notes to a code example, you can highlight areas that users need to change, that require additional explanation, or that have been added or changed since a previous release. One way to draw attention to particular parts of a code example is to use a retrievability aid such as numbers on certain lines and corresponding numbered notes at the end of the example, as shown in the following code example:

> ```
> <task> 1
> <title>Configuring InfoPlayer</title> 2
> <taskbody> 3
> <steps>
> <step> 4
> <cmd>Start InfoPlayer.</cmd>
> </step>
> <step>
> <cmd>Click Options > Preferences.</cmd>
> </step>
> .
> .
> .
> ```
>
> **1** The <task> element is the highest-level container for this topic.
>
> **2** Every topic has a title.
>
> **3** The <taskbody> element contains steps, but it can also contain prerequisites or the task and postrequisites.
>
> **4** Each step must contain the <cmd> element, but it can also contain a step result.

## Set the context for examples and scenarios

Be explicit about what's important in an example or scenario so that users will recognize its value. Although you might think that the purpose for including an example or scenario is obvious, your users might not. Be especially aware of the importance of setting the context when you've become so familiar with a product that you've lost the objectivity that is required to write content that is based on the users' perspective.

The introduction to an example or scenario should give users enough information to understand and apply the example or scenario to their own situation.

**Original**

> The following example creates a sequential data set in USER format.
>
> ```
> UNLOAD TABLESPACE DBNAME.TSNAME DB2 NO
> SELECT FNAME, NAME, ADDRESS, DATE_B, SALARY
>    FROM PERSONAL
> OUTDDN (SYSUT1)
> FORMAT USER           (
>    COL FNAME NULLID YES        ,
>    COL 3 TYPE CHARACTER(100)   ,
>    COL SALARY TYPE CHARACTER(7)
>    JUST RIGHT                  ,
>    COL 004 TYPE DATE_A )
> ```

The introduction to the original example does clearly state the results of the code example, but it fails to provide the details that will help users understand why this example is relevant to them.

# Easy to understand

**Revision**

The following SQL example creates a sequential data set in USER format. The new data set is based on data in the PERSONAL table. Refer to this example when you want to create a new data set that is based on data in an existing table.

This example changes data in the following three columns:

- The ADDRESS column is converted to fixed format and its length is increased to 100 characters.
- The SALARY column is converted from decimal to display characters and it is aligned on the units position with the sign first.
- The DATE column is formatted to display dates in MM/DD/YY format.

```
UNLOAD TABLESPACE DBNAME.TSNAME DB2 NO
SELECT FNAME, NAME, ADDRESS, DATE_B, SALARY
   FROM PERSONAL
OUTDDN (SYSUT1)
FORMAT USER              (
   COL FNAME NULLID YES           ,
   COL 3 TYPE CHARACTER(100)      ,
   COL SALARY TYPE CHARACTER(7)
   JUST RIGHT                     ,
   COL 004 TYPE DATE_A )
```

The revision includes an explanation of how this example can be used in other similar situations. It also describes the specific changes that are caused by running this example. By reading these descriptions, users can start to understand how they can use this example in their own work.

# Use similes and analogies to relate unfamiliar information to familiar information

Comparing the unfamiliar with the familiar helps users easily understand new information.

Similes and analogies are figures of speech that explicitly describe one thing in terms of another. Words such as *like*, *as*, and *as if* usually introduce similes and analogies. These figures of speech add color and interest as well.

Some users, for example, might have difficulty understanding the concept of defragmenting a hard drive. However, if you explain that a hard drive can become like a set of encyclopedias that have been put back in the wrong order, and that defragmenting is the process that puts volume A at the beginning, followed by volume B, and so on, users can understand the concept (provided that they are familiar with encyclopedias, of course).

When you use similes and analogies, make the basis of the comparison explicit. Avoid using metaphors, which are implied descriptions of one thing in terms of another.

The following paragraph tries to explain a concept by using a simile:

**Original**

> A *class hierarchy* is a ranking of classes that shows their inheritance relationships. It looks like a tree.

However, a vague assertion about the similarity of a class hierarchy to a tree isn't all that helpful. Various pictures of trees are likely to come to mind, but you don't know why they might look like a class hierarchy.

Easy to understand

**Revision**

> A *class hierarchy* is a ranking of classes that shows their inheritance relationships. It resembles a tree because each class is represented by a single leaf, or *node*. Related classes are connected to each other by lines, which are referred to as *branches*. The single point into which all classes terminate is called the *root node*.

The revised passage gives a reason for the suggested likeness. This situation could also be a good place to use a graphic to support the simile.

Analogies are similar to similes but are usually longer. Rather than relying on one area of similarity, they tend to involve a larger relationship. Consider how an analogy might improve the following passage:

**Original**

> The Status view shows the current activity based on the amount of resource that each app that is running is using. Usage is expressed as one of three status codes.

The original passage states the facts but lacks a framework that users can relate to. This description might be enough for experienced users, but novice users might need more explanation to fully understand the significance of the Status view.

**Revision**

> The amount of concurrent activity on your device affects the performance of an app in the same way that traffic conditions affect the speed of cars. When traffic is light, cars reach their destinations quickly. Conversely, when traffic is heavy, cars take longer to reach their destinations. The Status view shows the traffic conditions for each app that is currently running. These conditions are expressed as the following status codes:
>
> **Status code 1**
> Traffic is light. The app is not experiencing any warning or severe conditions for network resources.
>
> **Status code 2**
> Traffic is heavy. The app is experiencing one or more conditions that are slowing it down. If you need this app to run more quickly, consider stopping any apps that are not critical.
>
> **Status code 3**
> Traffic is stopped. The app is experiencing one or more severe conditions that prevent it from running at an acceptable level. For this app to run, you must stop some of the other apps that are currently running.

The revised passage uses an analogy that many people can visualize and understand.

Often technical information deals with subjects that are new to at least some users. If a subject is well understood by most users, no gap exists in their understanding, but you could choose to use similes or analogies to enliven the information.

When you use similes and analogies, make sure that the items that you use in your comparison are familiar to your entire audience. For example, the simile "The numeric identifier is like a social security number" is relevant only to people who live in the United States. Also, avoid using idioms and colloquialisms because they are, by definition, particular to a discrete set of users. For information about using the right language to address different audiences, see the style guideline "Convey the right tone" on page 267.

# Use specific language

Technical discussions include both general language and specific language. General language is typically appropriate in overviews, summaries, and topic sentences, in which concepts and procedures are introduced. However, in detailed usage information, you need to weed out abstractions in favor of specific language.

The following table provides just a few examples of abstract and general terms and phrases that, depending on the context, should be replaced by more concrete alternatives:

Table 7.1  Abstract words and phrases and concrete alternatives

| Abstract or general word or phrase | Concrete word or phrase |
|---|---|
| data, objects, or items | Documents, records, files, email, applications, spreadsheets, and so on |
| exceeds the size limit | exceeds a specific unit of measurement, for example, exceeds the limit of 16 MB |
| large files | files that are larger than some number, for example, files that are larger than 100 MB |
| problem | can cause the application to stop, can force users off of the system, can result in loss of data, and so on |
| machine | computer, notebook computer |
| resources | CPU resources, amount of disk space, number of people who are available to work on a project, amount of money that is required to develop a product feature, and so on |
| server | administration server, application server, connections server, and so on |
| storage | disk, disk space, memory |
| tools or utilities | migration, installation, reporting, archiving, cataloging, compression tools or utilities |

The use of specific language is essential in user interfaces. When users work with an interface, they are trying to accomplish a task. By using specific language in the interface, you can prevent users from having to stop in the middle of the task to obtain needed information. The following window is a typical login dialog:

**Original**

Server: [               ]

User name: [               ]

Password: [               ]

Depending on the context, the experience of the user, and possibly other factors, the field labels **Server** and **User name** might not be specific enough. Perhaps users have several types of servers in their environment as well as unique user names for those different servers.

**Revision**

Collaboration server: [ accounting01 ▾ ]

Server administrator user name: [               ]

Password: [               ]

Providing labels that are as specific and as meaningful as possible helps users to determine the appropriate values to enter. In the revision, users now have a much better idea about the server and user name that they need to enter. Note that in addition to providing a more precise name for the **Server** field, this field has been converted to a drop-down list of all available collaboration servers, which makes it even easier for users to complete their task. Changing the

Easy to understand

title of the **User name** field to **Server administrator user name** makes it clear that only the server administrator can log in and use this application, which prevents users who aren't administrators from wondering why they are unable to log in.

Another way to increase specificity in a UI is to include *input hints* directly in the UI, like the ones that are provided for the following input fields:

| | |
|---|---|
| Your name: | |
| SJPL ID number: | AB12-3456-78 |
| PIN: | AB1234 |

These hints depict the specific format for the values that users enter in the **SJPL ID number** and **PIN** fields, so users have a pretty good understanding of the values that they need to provide. Also, note that no description is provided for the **Your name** field because it's safe to assume that all users will understand what they need to type in this field.

# Concreteness checklist

When you document a procedure, a concept, an input field in a user interface, or any other technical subject, you can make the information that you write more concrete by relating it to an experience, skill, or knowledge that users already possess.

By using examples, scenarios, programmatic assistance, similes and analogies, graphic elements, and precise terminology that your users can relate to, you are increasing their chances for success. Users are much more proficient when they have a solid understanding of the information that they are using.

You can use the following checklist in two ways:

- As a reminder of what to look for, to ensure a thorough review
- As an evaluation tool, to determine the quality of the information

| Guidelines for concreteness | Items to look for | Quality rating<br>1=very dissatisfied,<br>5=very satisfiedt |
|---|---|---|
| Consider the skill level and needs of users | • Examples, samples, and scenarios are realistic and relevant to your audience.<br>• High-level concreteness elements are provided for novices; advanced concreteness elements are provided for experienced users.<br>• Concreteness elements address the real needs of your audience.<br>• Concreteness elements take into account a culturally diverse audience. | 1 2 3 4 5 |
| Use concreteness elements that are appropriate for the information type | • Embedded assistance includes appropriate programmatic assistance.<br>• User interface labels and control-level assistance is specific and descriptive.<br>• Concepts have appropriate examples, graphics, and scenarios.<br>• Tasks and task steps have appropriate examples and scenarios.<br>• Reference information has appropriate examples and samples.<br>• Troubleshooting information describes the error conditions and provides concrete actions to correct the problems. | 1 2 3 4 5 |

# Easy to understand

| Guidelines for concreteness | Items to look for | Quality rating<br>*1=very dissatisfied,*<br>*5=very satisfied†* |
|---|---|---|
| Use focused, realistic, and up-to-date concreteness elements | • Each concreteness element meets a particular need.<br>• Concreteness elements are realistic.<br>• Concreteness elements are up to date. | 1 2 3 4 5 |
| Use scenarios to illustrate tasks and to provide overviews | • Scenarios are used appropriately to provide overviews or illustrate complex tasks.<br>• Scenarios are based on fictitious but realistic circumstances. | 1 2 3 4 5 |
| Make code examples and samples easy to adapt | • Comments include clear instructions for using examples and samples.<br>• Variables are explained and are easy to identify.<br>• Code samples are provided online or with the product so that users can easily modify them for their own purposes. | 1 2 3 4 5 |
| Set the context for examples and scenarios | • The important or relevant aspect of examples and scenarios is clear.<br>• Introductions match examples and scenarios. | 1 2 3 4 5 |
| Use similes and analogies to relate unfamiliar information to familiar information | • Analogies and similes are provided where needed.<br>• The basis for comparison is clear and doesn't rely on culturally specific situations. | 1 2 3 4 5 |
| Use specific language | • General language is limited.<br>• Specific language addresses users' needs. | 1 2 3 4 5 |

To evaluate your information for all quality characteristics, you can add your findings from this checklist to Appendix A, "Quality checklist," on page 545.

CHAPTER 8

# Style

*The best style is the style you don't notice.* —W. Somerset Maugham

Style is the correctness and appropriateness of writing according to conventions. Like visual effectiveness, style is an expression of the "look and feel" of information. Maintaining a style for technical content means following standards and rules to ensure consistency and correctness.

Style and content are inextricably linked: consistent style helps users focus on the content and not on why one term is highlighted in bold in one place but not another. However, style isn't simply decoration that you apply to information—style helps users understand the information.

Generally, editors create style guidelines for the organization to ensure consistency across sets of information or across product interfaces. Style guidelines help writers create consistent information by providing quick answers to questions about highlighting, formatting, spelling, punctuation, tone, and markup language tagging. Not only does this consistency lend credibility and instill a sense of corporate identity to a company's content, it also helps teams, writers, and organizations deliver content that users expect from a product or line of products.

Many organizations also use specific templates, document type definitions (DTDs) for markup languages, reusable software code for user interfaces, cascading style sheets (CSSs), or Extensible Stylesheet Language (XSL) to

## Easy to understand

structure and format various types of content. Templates and consistent tagging help ensure that content has consistent appearance, organization, and required standard information, such as prerequisites in task topics.

Style entails making choices about tone, voice, cultural references, terminology, and presentation. To make information stylistically appropriate, correct, and consistent, follow these guidelines:

- Use active and passive voice appropriately
- Convey the right tone
- Avoid gender and cultural bias
- Spell terms consistently and correctly
- Use proper capitalization
- Use consistent and correct punctuation
- Apply consistent highlighting
- Make elements parallel
- Apply templates and reuse commonly used expressions
- Use consistent markup tagging

# Use active and passive voice appropriately

To make your sentences clear and engaging, use the active voice most of the time. Passive voice can be effective in some situations, but it also makes sentences longer and somewhat more cumbersome to read. If you want to emphasize who or what performs some action, use active voice. If you don't want to emphasize who or what does the action, use passive voice.

Active voice puts the agent of an action at the beginning of a sentence whereas in a passive sentence, the agent comes at the end of the sentence or isn't mentioned at all.

The following statement about how database administrators maintain catalogs is written in both active and passive forms:

| Active | The system administrator maintains the database catalogs. |
| --- | --- |
| Passive | The database catalogs are maintained by the system administrator. |

Passive voice requires a longer verb phrase and places the agent after the verb, or it doesn't mention the agent. In the sample passive sentence, the verb is *are maintained*, which is longer than the active verb *maintains*. The agent of the action is *administrator*, which comes at the end of the sentence.

Passive voice can be especially confusing when it's used in consecutive sentences in a paragraph.

**Original**

> To minimize the effects of high-frequency interference and other undesired electrical signals, a signal cleanup system (SCS) can be recommended. An SCS can be made up of a signal reference module, which is also known as a zero signal ground. Regardless of the name used, the intent is to provide an equal point of reference for equipment that is installed in a contiguous area for a wide range of frequencies. This is accomplished by installing a network of low impedance conductors.

In the original paragraph, every sentence contains a passive verb. The paragraph plods along as if it were stuck in deep mud.

# Easy to understand

**Revision**

> To minimize the effects of high-frequency interference and other unwanted electrical signals, you can use a signal cleanup system (SCS). You can create an SCS that includes a signal reference module, which is also known as a zero signal ground. The SCS, or zero signal ground, provides an equal point of reference for equipment that is installed in a contiguous area for a wide range of frequencies. To configure the equal point of reference, install a network of low impedance conductors.

Because the revised paragraph uses mostly active voice, users will understand who or what does the actions. They can also read and retain the information more effectively.

Use active voice in most situations, but use it especially when you write instructions. Passive voice in a step of a procedure can be confusing. In the following example, step 2 is written in passive voice:

**Original**

> 1. Connect the printer cable to your computer.
> 2. The printer should now be turned on.
> 3. Allow the computer to recognize the printer.
> 4. When the printer is successfully installed, print a test page.

What does step 2 really mean? Does the printer somehow turn itself on, or should it have been turned on at some earlier time? Users might be left wondering who or what turns on the printer.

**Revision**

> 1. Connect the printer cable to your computer.
> 2. Turn on the printer.
> 3. Allow the computer to recognize the printer.
> 4. When the printer is successfully installed, print a test page.

The revised step 2 makes it clear that the user must turn on the printer and when that action must be done.

You should use active voice when the agent of the action is important. Which of the following sentences is more appropriate for a marketing brochure?

| Active | The new InfoDBase software simplifies backup and recovery. |
|---|---|
| Passive | Backup and recovery is simplified by the new software. |

Although users likely won't be confused by reading either sentence, the active sentence focuses on the new InfoDBase software, which is what you might want users to buy or upgrade.

In some cases, you should use passive voice if the agent of the action isn't important or isn't obvious. Passive voice sentences are often more appropriate in reference information or formal scientific and engineering papers. In these types of information, the agent of the verb isn't relevant.

| Active | During a tonsillectomy, surgeons remove both tonsils from recesses on the sides of the pharynx. |
|---|---|
| Passive | During a tonsillectomy, both tonsils are removed from recesses on the sides of the pharynx. |

The active sentence causes users to focus on the noun *surgeons* rather than on *tonsils*. Although the original sentence is shorter, the passive sentence emphasizes the more important point of the sentence.

Passive voice can also be appropriate in error messages. In the following error message, users don't need to know that the system is doing some action:

**Original**

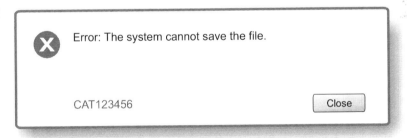

# Easy to understand

**Revision**

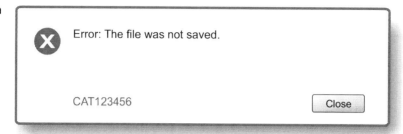

The passive voice is more appropriate in this case because what's important is that the file wasn't saved, not who or what is trying to save the file.

You might also use passive voice in error messages to avoid blaming users for making mistakes.

**Original**

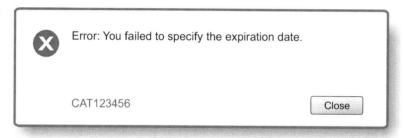

**Revision**

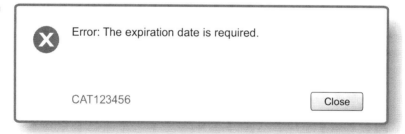

No one wants to be seen as failing anything, so using passive voice for such a message avoids blaming users by focusing on the problem instead of the user.

In general, use active voice unless the content and audience are better served by passive voice. Too much passive voice can make your writing more difficult to read, less engaging, and potentially misleading.

# Convey the right tone

Tone conveys how a piece of writing "sounds" or how it feels to the user. Tone also expresses your attitude toward the subject and the audience. An appropriate and consistent tone helps users do the task, learn the concepts, or make the right decisions without needing to interpret the rhetoric.

The tone of most technical information must be:

**Helpful**
Concrete and accurate information establishes a helpful tone.

**Direct**
Writing clear information establishes a direct tone.

**Authoritative**
Writing task-oriented, accurate, and complete information establishes an authoritative tone.

Many different tones are possible in technical writing. To achieve the tone that works best for a particular situation, you must understand the audience and the purpose for the information. For example, if you're writing a marketing brochure, you probably want to use a different tone than you would for a reference manual. Or if you're writing embedded assistance for a mobile app targeted to younger users, you might use a more playful, informal, and possibly humorous tone.

Easy to understand

## Informal tone

To achieve an informal tone, you can use contractions and a more conversational style. For example, the following web page uses an informal tone for a form that users complete if they have problems with the site:

The tone conveyed in this form is informal without being annoying. If you know that most of your users won't be confused or annoyed by the informal phrases, such as *What's going on*, and you think that your users will appreciate the friendly tone even when they're having problems, you can use the informal language.

Sometimes, you might be tempted to use slang, colloquial expressions, or idiomatic expressions to create a friendly, less intimidating tone. An *idiom* is a group of words that is established by usage as having a meaning that cannot be deduced from the individual words in that expression. For example, *at this stage*, *pass the time*, and *piece of cake* are idioms. Most idioms are colloquial, which means that they are used only in speech. Slang can be idiomatic, but slang terms and expressions are common to a specific group of people, time, or culture. For example, *cool* (to mean good or excellent) and *snail mail* (to mean physical mail as opposed to electronic mail) are terms used as slang.

Because language that is used to make information sound informal or friendly, such as humor, slang, or colloquialisms, can change over time, and because it can be misinterpreted, ensure that you reevaluate your word choice regularly. Update the text as necessary so that your tone doesn't seem dated, offensive, or misleading.

If you think that your tone is too casual or possibly offensive, be sure to test it with internal users from your own organization or in closed test sessions before you make the information public. If you translate, you should also ask your translation coordinator whether the information can be easily translated.

When you choose an informal tone, be sure that the tone isn't sarcastic or childish. Make sure you know under what conditions users might see that informal language. For example, don't be too casual when users just want to get something done quickly, for example, when they book flights or hotels. Or don't be flippant in an error message when users are already frustrated.

**Original**

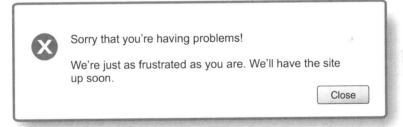

**Revision**

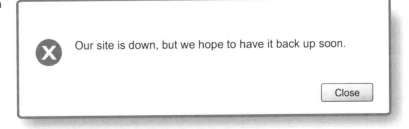

# Easy to understand

The revised message, although still informal, removes the patronizing and flippant tone by using more direct and neutral language.

## Pretentious tone

A pretentious tone is in stark contrast to a humorous, informal, or playful tone. Unfortunately, technical information is sometimes written with a pretentious tone. To be seen as credible and authoritative, writers might be tempted to use complex sentences, abstract technical terms, passive voice, and wordy phrases. This tone can be confusing and even intimidating.

**Original**

> One must recognize that regional expenditures could be diminished by the new accounting application.

The original sentence is not only confusing and wordy, but it is also pretentious. Using the pronoun *one*, passive voice, and overly formal words such as *diminish* and *expenditures* gives the sentence a pretentious tone. Such a tone tends to alienate users rather than engage them.

**First revision**

> Everyone must know that the new accounting application can reduce travel expenses.

If the idea that "everyone must know" is not important, you can omit it from the sentence:

**Second revision**

> The new accounting application can reduce travel expenses.

The last revised sentence conveys a neutral tone by using shorter words and active voice.

## Neutral tone

Although an informal tone can be effective under certain circumstances, using a more neutral, formal tone is generally easier to write, maintain, and translate. For example, in the following progress message, not only is the tone sarcastic, but some of the language will be difficult to translate:

**Original**

> **Searching...**
>
> We hope you can wait until we search our really huge warehouse. Please hang with us...
>
> Sorry...still lookin...

The original sentence tries to convey an informal tone by using the slang expression *please hang with us*. That language can quickly become passé and silly. The tone might also be especially frustrating to users who must already wait for the search to complete. The informal, somewhat flippant tone might even be offensive in some cultures.

**Revision**

> **Searching...**
>
> Please wait while we search for your items...

The revised message uses direct, formal English, and is therefore less likely to offend users or confuse translators.

Marketing or sales content is likely to have an informal tone. Sometimes, writers will try to impress users by using colloquialisms to create a friendly tone and make the product more appealing. However, be sure that you know your audience. If your content will be seen by nonnative English speakers, avoid colloquialisms, slang, and humor. Use a more neutral tone that won't be confusing, misleading, or insensitive to nonnative speakers of English.

# Easy to understand

**Original**

> InfoDBase will bend over backward to store all of your critical data.

Although *bending over backward* is an idiom that most native English speakers have heard and understand, it is difficult to picture it applied to a software application. Even native English speakers might have different interpretations as to what that phrase means.

**Revision**

> InfoDBase can quickly store all of your critical data.

The revised sentence uses a neutral tone and makes a more direct, concrete point than the original sentence without making users interpret an idiomatic expresssion.

## Avoid gender and cultural bias

Avoid making assumptions about or inappropriate references to gender, lifestyle, culture, or race in your writing. Pay special attention to the names of people and places, phone numbers, email addresses, the use of color, and other potentially culturally sensitive references.

**Original**

| Last name | First name | Phone number | Email address |
|---|---|---|---|
| Carter | Adam | (905) 954-6833 | carter@domain.com |
| LeBlanc | Paul | (301) 266-2110 | pleblanc@oursite.net |
| Taylor | John | (210) 454-3241 | taylorj@thissite.com |
| Smith | Fred | (408) 555-4444 | smithf@site.com |

In the original table, only male Anglo-Saxon names and US phone numbers are used.

**Revision**

| Surname | Given name | Phone number | Email address |
|---|---|---|---|
| Luchsinger | Sabine | 0041 44 101 12 51 | sabinel@oursite.net |
| Nguyen | Tom | (08) 3898 2083 | nguyentom@thissite.com |
| Carey | Linda | (408) 555-4444 | lcarey@domain.com |
| Rodrigues | Gary | 00351 21 206 08 02 | grod@site.com |

Even though the names in the table don't refer to real people, the revised names to some extent reflect the diversity of an international audience and avoid the sexism of an all-male list. The original table also reflects a bias toward the construction of names. For example, many cultures show their surnames first. The revised table uses the terms *Surname* and *Given name* to accommodate names from other cultures.

Include references to gender or race only if they are relevant. For example, the following sentence is acceptable in scientific information:

> The clinical study included 10 female patients and 12 male patients.

# Easy to understand

In addition, be careful with references to occupations. Don't write a sentence that implies that all nurses or administrative assistants are women or that all doctors and software programmers are men. The following sentence makes an assumption about the gender of an administrative assistant:

**Original**

> When errors occur, the administrative assistant can contact her manager.

You can revise the original sentence by changing the nouns and pronouns to plural:

**Revision**

> When errors occur, administrative assistants can contact their managers.

When you correct sexist language, avoid using *he or she* as in the following sentence:

> If the programmer tracks his or her tasks, he or she can determine his or her productivity.

The use of *he or she* or *his or her* quickly becomes awkward and stilted. Instead of using *he or she*, use plurals, the imperative mood, second person, or proper nouns, depending on the context of the particular sentence.

| Revision | Comments |
|---|---|
| If programmers track their tasks, they can determine their productivity. | Uses a plural noun (*programmers*) and a plural pronoun (*their*). |
| | Be careful not to mix singular nouns with plural pronouns. For example, this sentence is ungrammatical: "If a programmer tracks their tasks, they can determine their productivity." |
| | Although many people speak this way, the noun *programmer* does not agree with the pronouns *their* and *they*. |
| If you track your programming tasks, you can determine your productivity. | Uses an indicative verb and the second person (*you* and *your*). |
| Track your programming tasks so that you can determine your productivity. | Uses an imperative verb (*track*) and the second person (*you* and *your*). |
| For example, if Maria tracks her programming tasks, she can determine her productivity. | Uses a person's name, which can indicate that person's gender. |

Use of references to flags, countries, states, nations, or geographical borders can lead to unwanted political or legal problems for your company or organization. Even the use of colors, symbols, body parts, and animals can have culturally specific connotations. For more information about designing graphics for a diverse audience, see the visual effectiveness guideline "Ensure that visual elements are accessible to all users" on page 478. Also, see the topic "Writing for diverse audiences" in the book *The IBM Style Guide*.

# Easy to understand

## Spell terms consistently and correctly

Many industries and disciplines frequently invent terms, and those terms can take several months or years to be defined by established commercial dictionaries. For example, primitive email systems were invented as early as 1961, but the term *email*, or *e-mail*, was not added to a commercially available dictionary until 1982, according to Merriam-Webster dictionary. You must ensure that spellings are consistent and correct within topics, libraries, and user interfaces.

The high-tech industry continues to invent new terms that have multiple spellings, for example:

| | |
|---|---|
| addon, add-on, add on | read-write, read/write, read, write |
| blog, weblog | real-time, real time |
| classpath, class path | runtime, run-time, run time |
| client-server, client/server, client and server | sign in, signin, sign on, signon |
| combo box, combination box | signout, sign out |
| ebook, eBook | single sign-on, single signon |
| epub, epublication | standalone, stand alone, stand-alone |
| filename, file name | tablespace, table space |
| login, log in, logon, log on, log into, log in to | userID, userId, user ID |
| logoff, log off of, log off from | username, user name |
| logout, log out | website, web site, Web site |
| offload, off-load, off load | web, Web |
| online, on-line | vlog, vid-blog, movie blog |

In some cases, a simple spelling error might be annoying to the users of your product, but in other cases, a misspelling can be confusing and lead your users or translators to mistrust or misinterpret your information.

In the following passage, two different spellings are used for a *vlog*, which is a weblog that includes video clips:

**Original**

> You can upload your video log to any of the available servers. The vlog remains on the server for only 30 days.

**Revision**

> You can upload your video log to any of the available servers. The video log remains on the server for only 30 days.

This spelling inconsistency might lead users to think that a *vlog* and a *video log* are two different things, especially if those users are new to this technology. A vlog might be misinterpreted as some type of log file. The revised passage uses one spelling: *video log*.

Even subtle differences in spelling, such as between *toward* and *towards*, can be distracting if you don't use one version and use it consistently. For example, many words in English have variable spellings; in some cases, the differences stem from the dialect of English that is spoken. Words such as *recognize* and *recognise*, *color* and *colour*, *catalog* and *catalogue*, and *license* and *licence* are spelled differently depending on whether they are used in the United States or Great Britain.

Other terms have developed multiple spellings over time. For example, the words *cancelation* and *labeled* can be spelled with one *l* or two.

Another area of spelling that often troubles writers is knowing the correct spelling of hyphenated terms. Spelling of English words can be closed, open, or hyphenated:

### Closed
A word with two or more parts, such as a prefix and main word, is spelled as one word, for example, *nonlinear* or *microchip*.

### Open
Words are spelled as two or more words, for example, *mid Atlantic*, *half century*, *gray blue*, or *call out* (when used as a verb).

### Hyphenated
Word parts are connected by a hyphen, for example, *anti-inflammatory* or *bi-level*.

# Easy to understand

Often, a spelling might start as hyphenated and then evolve to be one word. For example, the word *email* used to be spelled with a hyphen (*e-mail*), but most style guides and dictionaries now mandate a closed spelling: *email*.

Be careful with the spelling of words that have prefixes and suffixes. For example, most words spelled with the prefixes *non*, *pre*, and *post* are rarely hyphenated unless they are prefixed to proper nouns or have unusual spellings (according to the *Chicago Manual of Style* and many other US style guides). Words such as *nonprofit*, *preassemble*, and *postrequisite* are never hyphenated, but *non-European* is hyphenated because the prefix occurs before a proper noun.

Spellings are further complicated by the part of speech of the word in question. Often, a spelling is closed or hyphenated when the word is an adjective or noun and open (two words) when the word is a verb. For example, you should spell the word *log in* as two words when it's a verb and one word *login* when it's an adjective or noun.

Never hyphenate or change a trademark in any way because a trademark is a type of intellectual property and is protected by law. For example, never add *non* or *'s* to a trademark such as *Java*. In many cases, usage guidelines for trademarks are provided on a legal web page by the owner of the trademark.

The following list shows the preferred spelling of some common technical terms.

| Noun or adjective | Verb |
|---|---|
| add-on | add on |
| backup | back up |
| callout | call out |
| fallback | fall back |
| login or logon | log in or log on |
| offload | offload |
| plug-in | plug in |
| setup | set up |
| sign-on | sign on |
| timeout | time out |

In addition to the problems of spelling technical terms, check the spelling of often confused terms, such as *principal* and *principle*. For example, in the following sentence, the word *recreate* is used incorrectly:

**Original**

> You must recreate the schema after you upgrade the application.

The word *recreate* means to give new life to or take recreation. It doesn't mean to create something again.

**Revision**

> You must re-create the schema after you upgrade the application.

The revised sentence uses the correct term *re-create*, which means to create something again. You must be careful with often confused or similarly spelled terms because most spell-checking tools cannot determine whether words such as *affect* or *effect* are used correctly in a particular sentence.

The following words are often confused or used incorrectly:

| | |
|---|---|
| accept, except | its, it's |
| access, assess | lay, lie |
| advice, advise | loose, lose |
| affect, effect | principal, principle |
| capital, capitol | recreate, re-create |
| cite, site, sight | than, then |
| compliment, complement | to, too, two |
| disc, disk | whose, who's |
| ensure, insure | your, you're |

To ensure that terms are spelled correctly, invest in spell-checking tools that can reference an accepted list of terms and their correct spellings.

# Use proper capitalization

Incorrect or inconsistent capitalization can cause confusion and affect translation. To make matters more confusing, the rules for capitalization vary from one country to the next. For example, in English and German, you capitalize the days of the week, but in French and Spanish, you don't. If you work in a global organization, you must decide which capitalization conventions to follow.

Capitalization rules can also vary for technical terms and spelled out abbreviations. Decisions about whether to capitalize names of technologies, components, and products can affect the way those terms are translated. When translators see a capitalized term, some might think it's a proper noun and therefore not translate the term.

For example, the following sentence unnecessarily capitalizes the term *Reminders*:

> You can configure Reminders to notify others when a record will be deleted.

A reminder is simply a notification sent through email and is used by users who work in a typical business, such as an insurance office. Capitalizing the term might give it unneeded attention or make users think it's something other than a simple reminder.

If your organization stores translated content in a database, you can incur extra cost if you have inconsistencies in capitalization. Those inconsistencies can cause the translator to translate the same nearly identical sentence twice because of the way that matching works in translation memory databases.

For example, in some languages these two sentences can be translated differently and will be stored as two different sentences in translation memory:

> The database file is stored in the Configuration folder.
>
> The database file is stored in the configuration folder.

Even if you do want to capitalize such terms, be sure that the capitalization is consistent.

For the text in a user interface, decide how to capitalize items such as table column headings, hover help text, field labels, and button labels. Inconsistent capitalization can cause problems with translation and can confuse users.

For example, all the field labels except one in the following interface use sentence-style capitalization, which means that only the first word is capitalized. Users might think that the *Office of Record* is somehow special or different from the other values.

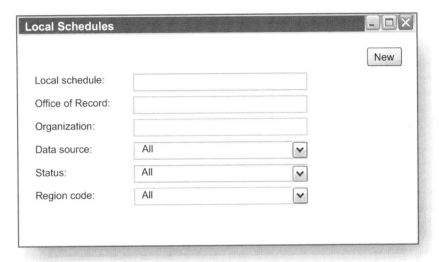

Users might be further confused when they see the same text in a different area of the interface with different capitalization:

## Easy to understand

Ensure that the elements in user interfaces are consistently capitalized. To help with that task, you can follow these capitalization rules:

| Capitalization style | Interface element |
|---|---|
| Sentence style | Check box labels |
| | Field labels |
| | Hover help text |
| | Inline (embedded) help text |
| | Input hints |
| | List boxes |
| | Message text (information, warning, and error) |
| | Radio button labels |
| | Status bar text |
| Headline style | Icon labels |
| | Links that are used for actions |
| | Menu choices, menu titles, and drop-down lists |
| | Navigation links |
| | Push button labels |
| | Section headings |
| | Table column headers |
| | Tab titles |
| | Table titles |
| | Window and dialog titles |

For example, if you applied the capitalization rules from the previous list, your user interface would look like this:

| Window Title |
|---|
| Product Name |
| Menu Item 1    Menu Item 2 |
| Tab Title 1   Tab Title 2  <br><br>Section Heading  <br>Page instructions. <u>Navigation Link</u>.    Table Title  <br><br>Field label: [       ]    \| Column Header \| Column Header \|  <br>[▼] Drop-Down List  <br><br>☐ Check box label  <br>⦿ Radio button label        [ Push Button Label ]  <br><br>Error message. |
| Status bar text |

For more capitalization guidelines, see "Capitalization" in *The IBM Style Guide*.

Easy to understand

# Use consistent and correct punctuation

Most organizations must create or adopt guidelines for punctuation because no single authority for English punctuation exists that provides rules for every situation. Just as writing styles change over time, some of the guidelines that govern the use of punctuation also change. For example, the newspaper industry currently does not use a serial comma before the last item in a list whereas in technical writing, this comma is often mandatory.

Technical style guides differ on the rules for punctuation. For example, some style guides require quotation marks around new terms while other guides mandate bold or italic highlighting. Or some guides instruct writers to use periods at the end of tooltips and others don't. If your organization doesn't have guidelines, compile a list of punctuation rules and ensure that those rules are appropriate for the industry or technology that you write for.

Although this book is not meant to replace a comprehensive set of punctuation rules, the advice for the following types of punctuation is applicable in most technical information:

- Commas
- Colons
- Semicolons
- Em dashes and hyphens
- Combinations of punctuation marks

In addition to these punctuation marks, consider how you'll apply other types of punctuation, such apostrophes, parentheses, quotation marks, and slashes. And decide how rules for punctuation might differ when you write embedded assistance in user interfaces. For example, decide whether to use periods in hover help text or tooltips.

For more information about the use of punctuation, see *The IBM Style Guide*.

## Commas

Commas are perhaps the most misused punctuation mark because they are used in so many ways. However, the following guidelines cover only the most common rules for commas.

### Separating essential and nonessential phrases and clauses

The most confusing comma rules are those for essential and nonessential (sometimes called *restrictive* and *nonrestrictive*) phrases and clauses. A phrase or clause is essential if it cannot be removed from a sentence without altering the meaning of that sentence. Nonessential means that the information can safely be removed from the sentence without affecting its meaning.

Enclose nonessential phrases and clauses in commas, but don't use commas around essential clauses and phrases. For example, the following *who* relative clause is nonessential and is correctly surrounded with commas:

> The InfoCorp IT administrator, who typically works with Windows servers, will install the new AIX servers.

The main point of the sentence is that the administrator will install the servers. Whether she works with Windows servers, though possibly interesting, is nonessential information.

The following subordinate clause, which starts with *although*, is nonessential and therefore must be preceded by a comma:

> Arthroscopic procedures can reduce patient risks and recovery time, although all invasive procedures create some risks.

The subordinate clause might be interesting, but it could be removed from the sentence.

A common mistake is to always place a comma before a clause that starts with *because*. Most of the time, the information in a *because* clause is essential. In this example, the sentence correctly omits the comma:

> You cannot check in the document because it is not checked out.

Easy to understand

## Separating sentences that use different types of conjunctions

Conjunctions are words that connect words, phrases, and clauses. There are three types of conjunctions:

**Adverbial**
  Includes words such as *however*, *finally*, and *therefore*.

**Coordinate**
  Includes *for*, *and*, *nor*, *but*, *or*, *yet*, and *so*; you can remember the coordinate conjunctions with the mnemonic *FANBOYS*.

**Subordinate**
  Includes words such as *because*, *even though*, *so that*, *whereas*, and *while*.

The general rules for each conjunction type are as follows:

### Joining sentences with adverbial conjunctions

When two or more sentences are joined by an adverbial conjunction, use a period or semicolon before the conjunction:

| | |
|---|---|
| Incorrect | The task was completed, however, some data cannot be processed. |
| Correct | The task was completed. However, some data cannot be processed. The task was completed; however, some data cannot be processed. |

The error in the incorrect sentence is called a comma splice, which is described in "Avoiding comma splices and run-on sentences" on page 287.

### Joining sentences with coordinate conjunctions

When two or more sentences are joined by a coordinate conjunction, use a comma before the conjunction:

| | |
|---|---|
| Incorrect | The task was completed yet some data cannot be processed. |
| Correct | The task was completed, yet some data cannot be processed. |

If the second half of the sentence doesn't have a subject and complete verb, don't use a comma:

| | |
|---|---|
| Incorrect | The configuration manager deploys the application, and removes cookies. |
| Correct | The configuration manager deploys the application and removes cookies. |

## Joining clauses with subordinate conjunctions

When you start a sentence with a subordinate clause (also called a *dependent clause*), which is a clause that starts with a subordinate conjunction, place a comma at the end of the clause:

| Incorrect | Even though some data cannot be processed the task was completed. |
|---|---|
| Correct | Even though some data cannot be processed, the task was completed. |

If a subordinate clause appears after an independent clause, use a comma between the clauses only if the second clause (subordinate) is nonessential. In the following example, the clause that starts with *because* is essential, which means that the information is necessary for users to understand the sentence:

| Incorrect | Damaged tendons require more time to heal than other types of tissue, because they receive little blood flow. |
|---|---|
| Correct | Damaged tendons require more time to heal than other types of tissue because they receive little blood flow. |

By knowing the difference between the three types of conjunctions, you can understand when and where to place the comma.

## Avoiding comma splices and run-on sentences

A *comma splice* occurs when you join two or more sentences (independent clauses) with just a comma. A *run-on sentence* is nearly the same except that no comma is used:

| Comma splice | You must install the application on the P33 server, that server is not yet available. |
|---|---|
| Run-on | You must install the application on the P33 server that server is not yet available. |

The clauses on each side of the comma can stand alone as a complete sentences. Therefore, the clauses cannot be joined with only a comma. To fix the problem, you can insert a period or semicolon between each clause, or you can insert a subordinate or coordinate conjunction. If you want to use an adverbial conjunction, you must use a period or semicolon. For example, here are several easy ways to fix comma splices and run-on sentences:

| Punctuation | Example |
|---|---|
| Period | You must install the application on the P33 server. That server is not yet available. |
| Period with an adverbial conjunction | You must install the application on the P33 server. However, that server is not yet available. |
| Semicolon | You must install the application on the P33 server; that server is not yet available. |
| Semicolon with an adverbial conjunction | You must install the application on the P33 server; however, that server is not yet available. |
| Coordinate conjunction | You must install the application on the P33 server, but that server is not yet available. |
| Subordinate conjunction | You must install the application on the P33 server although that server is not yet available. |

## Separating items in a series

Use commas to separate three or more nouns in a series. Whether you use the Oxford, or serial, comma depends on your style guide. To avoid ambiguities or misreading, you should add the serial comma before the last item in the series.

For example, the following two sentences have different meanings:

> Check the servers once per week for heat discharge, floor levels, connection status and security.
>
> Check the servers once per week for heat discharge, floor levels, connection status, and security.

In the first sentence, you might read the last phrase as *connection status* and *connection security*. Or you could read the last phrase as *connection status* and *security*. Without the comma, you can't be sure.

If you did intend the meaning to be *connection status* and *connection security*, you should insert the adjective *connection* before both nouns.

Depending on where you place the comma, users might interpret the sentences differently. If the information is translated, you might get different translations.

Commas are also used to separate adjectives in a series depending on whether each adjective modifies the noun individually. One easy way to check that each adjective separately modifies the noun is to try to insert the word *and* between each adjective. If inserting *and* makes sense, the comma is required.

In the following example, the adjectives *short* and *forged* separately modify the noun *pistons*:

| Incorrect | You can use short forged pistons in some two-stroke engines. |
| Correct | You can use short, forged pistons in some two-stroke engines. |

However, in some cases, the adjectives don't separately modify the noun:

| Incorrect | The new, record class will be available in the next release. |
| Correct | The new record class will be available in the next release. |

The adjectives are *new* and *record*, but *record* modifies the noun *class*, and *new* modifies *record class*. In addition, you cannot insert the word *and* between *new* and *record class*. The sentence "The new and record class will be available in the next release" doesn't make sense. Therefore, you don't use commas.

## Colons

Unlike commas, colons have only a few uses. They are most often used to introduce unordered or ordered lists or after field labels in user interfaces. Colons are also used to illustrate or exemplify what was stated before the colon.

### Introducing lists

When you introduce an unordered list, use a colon at the end of the introductory sentence. To make your text easier to translate, use a complete sentence before the colon:

| Fragment | You can install:<br>• Relational database<br>• Application server |
| Full sentence | You can install the following components:<br>• Relational database<br>• Application server |

One common problem with colons occurs when the colon is placed after a verb to introduce a list in a sentence. In the incorrect example, the verb *includes* is followed by a colon:

| Incorrect | A workflow management system includes: work items, activities, inboxes, and records. |
|---|---|
| Correct | A workflow management system includes work items, activities, inboxes, and records. |

This sentence doesn't require a colon because the verb adequately introduces the list of items. The colon is redundant.

### Explaining or illustrating ideas

Colons can also be used to illustrate, explain, or exemplify a point.

| No logical relationship indicated | The system programmers in region A can more effectively maintain the web application servers. The servers are housed in that facility, and the system programmers have the most up-to-date training. |
|---|---|
| Logical relationship indicated | The system programmers in region A can more effectively maintain the web application servers: the servers are housed in that facility, and the system programmers have the most up-to-date training. |

The second example clearly explains why the system programmers in region A are better suited to maintain the server because the colon helps to establish the logical relationship between the first sentence and the second. Although using a period is grammatically correct, the colon more effectively communicates the relationship between the two sentences.

You might be tempted to use an em dash between the two sentences, but the em dash is informal, and it's not typically used to illustrate or explain ideas.

### Emphasizing ideas

You can use colons before a word, phrase, or full sentence to emphasize an idea. You can also use a comma, but the colon adds more emphasis and formality. To be more emphatic, you can use an em dash, but em dashes are considered informal.

The following examples are listed in order of emphasis, with the em dash being the most emphatic:

| | |
|---|---|
| Comma | InfoDBase includes the latest technology for your web applications, SyncXML. |
| Colon | InfoDBase includes the latest technology for your web applications: SyncXML. |
| Em dash | InfoDBase includes the latest technology for your web applications—SyncXML. |

## Semicolons

Semicolons are used to separate two closely related sentences, but determining how closely related the two sentences are is somewhat subjective.

The following sentence correctly uses a semicolon between two closely related ideas:

> Trace amounts of condensation can accumulate on the condenser; this condensation must be removed periodically.

Unfortunately, semicolons are often used incorrectly to join an independent clause to a dependent clause. In most cases, you must use complete sentences on both sides of the semicolon.

The following incorrect example uses a semicolon before a dependent clause:

| | |
|---|---|
| Incorrect | Configuring retention policies can reduce the volume of archived files; although you must still provide enough storage for retained records. |
| Correct | Configuring retention policies can reduce the volume of archived files, although you must still provide enough storage for retained records. |

Because of the incorrect use of the semicolon, the part of the sentence that starts with *although* is a sentence fragment, or incomplete sentence. That clause can't stand alone, and therefore it can't be preceded by a semicolon.

The correct sentence uses a comma instead of a semicolon before *although* because the clause that starts with *although* is nonessential in this case.

Semicolons can also be used to separate phrases or clauses in a series when each phrase or clause includes commas. For example, the following sentence includes many commas and would be easier to read if semicolons separated each element:

| Incorrect | You install the server by using an interactive mode, which requires you to use the installation wizard, a silent mode, which requires you to use a response file, or a command-line mode, which does not require a response file. |
|---|---|
| Correct | You install the server by using an interactive mode, which requires you to use the installation wizard; a silent mode, which requires you to use a response file; or a command-line mode, which does not require a response file. |

However, in technical information, avoid creating such sentences with semicolons because the information is harder to scan. Instead, use an unordered list to make your information more retrievable.

## Em dashes and hyphens

In most authoring software, the em dash (—) and the hyphen (-) are different symbols and aren't interchangeable. Each is used for different situations.

### Em dashes for emphasis

Use em dashes, which are the width of the letter *m* in most fonts, to dramatically emphasize some part of a sentence. However, em dashes should be avoided in formal technical information if you want your tone to be neutral and direct. Because the em dash draws attention to the content that it sets off so strongly, it adds a level of informality to the text.

If you write less formal information, such as marketing information, you can use em dashes for emphasis. For example, this passage from marketing content effectively uses an em dash to highlight the product's strengths:

> The most scalable and powerful database on the market—InfoDBase Server Version 6—is now available for Linux environments.

If you do use em dashes, be sure to use them only for essential information. For example, the following sentence incorrectly uses an em dash around nonessential information.

| Incorrect | Additions to the project schedule—such as deadlines, requirements, and project codes—must be entered before the end of the month. |
| --- | --- |
| Correct | Additions to the project schedule, such as deadlines, requirements, and project codes, must be entered before the end of the month. |

The list of deadlines, requirements, and codes can be removed from the sentence without altering the main idea that additions to the schedule must be entered at a certain time.

Be careful not to overuse em dashes because they'll lose their effect.

If you do use em dashes, follow these guidelines:

- Use the correct symbol for the em dash with no spaces on either side of the em dash.
- Use the em dash either around a word or phrase or just before a word or phrase at the end of a sentence.

### Hyphens in spelling and compound words

Hyphens are used in some spellings in words with prefixes, suffixes, and medial elements (such as *in* in *mother-in-law*). For example, you hyphenate a prefix to a proper noun: *Indo-European*. Most common nouns that use prefixes and suffixes aren't hyphenated; for example, use *preinstallation*, not *pre-installation*.

You also use hyphens in compounds (two or more words) that are placed before a noun, such as *state-of-the art appliances* or *twentieth-century technology*. For example, *well-formed* is a compound adjective in the following sentence and requires a hyphen:

| Incorrect | The application requires well formed XML. |
| --- | --- |
| Correct | The application requires well-formed XML. |

You don't typically need the hyphen if the compound adjective comes after the verb:

| Incorrect | The XML must be well-formed. |
| --- | --- |
| Correct | The XML must be well formed. |

Be sure never to hyphenate words that end in *-ly* when they form a compound noun:

| | |
|---|---|
| **Incorrect** | Enter a fully-qualified path to the server. |
| **Correct** | Enter a fully qualified path to the server. |

## Combinations of punctuation marks

In addition to knowing the rules for each type of punctuation mark, you'll also need to determine the rules for using combinations of punctuation marks. Some style guides differ on these rules, so ensure that you stick to one set of rules for your information.

The most common combination that you need to clarify is how to use commas or periods with quotation marks. Both of the following sentences are correct, depending on the style guide that you adopt:

> For more information, see "Adding dimensions to cubes."
>
> For more information, see "Adding dimensions to cubes".

The period can go inside or outside the quotation marks depending on which style guide you follow. The same issue occurs with commas:

> New parameter strings, such as "start console," "link next," and "end process," must be added to the reference tables.
>
> New parameter strings, such as "start console", "link next", and "end process", must be added to the reference tables.

Again, both are correct, but you should choose one style and apply it consistently.

Although rules for how periods and commas are used with quotation marks are not universally accepted, other rules such as how to use periods with parentheses are.

Periods go inside parentheses when what's in the parentheses is a complete sentence:

| Incorrect | In the recipient list for the notification, select the administrator and click **Delete**. (The administrator can approve or ignore the notice). |
|---|---|
| Correct | In the recipient list for the notification, select the administrator and click **Delete**. (The administrator can approve or ignore the notice.) |

If only a part of a sentence is in the parentheses and that part can't stand alone grammatically, the period goes outside the parentheses:

| Incorrect | The brake fluid from the master cylinder is forced into a caliper, and the caliper presses against a piston, which squeezes the brake pads against the disc (rotor.) |
|---|---|
| Correct | The brake fluid from the master cylinder is forced into a caliper, and the caliper presses against a piston, which squeezes the brake pads against the disc (rotor). |

## Apply consistent highlighting

Highlighting refers to the emphasis given to an element by changing its visual attributes. When you use highlighting consistently and predictably, users can more easily focus on the content. They don't need to stop and think about what the difference is between entering a command that is displayed in lowercase and bold and one that is displayed in all capital letters in monospace, or why the same control on an interface is sometimes green and other times red.

**Text**

Consistent highlighting is important not only for technical documentation but also for user interfaces. Inconsistencies might lead users to think that the different highlighting means something different.

For example, in the following mobile device screen, the menu choices in the forefront of the screen use a different font and highlighting (italics) than is used in the initial set of menu choices:

Original

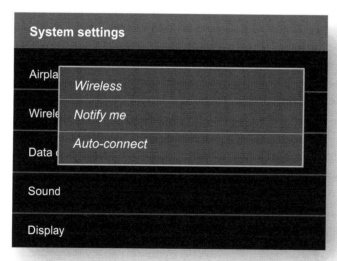

Some users might think that the different highlighting might indicate that the user interface control is disabled, meaning that it can't be tapped.

**Revision**

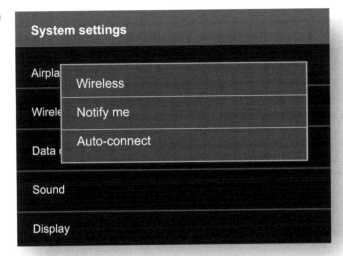

The revised second screen retains the same font and highlighting as the first screen and won't be confusing.

In technical information, even seemingly trivial differences in style can be confusing. For example, assume that each of the following instructions is from a different part of a product's information set:

**Original**

Press "ENTER."

Specify the `Search` command and press **Enter**.

At a command prompt, type `help` and press *Enter*.

The problems are obvious when you look at the instructions together. The word *Enter* appears in three different styles: double quotation marks, bold, and italics. These discrepancies are confusing and make the information look unprofessional. The information looks like it's not from the same product.

Easy to understand

**Revision**

> Press Enter.
>
> Specify the `Search` command and press Enter.
>
> At a command prompt, type `help` and press Enter.

The revised instructions use simple capitalization and no highlighting for the term *Enter*. Users won't stumble, as they would with the confusing highlighting in the original text.

To ensure that you use consistent highlighting, be sure that you and other writers on your team use a style guide that describes the rules for highlighting, such as what content gets bold, italics, all uppercase, initial capital letters, or monospace. Also, indicate what types of content should not be highlighted.

The following sample style guide describes how to apply different types of highlighting for product documentation:

Table 8.1  Highlighting style guide

| Item | Highlighting | Example |
| --- | --- | --- |
| API calls | None | Use the GetVersion method to retrieve the version number of your application. |
| Book and video titles | Italics | See the *InfoDBase Reference*... |
| Code samples | Monospace | Insert the following code:<br>`private static void iterateAddress`<br>`    Address mainAdd,`<br>`    int pageSizeFetch,`<br>`    int pageSizeDisplay)`<br>`{` |
| Commands in body or paragraph text | Bold | The **exit** command is not available. |
| Commands in instructions | Monospace | Enter the following command: `exit -server_name`. |
| Check boxes | Bold | Select the **Save my data** check box. |

**Table 8.1** Highlighting style guide (*continued*)

| Item | Highlighting | Example |
|---|---|---|
| Directory names | Monospace | The XML file is in the `C:\temp` directory. |
| Entered data | Monospace | Type `MyWorld` and press Enter. |
| Entry fields | Bold | In the **Table name** field, specify which table will store the records. |
| File names | Monospace | Open the `readme.txt` file. |
| Function names | Uppercase only | The CALC function calculates the length of the object. |
| Keyboard keys | None | Press Alt+F1. |
| Menu names and menu choices | Bold | To create a project, click **Project > New**.<br><br>From the **Properties** menu, click **New Property**. |
| Message text | Monospace | After you process the batch file, you might see the following message:<br><br>`The file cannot be found. Ensure that you entered the file name correctly and try again.`<br><br>If you see a `System resources not available` message, do not shut down the server. |
| New terms where defined | Italics | A *batch* is an accumulation of data to be processed. |
| Parameters | Bold monospace | Set the **`userid`** parameter for your Windows user name. |
| References to books | Italics | For installation instructions, see *Installing InfoDBase on AIX*. |
| Table and figure captions | Italics | Table 2. *Parts list for the AB29000* |
| Web addresses | None | You can find more troubleshooting information at info.com/appliances/support. |
| Window and dialog names | None | In the Project window, set the user roles.<br><br>You can add more files in the Add File dialog. |

For suggestions about what DITA XML elements to apply to file names, new terms, window titles, and so on, see "Use consistent markup tagging" on page 311.

## Lists

Lists that are inconsistently structured, capitalized, and tagged are distracting and potentially confusing.

Follow these guidelines for lists:

- Use a complete sentence for the list lead-in sentence to make translation easier unless your style guide specifies that you can use sentence fragments.
- Include at least two items in a list.
- Make list items grammatically parallel.
- Start each list item with a capital letter unless the first word is never capitalized.
- If one list item ends with a period because it is a sentence, use periods on all items. If none of the list items are sentences, don't use periods.
- Use a consistent style for lists (for example, bullet style, line spacing, and indentation).
- Don't nest lists more than two levels. For nested list items, use a different bullet type to indicate the different levels of nesting.
- Keep list items short so that they're easy to scan. If a list item is more than one or two sentences, consider using a different structure such as a definition list, sections, or headings.

The following list violates several of the previous guidelines:

**Original**

> The metadata catalog can contain:
> 
> - Dimensions
> - hierarchies that can be from any dimension
> - cubes
> - facts.

The original list doesn't use a complete sentence for its lead-in sentence, it doesn't use capitalization and punctuation consistently, and it's not parallel.

**Revision**

> The metadata catalog can contain the following objects:
> 
> - Dimensions
> - Hierarchies (from any dimension)
> - Cubes
> - Facts

In the revision, the introduction is a complete sentence, each list item starts with a capital letter, and all list items are punctuated consistently (no periods).

Easy to understand

# Make elements parallel

Parallel sentences and phrases are easy to scan because they set a predictable pattern and rhythm. Patterns and rhythm in language can make information easier to follow and remember.

For example, the following sentence isn't parallel:

**Original**

> By using records management tools, you can define legal holds, and you can also apply retention schedules and decide how to manage legal obligations.

The original sentence has two nearly parallel clauses (*you can define legal holds* and *you can also apply retention schedules*), but the last part doesn't match the same structure as the other two.

**Revision**

> By using records management tools, you can define legal holds, apply retention schedules, and manage legal obligations.

The revised sentence tightens up the longer clauses by using three grammatically parallel structures:

*define legal holds*
*apply retention schedules*
*manage legal obligations*

List items are sometimes difficult to make parallel when some of the list items are a single word or short phrase, but others are longer because they require more explanation.

# Style

**Original**

> If you want extension applications to retrieve data from your database, you must provide the following information:
>
> - Host name or IP address of the database server that stores your data
> - Port of the database server that stores your data
> - Name of the database or service that will serve the data
> - Name and password of the database user account that can be used to access the data. This name and password must also have operating system or root-level privileges. These credentials are typically created when InfoDBase is installed.

**First revision**

> If you want extension applications to retrieve data from your database, you must provide the following information:
>
> - Host name or IP address of the database server that stores your data
> - Port of the database server that stores your data
> - Name of the database or service that will serve the data
> - Name and password of the database user account that can be used to access the data
>
> This name and password must also have operating system or root-level privileges. These credentials are typically created when InfoDBase is installed.

One way to revise the list is to move the extra sentences from the last bullet into a separate paragraph outside of the list. However, be sure that this information is needed. It's likely that users might already know the information about the database name and password because they've probably already installed the product by the time they read this information.

**Second revision**

> If you want extension applications to retrieve data from your database, you must provide the following information:
>
> - Host name or IP address of the database server that stores your data
> - Port of the database server that stores your data
> - Name of the database or service that will serve the data
> - Name and password of the database user account that can be used to access the data

If the extra information isn't needed, the second revision is the better choice.

# Easy to understand

Ensure that the information in user interfaces, such as labels, menus, check boxes, and radio buttons, is parallel. For example, the text in the two radio buttons in the following dialog isn't parallel:

**Original**

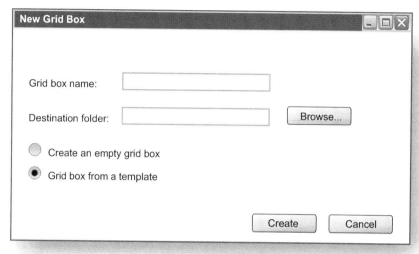

In the original dialog, the first radio button uses a verb phrase, but the second one uses a noun phrase.

**Revision**

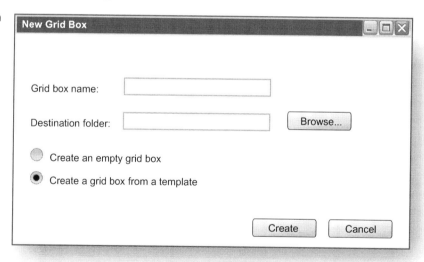

The revision uses two parallel verb phrases for the radio buttons: *Create an empty grid box* and *Create a grid box from a template*.

# Apply templates and reuse commonly used expressions

Templates and consistent phrasing guidelines relieve writers from having to create their own styles or worry about how to phrase text that is often repeated. To make your organization more efficient and your information more consistent, use templates and reuse common expressions.

For example, you can set up templates that include specifications for margin sizes, fonts, table styles, and even topic organization. You can go one step further and create a list of commonly used expressions for different situations, such as common phrases for error and warning messages. By reusing common or standardized text, you can reduce translation costs and distracting inconsistencies.

## Templates

A *template* is a collection of styles or tags that define the structure and appearance of a document. Templates or reusable code can also be used for interfaces. Some templates offer only a minimal number of styles, so writers, editors, and visual designers must define styles as needed. Other templates are more rigid and don't allow writers to redefine items such as fonts, tabs, character sizes, or line spacing.

Minimally defined templates might allow writers more flexibility to present information creatively and even to satisfy a particular user's needs, but writers run the risk of delivering information that is inconsistently presented, which could be confusing and distracting to users of the information.

Even though rigid templates keep writers from changing styles to fit particular needs, these templates also ensure consistent presentation. Because information might be written by multiple writers or maintained by different writers over time, a rigid template helps to ensure that writers don't add different styles to their content.

If you are involved in creating templates for your information, carefully consider the styles that you adopt and decide whether and where deviations are allowed. Some occasional deviations might be an option if the template doesn't meet the needs of specific users. For example, you might allow a particular type of two-page document to expand to three pages if that expansion makes sense for enough users.

Darwin Information Typing Architecture (DITA) is an XML open source data model for authoring technical content, and it provides templates for topic-based information. These templates help to ensure that writers across an organization create task, conceptual, and reference information in the same way.

The following example shows a DITA task topic that contains sections for a title, prerequisites, and steps:

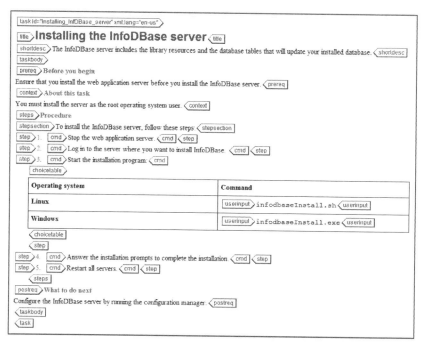

When this DITA topic is transformed to an HTML, PDF, or EPUB (electronic publication) format, the appearance of the topic is controlled by a cascading style sheet (CSS) or Extensible Stylesheet Language (XSL). The advantage to controlling styles with a CSS or XSL file is that if you want to make one change to, say, the appearance of tables in all of your topics, you change only the CSS or XSL file. These style sheet files relieve writers from having to maintain styles in the parts of the documentation or user interface that only they own. That centralized control also helps to ensure that all the information across the organization has the same structure and appearance.

For software products, encourage your development or engineering teams to use common interface templates or reusable code to ensure that the tools, components, and products that you work on have a similar look and feel. As a writer, you can help to ensure that your teams establish guidelines for interface elements, such as buttons, tabs, dialogs, and toolbars.

## Commonly used expressions

To make your information consistent across the organization and to reduce translation costs, you can create a list of commonly used words, phrases, and sentences. For example, should writers use the phrase "For more information about" or "For more information on"? Even subtle differences in English expressions can increase translation costs. In technical information, it's better to express the same idea in only one way.

Even if the expressions aren't always completely interchangeable, you should strive to say the same thing in the same way whenever possible. For example, you should have a common way to introduce lists, tables, commands, videos, and other elements.

Text in documentation or in user interfaces can benefit from reusing common expressions. For example, many software products require user names and passwords. However, simple reminders about what characters aren't allowed in a password are often written differently:

- The password cannot contain the following characters: @ # $ % * ; , () ! ^
- Passwords cannot contain the following characters: @ # $ % * ; , () ! ^
- The following characters are not allowed in passwords: @ # $ % * ; , () ! ^
- The characters @ # $ % * ; , () ! ^ are not permitted in a password.
- When you specify a password, be sure it does not have @ # $ % * ; , () ! ^.
- Do not use these characters in a password: @ # $ % * ; , () ! ^

The list could go on, but you can see how easy it is to say the same thing in many different ways.

As another example, video launch pages often include a statement about how long the video lasts, which is useful information, but there's no reason to state that fact in different ways:

- This video lasts 5 minutes.
- This video lasts five minutes.
- This video will last 5 minutes.
- This video takes 5 minutes.
- This video will be about 5 minutes long.
- The video should be complete in 5 minutes.
- Duration: 5 minutes.

In technical information, such variations aren't useful or practical. And if you translate your content, the variations add unnecessary cost to your organization.

Easy to understand

Error, warning, confirmation, and informational messages provide an excellent opportunity for you to reuse common expressions.

For example, in the following two messages from the same product, each mentions contacting an administrator:

**Original**

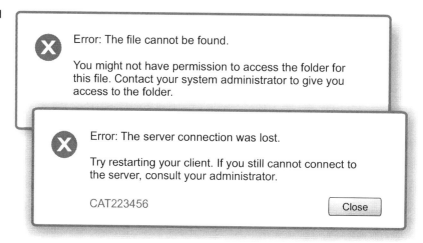

The phrases *contact your system administrator* and *consult your administrator* are unnecessarily inconsistent.

**Revision**

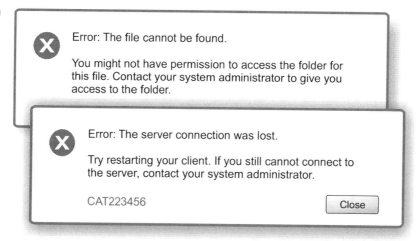

The revised messages both use the same phrase. Even if you don't translate your content, consistency in phrasing reduces distractions for users. And providing writers or engineers with a list of commonly used expressions can be helpful, especially for those who are new to authoring message text.

Many types of reference information also should reuse common expressions. For example, if you create many command topics for a software product, each topic should follow a template design and reuse common text. For example, the following two topics express the same ideas in slightly different ways:

**CHANGE_RECORD_TYPE command**

Use the CHANGE_RECORD_TYPE command to change a record type to electronic or physical.

**Syntax**

```
CHANGE_RECORD_TYPE volume
-electronic | physical
```

**Example of changing a record type**

This example command changes the volume to a type of electronic:

```
CHANGE_RECORD_TYPE ARG dod
electronic
```

**Command for CHANGE_RECORD_TYPE**

You can use the CHANGE_RECORD_TYPE command to change a record type to electronic or physical.

**Command syntax**

```
CHANGE_RECORD_TYPE volume
record_name owner_name password -
electronic | physical
```

**Example:** changing a record type

The following example command changes the record named dod1234 from the ARG volume to a type of electronic:

```
CHANGE_RECORD_TYPE ARG dod1234
g_wilder xyz5432 -electronic
```

One topic uses *CHANGE_RECORD_TYPE command* as its title and the other uses *Command for CHANGE_RECORD_TYPE*. The other headings are different, and the text under those headings is different.

The differences in the text might seem trivial, but if you want to lower translation costs, use the same text for the same things. What's important is that you decide which style of text you want to use for all topics with this type of content and stick to the same reused expressions.

Applying a template for this type of information is a good start, but you can also ensure that each section uses standardized, or common, text so that users get a predictable experience, you save on translation costs, and writers can get their jobs done more quickly. For example, you might decide that the common text in the first paragraph of your command reference topics must always start with "Use the *command* to *do this task*."

Discovering the expressions that are used across a large organization can be challenging. You might be able to use or create tools that can comb through documentation or websites and flag sentences that have similar words and similar structure. You might also form workgroups of writers and editors to start a list of common expressions and evaluate which expressions should be used throughout the organization.

To find areas that might have reusable text, you can start by assessing the following types of information:

| Information type | Items that have potentially reusable text |
|---|---|
| User interface text | • Interface text such as button labels, menu selections, and window titles<br>• Error, warning, informational, and status messages<br>• Auto-generated email notifications<br>• Warning or caution labels for software and hardware products and appliances |
| Product documentation | • Reference information such as introductory text for entering commands, descriptions of syntax, or parts lists<br>• Introductions to tables and illustrations<br>• Introductions to steps or substeps<br>• Accessibility information<br>• Welcome or home pages for websites<br>• Other often repeated phrases such as phrases that describe default values<br>• Troubleshooting information<br>• Section titles for readme files<br>• Contact information for an organization or company |
| Multimedia | • Audio-enabled navigation systems, for example, giving directions (*turn right* versus *go right*)<br>• Introductions, summaries, or other common text used in videos |

You can build the list over time and make that list available to other writers and engineers who will need to reuse these expressions.

For information about reusing standardized legal text, see the accuracy guideline "Reuse information when possible" on page 86.

# Use consistent markup tagging

If you author information in a markup language such as HTML or XML, ensure that you tag your content consistently. This consistency makes it easier to make global changes to your tagging. For example, if one team uses the <cmdname> DITA element for commands that users enter in an application and another team uses the <userinput> element for the same type of content, you can't easily make a style change to the same content. You'll have to fix the inconsistencies first.

Consistent tagging also prevents your users from becoming distracted by differences in highlighting. The following example lists applications, their default location, and their startup commands. However, the information in two of the rows uses different tagging than the information in the other rows.

**Original**

| Application | Default location | Start command |
|---|---|---|
| DB Access | InfoDBase/apps/users/DBaccess | db access start |
| DB Accounts | InfoDBase/apps/accounts/DBaccounts | db accounts start |
| DB Scale | InfoDBase/apps/scale/DBscale | db scale start |
| DB Storage | InfoDBase/apps/tiers/storage/DBstorage | db storage start |

Table 1. InfoDBase application locations and start commands

In the third row, a <uicontrol> element was used instead of the <cmdname> element. And even though the <tt> and <filepath> elements in the fourth row can render nearly the same appearance in HTML (monospace), using two different elements can lead to problems later if you or your team wants to change the way those elements appear in output.

# Easy to understand

**Revision**

*Table 2. InfoDBase application locations and start commands*

| Application | Default location | Start command |
|---|---|---|
| DB Access | InfoDBase/apps/users/DBaccess | `db access start` |
| DB Accounts | InfoDBase/apps/accounts/DBaccounts | `db accounts start` |
| DB Scale | InfoDBase/apps/scale/DBscale | `db scale start` |
| DB Storage | InfoDBase/apps/tiers/storage/DBstorage | `db storage start` |

(Table cells show DITA tags: table, title, filepath, cmdname)

With semantic markup languages such as DITA XML, it's even more important to use consistent tagging across your information sets. Semantic markup requires that you tag a piece of content for what it is and where or how it's structured, not for how you want it to appear on a web page, user interface, or PDF.

For example, if you're providing information about application programming interfaces (APIs) and you're describing a method such as GetCode, you should tag that term with the <apiname> element. The CSS or XSL will control how that element appears in HTML or PDF.

Using DITA elements consistently is sometimes more difficult because there are many more DITA elements than HTML elements and because DITA can be extended, which means you can create elements when you need them.

You can use the following example style guide to govern the use of DITA elements:

Table 8.2    DITA tagging style guide for InfoDBase documentation

| Item | DITA element | Example |
|---|---|---|
| Book and video titles | <cite> | See the <cite>InfoDBase Reference</cite>. |
| Commands | <cmdname> in body text; <userinput> in steps | The <cmdname>exit</cmdname> command is not available.<br><br>Enter the following command: <userinput>exit -<varname>server_name</varname></userinput>. |
| Check boxes | <uicontrol> | Select <uicontrol>Save my data</uicontrol>. |

Table 8.2  DITA tagging style guide for InfoDBase documentation *(continued)*

| Item | DITA element | Example |
|---|---|---|
| Entered data | &lt;userinput&gt; | Enter the following application name: &lt;userinput&gt;MyWorld&lt;/userinput&gt;. |
| File and directory names | &lt;filename&gt; | Open the &lt;filename&gt;readme.txt&lt;/filename&gt; file. |
| Keyboard keys | None | Press Alt+F1. |
| Menu names and menu choices | &lt;menucascade&gt;, &lt;uicontrol&gt;, or both | To create a project, click &lt;menucascade&gt;&lt;uicontrol&gt;Project&lt;/uicontrol&gt;&lt;uicontrol&gt;New&lt;/uicontrol&gt; &lt;/menucascade&gt;. |
|  |  | From the menu, click &lt;uicontrol&gt;New Project&lt;/uicontrol&gt;. |
| Message text | &lt;msgblock&gt; or &lt;msgph&gt; | After you process the batch file, you might see the following message: &lt;msgblock&gt;The file cannot be found. Ensure that you entered the file name correctly and try again. &lt;/msgblock&gt; |
|  |  | If you see a &lt;msgph&gt;System resources not available&lt;/msgph&gt; message, do not shut down the server. |
| Parameter or command variables | &lt;parmname&gt; | Set the &lt;parmname&gt;userid&lt;/parmname&gt; parameter for your Windows user name. |
| Window, dialog, and page names | &lt;wintitle&gt; | In the &lt;wintitle&gt;Project&lt;/wintitle&gt; window, set the user roles. |
|  |  | You can add more files in the &lt;wintitle&gt;Add File&lt;/wintitle&gt; dialog. |

For information about how to use DITA elements effectively and consistently, see the book *DITA Best Practices*.

Easy to understand

# Style checklist

A consistent and appropriate style helps users understand information. Effective style is transparent and doesn't hinder users' understanding of the information. Users notice style only when highlighting and formatting is inconsistent; when the tone is inappropriate; when punctuation, capitalization, or spelling is incorrect; or when templates are not applied properly. When users pay attention to the style, they probably don't pay attention to the meaning of the information.

You can use the following checklist in two ways:

- As a reminder of what to look for, to ensure a thorough review
- As an evaluation tool, to determine the quality of the information

| Guidelines for style | Items to look for | Quality rating<br>*1=very dissatisfied,*<br>*5=very satisfied* |
|---|---|---|
| Use active and passive voice appropriately | - Active voice is used in most cases.<br>- Passive voice is used only when the actor or agent isn't known or isn't important.<br>- Passive voice is used in messages to avoid blaming the user. | 1 2 3 4 5 |
| Convey the right tone | - The tone for formal information is helpful, direct, and authoritative.<br>- The content is free from a pretentious tone.<br>- Content that uses a friendly, humorous, or informal tone is helpful but not sarcastic or patronizing.<br>- If idioms or slang is used, the information is tested and checked periodically to ensure that the text is not dated or inappropriate. | 1 2 3 4 5 |
| Avoid gender and cultural bias | - Sample names, phone numbers, email addresses, and other sample data are appropriate for an international audience.<br>- References to gender, occupation, or position are free from bias. | 1 2 3 4 5 |

# Style

| Guidelines for style | Items to look for | Quality rating<br>1=very dissatisfied,<br>5=very satisfied |
|---|---|---|
| Spell terms consistently | • Technical terms are spelled only one way.<br>• Terms with prefixes and suffixes are spelled correctly.<br>• Terms that are trademarks are never changed, such as by adding a prefix or 's.<br>• If you author in English, terms with variable spellings, such as with British and US English spellings, are spelled in only one way.<br>• Commonly confused terms such as *affect* and *effect* are used and spelled correctly. | 1 2 3 4 5 |
| Use proper capitalization | • Technical terms use consistent case.<br>• Technical terms aren't unnecessarily capitalized.<br>• User interface controls are correctly capitalized. | 1 2 3 4 5 |
| Use consistent and correct punctuation. | • Punctuation is used correctly and consistently.<br>• Guidelines are established and followed for controversial punctuation such as the serial comma and the use of quotation marks with periods and commas. | 1 2 3 4 5 |
| Apply consistent highlighting | • Style and highlighting of items such as button labels, hover help text, windows, and messages are consistent in user interfaces.<br>• Commands, check boxes, labels, file names, headings, titles, and other elements are highlighted correctly in the documentation according to the style guide.<br>• Lists are short and parallel, consistently punctuated, include at least two items, and aren't nested beyond two levels. | 1 2 3 4 5 |
| Apply template designs and reuse commonly used expressions | • Templates are used appropriately and consistently for information such as task, concept, and reference topics.<br>• Common expressions are reused when possible. | 1 2 3 4 5 |
| Use consistent markup tagging | • Markup language tagging is correct and consistent according to the style guide.<br>• DITA elements are semantically correct and not applied simply for appearance in output. | 1 2 3 4 5 |

To evaluate your information for all quality characteristics, you can add your findings from this checklist to Appendix A, "Quality checklist," on page 545.

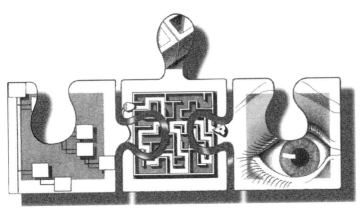

# PART 4

# Easy to find

When users struggle to find information, they can become frustrated and dissatisfied with the product. Information must be designed and structured in a way that makes it easy to access. Information that is easiest to find is often the information that users don't need to look for at all because it's already right where when they need it.

| | |
|---|---:|
| **Chapter 9. Organization** | **319** |
| Put information where users expect it | 322 |
|     Separate contextual information from other types of information | 324 |
|     Separate contextual information into the appropriate type of embedded assistance | 332 |
|     Separate noncontextual information into discrete topics by type | 337 |
| Arrange elements to facilitate navigation | 345 |
|     Organize elements sequentially | 350 |
|     Organize elements consistently | 354 |
| Reveal how elements fit together | 360 |
| Emphasize main points; subordinate secondary points | 366 |
| Organization checklist | 376 |
| **Chapter 10. Retrievability** | **379** |
| Optimize for searching and browsing | 381 |
|     Use clear, descriptive titles | 381 |

| | |
|---|---:|
| Use keywords effectively | 384 |
| Optimize the table of contents for scanning | 389 |
| Guide users through the information | 394 |
| Link appropriately | 399 |
| Link to essential information | 400 |
| Avoid redundant links | 405 |
| Use effective wording for links | 409 |
| Provide helpful entry points | 413 |
| Retrievability checklist | 420 |
| **Chapter 11. Visual effectiveness** | **421** |
| Apply visual design practices to textual elements | 424 |
| Use graphics that are meaningful and appropriate | 431 |
| Illustrate significant tasks and concepts | 431 |
| Make information interactive | 441 |
| Use screen captures judiciously | 448 |
| Apply a consistent visual style | 460 |
| Use visual elements to help users find what they need | 467 |
| Ensure that visual elements are accessible to all users | 478 |
| Visual effectiveness checklist | 483 |

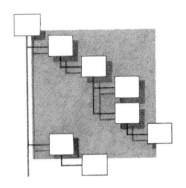

CHAPTER 9

# Organization

*A place for everything, and everything in its place.* —Charles A. Goodrich

Organization is how elements of information fit together to form a coherent arrangement of parts that makes sense to users. Organization applies both to the arrangement of two or more parts and to the arrangement of the information within each part. The parts that you organize range in size from the largest parts of the information set, such as an installation guide or a support website, to the smallest elements of information, such as field-level assistance in a user interface or an example in a step.

Easy to find

Figure 9.1  The elements of an information set fit together to form larger organized parts that are presented to users on various devices or media

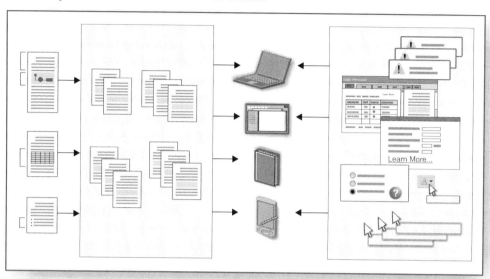

When you plan your writing at a high level, consider the organization of what you will produce from the top down. The *top-down perspective* is like an outline that shows how the main parts will be structured and how they relate to each other. You build your outline or topic model to ensure that you are planning to deliver the right content and that you have an idea of where each part goes. The completeness guideline "Cover all subjects that support users' goals and only those subjects" on page 115 provides guidance about planning the content that you need to write.

The main parts should be laid out in a way that makes sense to users. For example, for a set of topics about installing a product, you need topics for planning, for prerequisites, and for installing, and the topics for planning need to be before the topics for installing.

You organize the main parts into smaller elements, such as topics. In turn, topics contain organized paragraphs, lists, graphics, and other elements. Similarly, user interfaces contain organized windows, dialogs, and messages, which in turn contain organized tabs, fields, buttons, text, and labels. All of these elements are arranged to form the whole information set.

The elements within each part must be arranged so that the structure of the part is clear to users. Well-organized content gives users an implicit sense of the structure of the whole and of the parts. Users can then build expectations about where elements will be so that they can find what they need.

To plan and write information that is effectively organized, be sure that each part, whether large or small, has its own coherent structure and fits well in the whole. To organize information well, follow these guidelines:

- **Put information where users expect it**
- **Arrange elements to facilitate navigation**
- **Reveal how elements fit together**
- **Emphasize main points; subordinate secondary points**

# Put information where users expect it

An element of information, whether large or small, is in the right place when it's in the first place that users look, or even better, when it's right in front of users before they realize they need it.

As discussed in Chapter 1, as technology evolves, technical information must adapt to keep up with the expectations of busy users and with changing information delivery mechanisms. The once-traditional set of bulky reference manuals and user guides gave way a long time ago to web-based information, which in turn has given way to information that is shorter and delivered right where it's needed. Information might be delivered in many ways, such as:

- As part of users' interactions in the user interface
- In videos
- On the box that a product is packaged in
- On the installation screens
- In a mobile device app
- In topics on a website

To exploit these methods of delivery and put information where users expect it, you need to organize information so that it can be delivered in smaller pieces.

**Figure 9.2** Information can be accessed from a multitude of devices or media, which are fed by an organized set of information elements

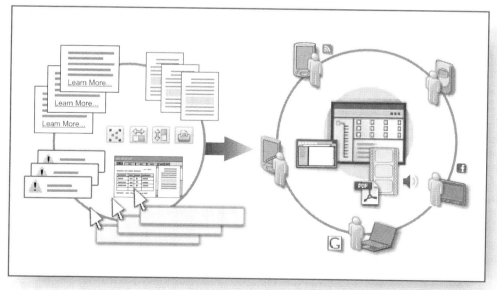

An element of information might need to be small enough to fit on a mobile screen or in hover help, or it might be as large as an installation guide. Regardless, the element should be delivered where users expect it and provide the right information based on users' needs. Any element or set of elements that is delivered in one place, such as an administration guide, a field on a website, or a message in a user interface, must have sufficient content to make sense in the context where users read it. Ensuring that the content is small enough to fit where it's needed while still providing enough information for users is a bit of a balancing act.

If you consider the number of elements that you might need to deliver the right content in the right places, the task of organizing your information into the necessary elements can seem overwhelming. A good way to start is by identifying which content is contextual and which is not contextual.

**Contextual information**
> Information that is delivered in the user interface. Contextual information can be further separated into elements such as messages, labels, or contextual help topics based on where users need the information.

**Noncontextual information**
> Information that is delivered outside the interface. This information includes content about the product overview, installation, and high-level tasks. Noncontextual information should be organized into topics that provide only what is needed. Each topic should be distinct from other topics so that it can be read by itself.

As you write and organize the smaller elements of your information set, remember that:

- Elements must be small enough to be delivered where users expect to find them.
- Elements must have sufficient content to be meaningful when read alone.
- Elements must be presented in a coherent arrangement of parts so that users know where to expect each element.

An element is considered *discrete* if it is individually distinct and makes sense when delivered to users by itself. Topics and error messages are discrete elements because they can be read alone. Discrete elements can be delivered with other elements to form a larger part, but they are meant to be read one element at a time. In most cases, a paragraph is not a discrete element because it needs other paragraphs or context to make sense. Elements that are not meant to be delivered alone, such as paragraphs, must flow coherently with surrounding elements to convey ideas clearly.

To organize your content into discrete elements that are delivered where users expect, follow these guidelines:

- Separate contextual information from other types of information
- Separate contextual information into the appropriate type of embedded assistance
- Separate noncontextual information into discrete topics by type

## Separate contextual information from other types of information

Contextual information explains the piece of the product that users are working with. For example, the arrows on the lid of a medicine bottle are contextual because they explain how to use the bottle lid. For software products, contextual information might explain product windows, fields, and controls. This information might include the syntax for a field, default values, conditions for when a control is available, what a control does, or what to do in a product window.

Contextual information makes sense only "in context." Contextual information that strays out of its context is in the wrong place. For a medicine bottle, the arrows are of little help if they are printed on the bottom of the bottle or shown on a website.

If you put contextual information with other types of information, such as user task topics or overview information, you waste resources because that information is not where users expect it. For example, the following explanation of a window and table of icons might add visual interest to a user's guide, but it's not helping users:

Original

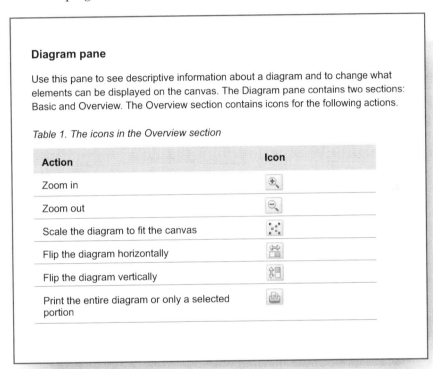

Information that is in the wrong place or repeated unnecessarily is useless to your users. The information in the table is out of context in a user's guide because it is useful only when users are working in the Overview section of a window. The information above the table is also contextual and arguably even more useless because users probably don't want to memorize the names of the sections of the window they will use.

# Easy to find

**Revision**

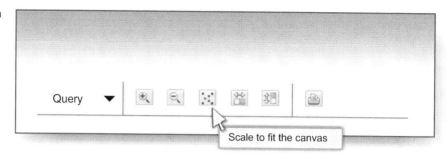

Users who want to know what a button does are likely to either hover over the button and look for a tooltip or try clicking the button to see what happens. They are much less likely to stop what they are doing and look for the button descriptions in the user's guide. Because the table and explanation are not needed or expected in the user's guide, they are removed in the revision. The contextual information is now available only in the user interface, where it belongs.

When you add contextual information to noncontextual topics, you write a lot more content than is needed and put that content where it does users little good. In particular, avoid creating task topics that walk users through the elements of the windows or dialogs that are part of the task. As explained in the task orientation guideline "Focus on users' goals" on page 32, task topics should be higher level and focused on goals, not features or functions. If task topics contain contextual rather than goal-focused information, you get three unwelcome results:

**Information that is in the wrong place**

    Contextual information for windows, dialogs, buttons, and fields is buried in task topics instead of provided in the interface. Users who try to use the interface don't get the help they need where they need it.

**Information that is repeated unnecessarily**

    Some windows might be used in multiple tasks, and low-level contextual steps for those windows are repeated in several user task topics. For example, suppose that users of your product log in infrequently and so often need to reset their passwords. If you add a step about how to reset a password to every user task for which users might need to log in, you will repeat that step many times out of context. A better approach is to add a **Reset your password** link to the login dialog.

**Cluttered tasks**

> Contextual information causes task topics to become cluttered. Many controls are just features and not required parts of any user tasks. In the previous example, the icons are features of a window in the product, but they aren't really steps of a user task. When task topics serve as the only method of explaining the user interface, these topics will have extraneous steps for minor or unnecessary controls.

Separating contextual information and keeping it separate is key to providing the right information in the right place. In practice, however, it's not as easy as you might think. Time and again, contextual information will sneak into the wrong places, even when you have defined a clear organization scheme. Your organization scheme will likely be challenged by some typical situations:

**Application developers are unable to fix a defect, and the writer is asked to handle it in the documentation**

> Suppose a mobile app shows wait times for attractions at an amusement park. Many users are frustrated because the app does not refresh the wait times frequently enough. A good solution is to have the app show the latest wait-time data or at least provide a prominent refresh button on the screen.
>
> A less desirable, but all too common, solution is to add a note to the app's download page that reminds users about the hidden refresh button. This solution pretends to fix the problem, but instead it makes the problem worse. Because the download page is not where users expect to find information, users will continue to have the problem with the refresh button; however, now the real defect might never be fixed.

**Other teams, such as support or development, have assumptions about where content should be**

> Removing contextual information from noncontextual places in your information set might initially be disruptive to support, development, or test teams. Suppose the support or development team is used to looking up connection properties in user's guides for all products. These team members will be affected if those properties are now available only as choices on the connection dialog. The writer might be pressured to put those properties back in the user's guide even though users need the properties only when they use the connection dialog.
>
> If the writer puts the content back in the user's guide, the support team will be happier; however, the content in the user's guide does not help users and is likely to become inconsistent with the properties in the dialog.

**The marketing team wants to highlight new features in the documentation**

Suppose the marketing team wants to highlight a new import feature for a new release of an established product. The writer is asked to mention the new import feature everywhere possible, such as in high-level task topics, scenario and overview topics, and help topics for related windows. Although cluttering the documentation with advertisements about this import feature might satisfy the marketing team and would slightly improve the chances that users will learn about the existence of this feature, all of this extra clutter does not add value because the information is not where users expect to see it.

The best way for the writer to help users with the new feature is to ensure that the feature works as expected and that any necessary contextual information is provided only where it is needed. In many cases, the new features list is probably the only noncontextual place where users would expect to see the new feature mentioned.

Adding a note, step, paragraph, or sentence in the wrong place never helps users. Keep contextual information in context.

The following web page from a bank website provides contextual information inside the steps of a task:

**Original**

> ### Go Mobile: Get started with mobile banking
>
> If you have an account with InfoBank, follow these steps to set up the secure mobile bill payment service and start paying bills from a mobile device:
>
> 1. From a web browser, log on to your online banking account at example.infobank.com and set up your account for mobile bill payments:
>
>    a. Authorize your account for mobile bill payments by going to **Settings > Preferences > Mobile Settings** and selecting the **Yes, I want mobile** check box.
>
>    b. Set up your payees from the web browser. To set up payees, go to the Payees tab and click **New Payee**. Be sure to fill out the required fields, including the field for the account number. Do not enter spaces or hyphens in the account number field.
>
>    **Tip:** Specify short, meaningful nicknames for your payees so that the names will be easy to read on the smaller mobile device.
>
>    Click the **Save** button to save the payee. After payees are set up, you can pay them from your mobile devices.
>
> 2. Install the mobile banking app on one or more mobile devices.
>
>    From your mobile device, go to the app store and download the InfoBank mobile banking app. If prompted, specify the same user ID and password values that you use on the example.infobank.com website.
>
> 3. Start using your device for mobile banking. You can check balances, transfer money, or pay bills. To pay bills, tap the **Pay a Bill** button. The fields on the Pay Bills screen are:
>
>    **Payee**
>        Choose from a list of payees or nicknames that are set up for your account.
>    **Amount**
>        Specify the amount to pay in US dollars.
>    **Deliver by**
>        Specify when the payment should arrive.
>    **Pay**
>        Tap this button to send the payment now.
>    **Schedule Payment**
>        Tap this button to schedule recurring payments.

The original web page shows the contextual information for some of the controls on the New Payee window and all of the controls on the mobile Pay Bills screen mixed in with the task steps for getting started with mobile banking. But users who fill out the New Payee window won't be reading the web page with the steps, and users who look at their mobile device shouldn't need to view the web page while working with the app. The information is not in the right place. If any of the contextual information is needed by users, it must be presented where users need it.

**First revision**

> ### Go Mobile: Get started with mobile banking
>
> If you have an account with InfoBank, follow these steps to set up the secure mobile bill payment service and start paying bills from a mobile device:
>
> 1. From a web browser, log on to your online banking account at example.infobank.com and set up your account for mobile bill payments:
>
>    a. Authorize your account for mobile bill payments by going to **Settings** > **Preferences** > **Mobile Settings** and selecting the **Yes, I want mobile** check box.
>
>    b. Set up your payees from the web browser. To set up payees, go to the Payees tab and click **New Payee**.
>
>    After payees are set up, you can pay them from your mobile device.
>
> 2. Install the mobile banking app on one or more mobile devices.
>
>    From your mobile device, go to the app store and download the InfoBank mobile banking app. If prompted, specify the same user ID and password values that you use on the example.infobank.com website.
>
> 3. Start using your device for mobile banking. You can check balances, transfer money, or pay bills.

The first revision no longer includes contextual information about buttons and fields on the web interface and on the mobile app. Information about where to go on the web interface to set up the account and payees is provided to make sure users get to the right web interface pages and have an idea what to do, but the other details aren't provided. Users who go to the web interface pages and mobile app screens will expect to use the embedded assistance to figure out how to use them.

# Organization

**Second revision**

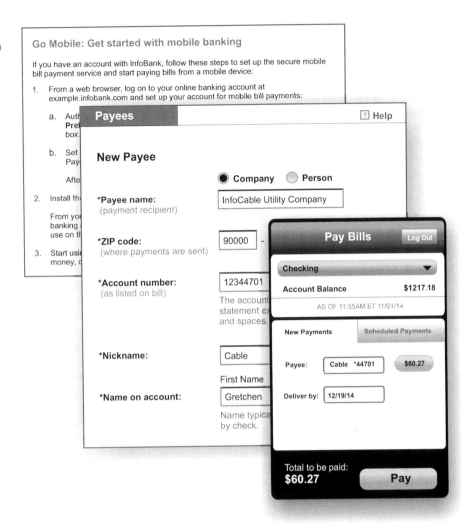

The second revision shows contextual information in the web interface page and mobile app screen, which is where users need it and expect it to be. When users need to create payees, they can read the labels and text on the New Payee page if they need information. When users need to pay a bill, they can rely on the embedded assistance in the mobile app screen to understand how to pay bills. You might notice that some of the labels do not match the labels that were mentioned in the original web page. Keeping contextual information in context provides an added advantage of requiring less maintenance for noncontextual content.

Determine what information is contextual and provide that information separately from other information, such as higher level task or conceptual topics. Ensure that contextual information is available from the interface and not scattered in places that users do not expect to find it.

## Separate contextual information into the appropriate type of embedded assistance

As explained in Chapter 5, "Completeness," contextual information is best presented in a consistent pattern that discloses information progressively to users. To develop a pattern for disclosing information, you need to take into account all of the information that users will need. This information, therefore, begins in the user interface as embedded assistance. The pattern should map a progression through the types of embedded assistance to assistance that is linked from the user interface and finally, if needed, to noncontextual information that you might deliver on the web.

**Figure 9.3** Application of a sample pattern with the following types of assistance shown:
**1.** Default value, **2.** Label, **3.** Input hint, **4.** Static text for a field, **5.** Static text for the dialog,
**6.** Hover help, **7.** Contextual help topic, **8.** Noncontextual topic

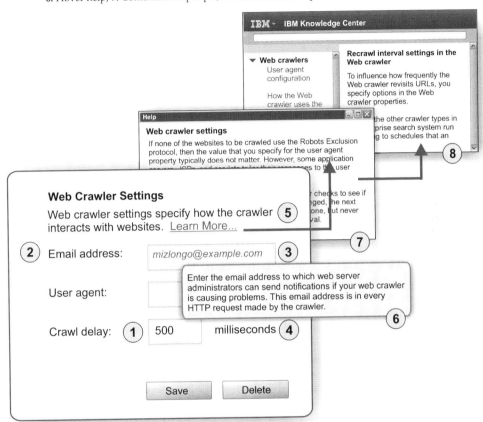

Each user interface has different capabilities, but all user interfaces, even hardware products, include labels and messages. Well-designed user interfaces include some programmatic assistance, such as default or detected values. Some user interfaces have space for providing static descriptive text. Some provide capability for hover help or F1 help on fields and controls. Most user interfaces offer one or more mechanisms for displaying help: a single **Help** link in a banner, a **Help** menu choice, a **Help** push button in windows, or text links within the user interface.

If you consider all of the possible mechanisms for providing information, you can then determine what types of information elements you can provide with each mechanism in your pattern. For example, the label and message elements are always included in an interface, and so your pattern must include mechanisms to provide the label and message elements. The additional, optional mechanisms can provide elements such as description, syntax, context, or examples. So you might have a pattern with the following mechanisms for information elements:

Figure 9.4  Sample pattern for using mechanisms to progressively disclose contextual information elements in the user interface

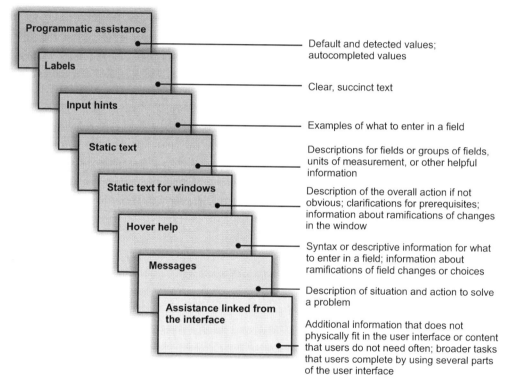

You can use the pattern to help you map where to put different types of content so that the content is delivered appropriately and users get what they expect. Some information elements work better when they are delivered in specific ways. For example, using hover help to present examples of long strings that users must enter is not effective because users cannot see the string as they type.

Follow the pattern consistently to ensure that the user experience is consistent across your set of products. If most windows and dialogs in an interface provide links to help topics from static text with the label **Learn more**, users are not likely to look for help links in other windows from a menu choice.

Original

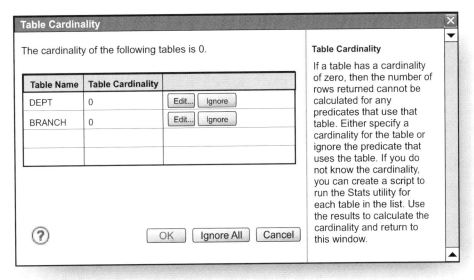

Users who use the original dialog must click the help button and read the explanation about what's happening and instructions for what to do.

First revision

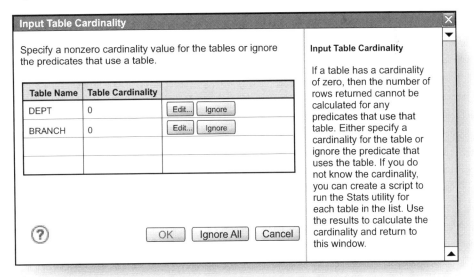

Easy to find

In the first revision, some information about what to do is removed from the help and put where users can see it. However, to learn why the cardinality is needed and how to determine the cardinality, users must still access the help.

**Second revision**

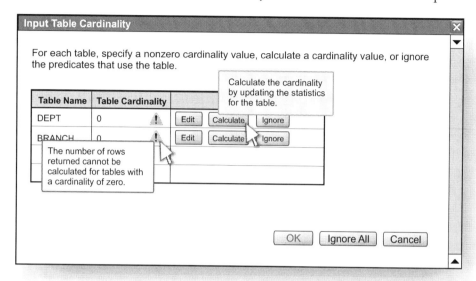

In the second revision, the embedded assistance in the dialog provides users with an explanation of what's happening and what to do. The warning icons indicate a problem with the cardinality, and the hover help explains why the cardinality is needed. In addition, the dialog is altered so that users do not need to leave the task to run a script to calculate the cardinality for each table. Now they can calculate the cardinality without leaving the task. The help icon is removed because the help pane is no longer needed.

When you create your pattern, keep in mind that each click that takes users away from the interface takes them further from their task. Always strive to keep the information that users need to complete a task as close to that task as possible. For example, the syntax for a complex value is probably most useful when it's included as static text next to the field. If you include it as hover help, it's slightly less useful, and if you include it in the help pane, it's even less helpful.

## Separate noncontextual information into discrete topics by type

Information is *noncontextual* if it can be understood outside the context of an interface. Noncontextual information includes overview, planning, installation, high-level task, concept, and reference information. This type of information might be delivered in HTML-based information sets, PDF or EPUB books, or other types of media. Although users might find noncontextual information by following a progressive disclosure path from an interface, they also expect to be able to find it by searching.

Noncontextual information is easier to find and consume when it's organized into distinct topics. Most users who search for a nugget of information, such as the syntax of a command, want to read about the syntax and move on. If you organize your noncontextual information into discrete topics, you will be able to:

- Provide each topic in a format that can be searched for and delivered separately
- Combine related topics to form larger structures when needed
- Reuse topics in different places or for different audiences

**Topics**

*Topics* are discrete, meaningful elements of information, each with a title and at least one paragraph that serves as an abstract or short description. A topic, also known as a chunk, article, or module, must be long enough to cover a subject adequately and make sense if read by itself.

Well-organized information is easier for writers to reuse effectively for different types of output. By writing discrete topics that do not require a specific structure or linear arrangement, you enable the topics to be organized or reused in different ways. A single topic can appear in multiple topic sets, in multiple information centers, in a PDF or EPUB document, in a printed book, and on a website. You can also share topics across topic sets for:

- Multiple product features
- Multiple operating systems
- Multiple audiences

Easy to find

Figure 9.5   Distinct topics contain paragraphs and other elements to adequately cover one subject. Topics can be organized into sets or delivered separately in different media.

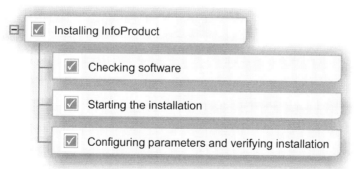

Organizing content into discrete topics gives you the flexibility to reuse the content or any portion of it for a variety of presentations. In addition, because topics are discrete and generally short, users who find the right topic will get just the information they need.

The following list of topics shows a hierarchy of tasks in which the task of verifying the installation is combined with the task of configuring parameters:

**Original**

- Installing InfoProduct
    - Checking software
    - Starting the installation
    - Configuring parameters and verifying installation

The final topic is not discrete because the task of verifying installation is not part of the task of configuring parameters. One user might verify the installation while another user configures the parameters. In addition, users who need to configure additional parameters at some later time do not expect to see steps for verifying the installation of a product that is already installed. The two tasks should not be combined into one topic.

**Revision**

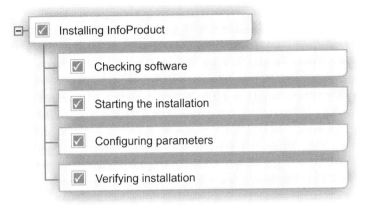

The revision shows verifying installation as the last discrete task at the same level as the other subtasks of installing the product. The revision allows for a logical separation of different tasks. Organizing the tasks into discrete topics might mean writing more topics, but it makes for a better user experience.

**Topic types**

Each topic should generally contain only one type of information. Most technical information falls into one of three types:

| Topic type | Definition | Example |
|---|---|---|
| Concept | Describes ideas, relationships, or processes. Concept topics provide information that users might need to know when they work with a product or do a task. | A concept topic called "Data consolidation" might contain several paragraphs and a graphic that explain what data consolidation is. |
| Task | Provides information to help users complete a task and often includes step-by-step instructions. Task topics can also provide rationale for the task, prerequisites, and examples. | A task topic called "Assigning authorizations" might contain a brief overview of assigning authorizations, why or when the task is needed, a list of prerequisites, steps to assign the authorizations, and a list of any postrequisite (follow-on) tasks. |
| Reference | Provides quick access to facts that users might look up, such as conversion tables, syntax rules, message explanations, and code samples. | A reference topic called "Environment variables" might contain a table that lists the operating systems, valid settings, and a default setting for each environment variable. |

If you organize noncontextual information into topics by type, you can more easily keep topics discrete, and you can provide a consistent structure so that users know what to expect.

For example, if all of the task topics contain a brief overview, prerequisites, steps, and postrequisites, users know what to expect in a task topic. If you add conceptual information to the steps, users who look for concepts will have a harder time finding that information, and users who are trying to do a task are interrupted by conceptual information. Similarly, users who look for syntax and examples of commands will have trouble locating the reference information if it is combined with concepts or tasks.

**Original**

### Finding the union of geometries

The union of multiple geometries is the logical OR of space. The source geometries must be of similar dimension.

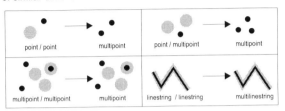

To find the union of multiple geometries:

1. Add a Union operator to the Query window canvas.

2. In the map window, find the first geometry. For example, select a city park.

3. *and so on...*

To find the union with the ST_Union function, use the ST_Union function in a query. The ST_Union function takes two geometries as input parameters and returns the geometry that is the union of the ST_GEOMETRY.

**Examples**
The following examples show how to use the ST_Union function. First, geometries are created and inserted into the SAMPLE_GEOMS table, and then ST_Union is used to find the union of those geometries.

```
infogeo CREATE TABLE sample_geoms (id INTEGER, geometry,ST_Geometry)
INSERT INTO sample_geoms VALUES (1, ST_Geometry( 'polygon ((10 10,
10 20, 20 20, 20 10, 10 10) )', 0))
INSERT INTO sample_geoms VALUES (2, ST_Geometry( 'polygon ((30 30,
30 50, 50 50, 50 30, 30 30) )', 0))
INSERT INTO sample_geoms VALUES (3, ST_Geometry( 'polygon
```

The original topic mixes conceptual, task, and reference information, which makes the different types of information harder to find.

Revision

### Union of multiple geometries

The union of multiple geometries is the logical OR of space. The source geometries must be of similar dimension.

The resulting
If it can be re
types is used
type is used.

**Related info**
ST_Union fu
Finding the u

### ST_Union function

The ST_Union function takes multiple geometries as input parameters and returns the geometry that is the union of the given geometries. The resulting geometry is represented in the spatial reference system of the first geometry.

If any of the multiple given geometries is null, then null is returned.

**Syntax**
>>--Infogeo.ST_Union--(--geometry1--,--geometry2--)--><

**Paramet**

**geometr**
A value
with ge

**geometr**
A value
with ge

**Return t**
ST_Geo

**Example**
The follo
geometri
then ST_

```
infoge
geometr
INSERT
'polygo
INSERT
'polygo
INSERT
'polygo
```

**Related**
Finding t
Union of

### Finding the union of geometries

The union of multiple geometries is the logical OR of space. The source geometries must be of similar dimension.

To find the union of multiple geometries:

1. Add a Union operator to the Query window canvas.
2. In the map window, find the first geometry. For example, select a city park.
3. Drag the first geometry to the Query window canvas and place it inside the Union operator.
4. Add another geometry.
   a. Return to the map window and find the next geometry. For example, select a nearby school.
   b. Drag the next geometry to the Query window and place it inside the Union operator.
5. Repeat step 4 for any additional geometries.
6. Click **Run** and give your new geometry a name, such as "Overlay of parks and schools in grid 9."

**Related information**
ST_Union function
Union of multiple geometries

The revision shows three topics:

- "Union of multiple geometries," a concept topic
- "ST_Union function," a reference topic
- "Finding the union of geometries," a task topic

Users can find what they need quickly and don't need to read through concepts to find tasks and syntax.

Not every situation will call for one task, one concept, and one reference topic. For example, if the task of finding the union of geometries is a simple task that a user can do easily from the user interface, then you might need only a concept topic and a reference topic in the previous example. In other cases, you might need multiple reference topics or no concept topics or multiple task topics. The number and types of topics depend on the content and the audience.

Organizing content by type provides a more flexible architecture that can accommodate change more easily. For example, if you create a separate topic for each task, you can easily add, remove, or rearrange tasks when the product changes or when you want to provide another organization scheme for a different set of users.

Your company might provide templates for the three basic topic types as well as troubleshooting topics, error message topics, or even contextual help content. If you use DITA, templates for some topic types are provided for you. You can read more about concept, task, and reference topics in *DITA Best Practices*.

## Topic sets

A *topic set* is a structured group of related topics that are delivered together and cover a user task or user goal. You might have a topic set for your installation instructions, a set for your troubleshooting content, and a set for each main user goal.

Figure 9.6  Topic sets can contain any number of related topics

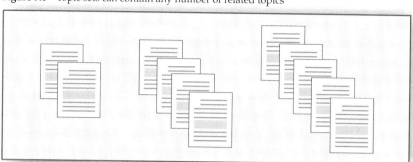

Although each topic should make sense on its own, most topics are delivered as part of one or more topic sets. Every topic set has a navigation scheme that indicates the hierarchy and relationships of the topics to each other. Suppose that you are documenting the task of creating custom reports. You know that users will want to read about the task as well as read the sample report reference topic and a set of concept topics about each type of report. You create a topic set that contains all of those topics and a hierarchy and order for the topics.

Figure 9.7  A hierarchy for a topic set

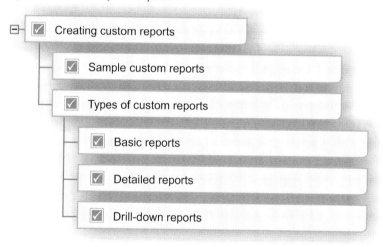

Make sure that all of the topics in the set make sense together. You can deliver each topic set alone or combine it with other sets as needed.

## Arrange elements to facilitate navigation

If your information has a confusing structure or no discernible structure, users will have trouble navigating and using it. Ideally, you want the elements of any set of information to be arranged so well that users do not spend time having to think about the organization.

When you arrange elements together, you help tell a story. Make sure that the right elements are presented together and that a hierarchy or structure is in place to help users find information and understand relationships.

Elements in a topic, such as paragraphs, steps, or tables, must be arranged together in a coherent way, which might be according to sequence, a progression from general to specific, or a pattern for consistency with other similar topics.

Elements in an interface should be arranged to facilitate understanding of user tasks and navigation through those tasks. The arrangement should also make clear the relationships of the smaller elements, such as fields and buttons, to a task and to each other.

The following dialog is not arranged to aid navigation. Even if good labels and assistance are provided, users will have trouble knowing where to start and what to do because the structure of the dialog does not provide any clues about how actions or items are related.

**Figure 9.8** A dialog design with an unclear navigation. Users have no obvious starting point or clear idea of relationships.

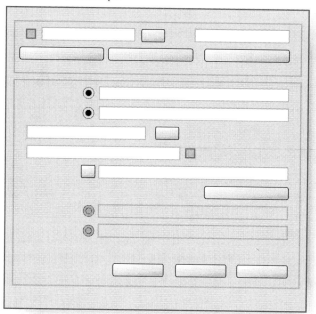

For topics, a topic map, navigation pane, or table of contents uses hierarchy and order to show relationships. The more topics you have, the more important it is to arrange them in a way that makes sense to users. When information is organized into unfamiliar, arbitrary, or ambiguous categories, users will feel lost and must work harder to find what they need.

The organization of the following set of reference topics is based on the complexity of the information; however, this organization scheme assumes that users will know which topics are suited for beginners and which are suited for advanced users.

# Organization

**Original**

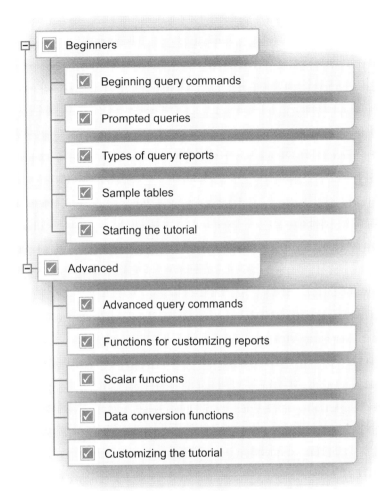

In the original structure, topics are divided between beginner and advanced levels, so users must guess which category applies to the information they want to know. The division between the two branches is subjective and depends on users' assessments of their own skill levels with the product. Some users might look for the topic "Prompted queries" under the Advanced branch, while other users could look for "Customizing the tutorial" under the Beginners branch.

**Revision**

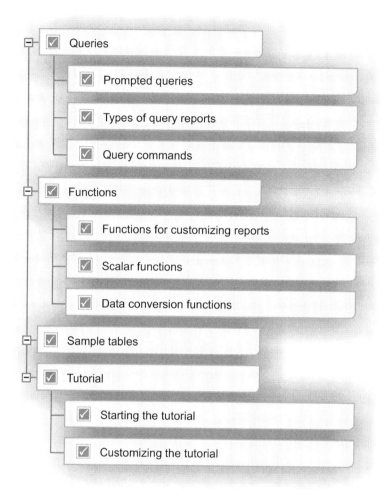

In the revised structure, the topics are arranged into categories that make sense to all users.

The order of topics within a category can indicate a logical progression based on various factors. For example, topics might be arranged in sequential order, in numerical or alphabetical order, from abstract to concrete, or even from most useful to least useful. If topics have no obvious sequential or categorical arrangement, you need to find other criteria for organization.

Suppose you have five topics that support the user task of monitoring performance. Your topic map might have a parent topic for the task and child topics that support the task or that contain subtasks. The following topic map shows a user task of monitoring performance with four subordinate topics: a concept topic, two reference topics, and one task topic.

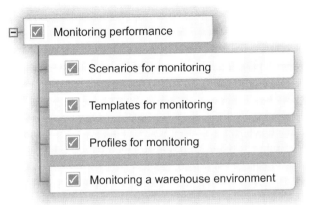

In this topic set, users might not care much about the order of the child topics because only four child topics exist and these topics don't seem sequential. Regardless of how these topics are arranged, users will have to scan through them to find what they want. In this case, although a different order might be just as reasonable, the writer arranged the topics in the order of usefulness for most users.

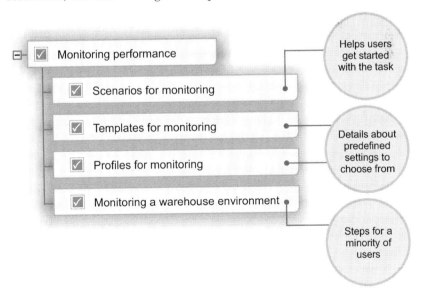

Easy to find

When an organization is obvious to users, information is easy to navigate. Just as a visitor to a home should expect to be able to find the dining room from the kitchen, users who browse any area of an information set should have a sense of where they are in the whole and how the current piece of information relates to the pieces around it.

Information arrangement applies to more than just topics. Arrange all elements, whether they are paragraphs in a topic, topics in a set, or embedded assistance elements in a user interface, in a way that supports your users. To arrange information to help users navigate your content:

- ❑ Organize elements sequentially
- ❑ Organize elements consistently

## Organize elements sequentially

The first experience users have with your content might be to evaluate the product for use and to gain an overview of the product's capabilities. Later, users will need content that supports the task of installing or assembling the product, and then users will need content to support key tasks. Your organization scheme needs to support these primary uses, and your content should be organized to reflect the sequences of tasks.

Because users generally complete key tasks in a certain order, the navigation should be arranged to reflect that order. If what users do in task Y depends on what they did in task X, the topics should be arranged so that task X precedes task Y.

**Original**

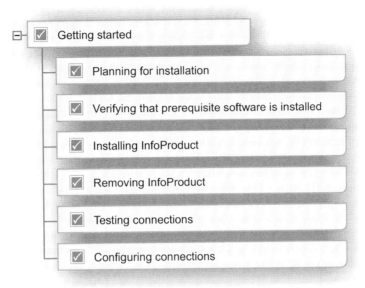

The original outline shows the task of configuring the connections following the task of testing the connections. In addition, the original outline shows the task of removing InfoProduct as a getting started task. Users are not likely to remove a product right after they install it, nor are they likely to look for the task of removing the product under the getting started branch.

**Revision**

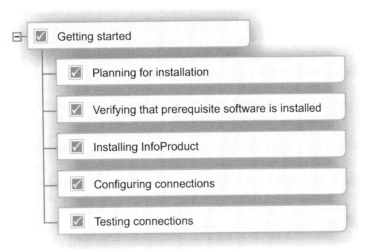

Users can't test the connections until after they configure the connections, as shown in the revised outline. The revised outline does not include the removing task with the other getting started tasks.

You also need to fit concept and reference topics into the outline in a way that makes sense to users. If concept and reference topics are associated with particular tasks, you can arrange these topics with the tasks and in an order that users would expect them or need to access them.

A set of conceptual or reference topics might not belong with any particular task or might not fit well into a task structure. In these cases, you can arrange the topics for quick retrieval, such as alphabetically or numerically. With an alphabetical arrangement, whether in a book or in an information center, users know that the topic "Infolist command" is before the topic "Store command" and after the topic "Grant command." Similarly, with a numerical arrangement, users know that message 120 is before message 395.

A notable exception to the guideline to organize sequentially is window-level assistance, or contextual help topics. These topics are typically not organized together in any sequence because they are not meant to be read sequentially.

**Figure 9.9**  Contextual help topics provide assistance for the current window when needed

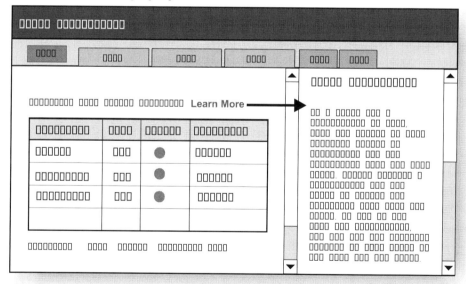

Organizing content sequentially is also important at the level of sentences and steps. If the information for a task is not presented in the order that users need it, users waste time and become frustrated.

**Original**

> **Installing InfoTax**
>
> To install the InfoTax software:
>
> 1. Run the setup.exe program. To access the setup.exe program, extract the InfoTax software files by using a file decompression tool. Before you install the software, ensure that you have enough storage space on your hard disk. You need 500 MB of free space.
>
> 2. Follow the instructions that are provided on the installation screens.
>
> System requirements:
> 5.2 GHz processor
> 2 GB RAM
> A directory in which to store the program files. The name of the directory must begin with the characters *tax*.
>
> You will need your 16-digit product registration key. You can find the key on the download page or in the license agreement email.
>
> **Note:** Shut down all other programs before you install InfoTax.

In the original task, information that users need before they install the InfoTax software is buried in the topic and is not in the order that users need it. Users might start the installation task and waste time before realizing that they need a registration key or a faster processor.

Step 1 instructs users to run the setup.exe program, but they have not yet created the directory from which to access the file. Those instructions are provided, but not in an order that users can relate to. And finally, the requirement to shut down all other programs before installing InfoTax is not mentioned until after the installation steps.

Easy to find

**Revision**

> **Installing InfoTax**
>
> Before you install InfoTax:
>
> - Locate your 16-digit product registration key, which is provided on the download page or in the license agreement email.
> - Ensure that your system meets the following requirements:
>   - 5.2 GHz processor
>   - 2 GB RAM
>   - 500 MB of free space
> - Stop all programs that are currently running.
>
> To install the InfoTax software:
>
> 1. Create a directory in which to store the program files. The name of the directory must begin with the characters *tax*.
> 2. Extract the program files by using a file decompression utility.
> 3. Run the setup.exe program. Follow the instructions that are provided on the installation screens.

In the revised task, all of the requirements that must be met before the installation are explained before the installation steps, and the steps are provided in the order that users must do them to avoid making mistakes or losing work.

The task orientation guideline "Group steps for usability" on page 53 provides more information about organizing steps.

## Organize elements consistently

Similar content should be organized consistently across a product or a set of products. The organization of embedded assistance, the topic navigation, and even the contents of similar topics should be familiar to users and follow a consistent pattern.

# Organization

For a topic set, inconsistent navigation can be caused by either inconsistent content in topics or an unclear or ill-advised navigation scheme. When a topic set is organized inconsistently, some topics might be hidden or too hard to find in the navigation. In addition, any obvious inconsistencies look sloppy and can leave users wondering if content is missing, redundant, or wrong. For example, in the following navigation, the child topics are organized differently for each tag:

Original

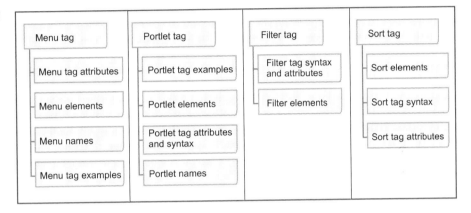

In the original navigation, it's difficult to know whether all of the content is there and how it is organized for any one tag based on the organization of other tags. For example, each tag has content to explain the attributes for the tag:

| Tag | Content about attributes | Are attributes provided with the syntax? |
| --- | --- | --- |
| Menu tag | In the first child topic | No |
| Portlet tag | In the third child topic | Yes |
| Filter tag | In the first child topic | Yes |
| Sort tag | In the third child topic | No |

Users who look closely might find that the attributes are provided with the syntax for the Portlet tag and for the Filter tag.

355

Easy to find

**First revision**

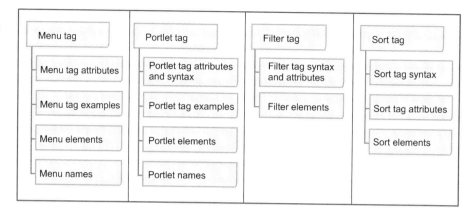

In the first revision, the child topics are arranged in the same order for each tag, which makes additional organization inconsistencies more obvious. Users can see that the problem is not just in the order of the topics, but in the content of the topics themselves.

Some tags are missing child topics or have combined child topics. For example, the syntax content is presented in three different ways:

| Tag | Content about syntax |
|---|---|
| Menu tag | Not shown; might be in the parent topic or another topic |
| Portlet tag | Provided in a topic with the attribute content |
| Filter tag | Provided in a topic with the attribute content |
| Sort tag | Provided in a distinct topic |

This inconsistency is now easy to spot in the navigation. Users can also see that not all tags have topics for their examples. It's difficult to tell from this navigation whether examples for the Sort and Filter tags are available at all. And if they are available, which topic are they in? Users will have a hard time finding what they need in this inconsistent topic organization.

**Second revision**

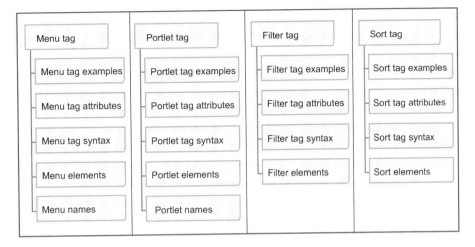

In the second revision, the topics are organized consistently for each tag. Information is now separated into distinct topics so that the organization scheme is consistent. In addition, the examples topic is now the first child topic for each tag because users like to start with examples of tags and then look for more detail if needed. In this revision, the Filter tag and the Sort tag are each missing a topic about names because neither tag had content to support a names topic. Consistency does not require creating topics where none are warranted.

In addition to these improvements, you can improve retrievability if you remove many of the redundant words from the child topic navigation titles. For example, the word *Menu tag* can be removed from navigation titles of the first three child topics under the Menu tag parent topic. See the retrievability guideline "Optimize the table of contents for scanning" on page 389 for information about facilitating scanning.

The organization of elements within topics should follow a similar pattern across a product and other related products. For example, if you include restrictions before the steps in some task topics and after the steps in other task topics, users might not notice some of the restrictions. Similarly, if a quick start card for one product contains a link to a download page in the first line of the card, but the quick start card for a related product puts a link to a support page there instead, users might take a few wrong turns before finding the download page.

When you define a pattern to follow, ensure that your organization scheme is not too rigid. Consistent organization is good, but only when the patterns can accommodate new or changed content. Sometimes writers will try to get around strict organization schemes by inserting paragraphs that are marked as notes. If you find that some appropriate information is not fitting into your pattern or has no home, then you likely need a more flexible organization scheme.

Similarly, if you create a pattern for your embedded assistance and do not follow the pattern consistently, users might not get the information that they need. Suppose that you typically use hover help to provide assistance for fields that users need help with, but then you decide to provide field-level assistance for a few fields in a separate help pane. Users who are used to getting help in hover help are not likely to look in the help pane for those few fields. Those users are more likely to hover over the fields and assume that no information is provided.

The following two screens of a mobile app show an inconsistent embedded assistance pattern. In one screen, the labels are embedded in the fields and descriptive text is above the labels; in the other screen, the field label is above the field, and descriptive text is in the field.

If you don't follow an embedded assistance pattern consistently, you will likely have several screens that are not consistent. Eventually, you might find that one or more of your patterns is not working well. For example, the right screen in the example includes descriptive text in the field. If that text is complex, users might need it as they type but not be able to refer to it.

Most people can relate to the frustration of not being able to find an account number on a statement or a phone number on a website because these appear in different places depending on the statement or the website. This type of inconsistency is less excusable within a product or related products.

Consistent organization helps users become familiar with the structure of information so that they can find what they need with confidence. When you present information consistently, users learn to predict what information they will find and where they will find it.

Easy to find

# Reveal how elements fit together

Because users do not read most information linearly, users need to understand where they are and where they can go. Like travelers, they need a map both at the beginning of a journey and as their trip progresses to keep their bearings in relation to the destination.

For topics, the navigation pane or site map provides a map to those users who want a sense of the breadth of the information. However, if a navigation structure cannot provide sufficient information to help users understand key relationships, you should provide a better map for your users:

- You can provide task overview topics or content in a parent topic that shows how topics relate to each other. Users can return to these parent topics as necessary to link to other topics. In this way, users get a sense of the structure of the information and can read some of it linearly if they want to.
- You can add effective transitions with links that convey the relationships of topics to other topics in a topic set.

These aids help users to verify their sense of the whole. If they can't verify it or if a piece doesn't fit, something is wrong or missing.

Providing the topics and a navigation alone is often not enough. The following task topic is organized according to the order that users should do the subtasks or read the supporting topics, but it doesn't convey how the pieces work together to make a whole.

**Original**

> To upgrade to version 8, you need to read these topics:
>
> * Hardware requirements
> * Software requirements
> * Applying the latest updates
> * Stopping InfoProduct processes
> * Backing up your infoproduct.inf file
> * Running the upgrade utility for Windows
> * Running the upgrade utility for UNIX
> * Verifying the upgrade
> * Setting up a new profile

Users who see the original task topic see a lot of links, but no real structure, and they are left to figure out where to start and which topics to read. In addition, users are not told what to do after the upgrade task is done.

**Revision**

> To upgrade to version 8:
>
> 1. Ensure that you have the correct hardware and software:
>
>     * Hardware requirements
>     * Software requirements
>
> 2. Apply the latest updates.
>
> 3. Stop all InfoProduct processes.
>
> 4. Back up your infoproduct.inf file.
>
> 5. Run the upgrade utility:
>
>     * On Windows
>     * On UNIX
>
> 6. Verify the upgrade.
>
> 7. Set up a new profile.
>
> After you finish upgrading InfoProduct, you can start the processes again.

The revised topic shows the relationship of the subtasks to each other and to the higher-level task. The revised topic also mentions what to do after completing the task, so users are not left stranded.

Easy to find

The following parent topic doesn't provide a "big picture" for the tasks that are involved in using the sample program:

**Original**

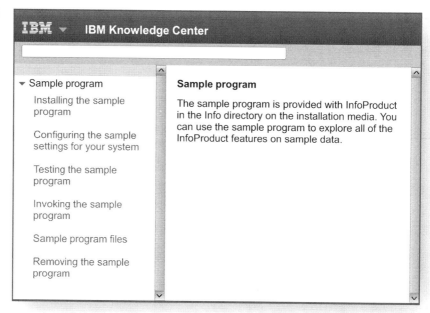

The original set of topics includes a parent concept topic about the sample program with several child task topics and one child reference topic. Users cannot tell from the navigation pane whether the order of the child topics is important or how the tasks relate to each other. The parent topic in the right frame does not explain the relationships of the subtopics. Users are given a paragraph on the right and a set of disjointed topics, or unconnected dots, on the left.

**Revision**

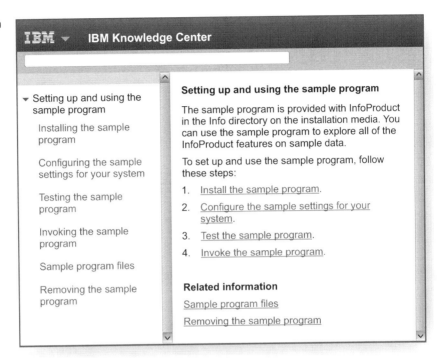

The revised set of topics includes a parent task topic that explains that the first four subtopics are all subtasks of the parent task and indicates the order in which to do them. The parent task topic provides information about the subtopic relationships that users cannot grasp from the navigation pane alone. In the revision, the dots are now connected to form a picture of the parent task.

Not all parent tasks require steps. Some parent tasks introduce the task, and users can choose the path that applies to them. Task overviews for a high-level goal might introduce and explain the goal in text format without using steps. The important thing is that these topics provide users with the big picture.

Because topics are not always read linearly, you cannot assume that users have read topic A before reading topic B or that users who read topic A will know to go to topic B next. Therefore, introductions to tasks such as "Now that you have configured the receiver, you must configure the amplifier" are not sufficient. In the "Configuring the amplifier" topic, you need to mention that "Configuring the receiver" is a prerequisite task, and consider providing a link to the "Configuring the receiver" topic.

# Easy to find

The same advice applies to other transitions between topics. Sentences such as "The next section explains how to configure the printer driver" assume that users see the sections in a certain order. A better way of mentioning a postrequisite task might be to provide a sentence like this in your topic: "After you complete this task, you must configure the printer driver."

The following passage shows the last step in a task topic that explains how to install a product:

**Original**

> 9. Press Enter. The InfoProduct installation process begins.

The original topic ends as soon as the installation process is started and doesn't point users to the next place to go in the information set. Users must figure out what to do next, if anything.

**Revision**

> 9. Press Enter. The InfoProduct installation process begins.
>
> **What to do next**
>
> Next, you can configure InfoProduct settings for your system or import them from another system:
> - Configuring InfoProduct settings manually
> - Importing shared InfoProduct settings

The revised topic explains to users that they can now configure their settings. The links to the related tasks are provided at the end of the topic for users who want to configure the settings now.

For user interfaces, good user interaction design helps users keep their bearings. Labels, placement, and groupings in an interface provide users with a sense of relationships. Users can also keep their bearings with navigation aids such as **Next** and **Back** buttons.

In the following figure, the pages of the installation wizard are shown on the left and check marks indicate which pages are complete. Users can tell where they are in the task and where they are going.

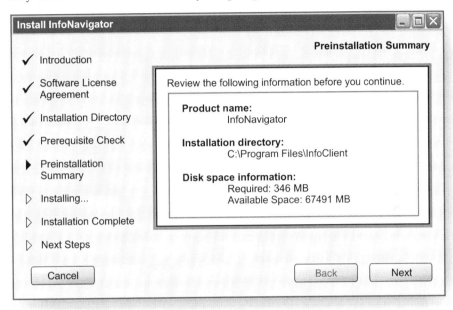

See the retrievability guideline "Guide users through the iformation" on page 364 for more ways to help users keep their bearings.

# Emphasize main points; subordinate secondary points

You can use emphasis and subordination to distinguish main points from supporting information. Users need a way to distinguish the highways from the back roads so that they can keep track of the main points without getting diverted off course by secondary points.

## Main points

Emphasizing a main point involves one or more of the following techniques:

### Placement

Placing key topics or words first is the best way to emphasize a main point: in the heading or title, in the introduction, at the top of a graphic, in the first sentence of a paragraph, in the first words of a sentence.

### Introduction

An introduction, or overview, should make clear to users why subsequent information is important or how that information is relevant. Such an orientation places the subsequent information in context and therefore makes it easier for users to learn or to understand the order of content to come.

For example, the main point of a topic should be made in the first paragraph. Users should be able to identify the main point immediately and determine if they are in the right topic, or in some cases, get all the information they need without having to read additional details.

### Repetition and reinforcement

Repetition keeps users' minds on the point throughout a topic or product interface. Effective reinforcement includes making a point in more than one way, such as by text and illustration, or by general statement and example.

### Details

Details help emphasize what's important. Users should be given more details about important information and fewer details about information that is less important.

The following paragraph seems to be about paragraphs at first:

**Original**

> Paragraphs can have their first lines indented or not; they can be set to the left margin or indented under headings; they can be separated by one blank line, two blank lines, or none. But, in the source, they all begin with a tag designating a paragraph. Similarly, a highlighted phrase can be underlined when it is printed on one device, appear in italic when printed on another, and display in red in a window. It is still the same kind of element, a highlighted phrase, identified by the same highlighting tag. Therefore, the generalized markup tag identifies the kind of element that a portion of a document is.

Users need to read the whole original paragraph to determine what the paragraph is about. Some users might not grasp the intended subject at all. The actual main point of the paragraph, how to use markup tags, is not introduced or reinforced.

**First revision**

> The generalized markup tag identifies the kind of element that a portion of a document is, not the way that the element is to be formatted. Paragraphs can be set to the left margin or indented under headings; they can be separated by one blank line, two blank lines, or none. But in the source, they all begin with a paragraph tag. Similarly, a highlighted phrase can be underlined when it is printed on one device, appear in italic when printed on another, and display in red in a window. It is still the same kind of element, a highlighted phrase, identified by a highlighting tag. Therefore, when marking a portion of a document, ask yourself, "What kind of element is this?" not "How do I want this to look?"

The first revised paragraph begins with the topic sentence. It uses repetition and contrast to emphasize the main point.

# Easy to find

**Second revision**

> The generalized markup tag identifies the kind of element that a portion of a document is, not the way that the element is to be formatted. For example:
>
> - Paragraphs are identified with a paragraph tag. They can be set to the left margin or indented under headings; they can be separated by one blank line, two blank lines, or none. But in the source, they all begin with a paragraph tag.
>
> - A highlighted phrase is identified by a highlighting tag. It can be underlined when it is printed on one device, appear in italic when printed on another, and display in red in a window. It is still the same kind of element, a highlighted phrase.
>
> Therefore, when marking a portion of a document, ask yourself, "What kind of element is this?" not "How do I want this to look?"

The second revision uses a list to set off the examples more clearly and make the information easier to scan. It also uses placement and reinforcement in the list items to keep the focus on the main point. The improved organization of the passage to emphasize the main point has improved the clarity of the passage as well. See the clarity guideline "Write coherently" on page 174 for a related discussion about using placement to emphasize ideas for better comprehension.

In websites or user interfaces, placement is often used to emphasize main points. The following welcome page shows one way that you can give more focus to some ideas and less to others:

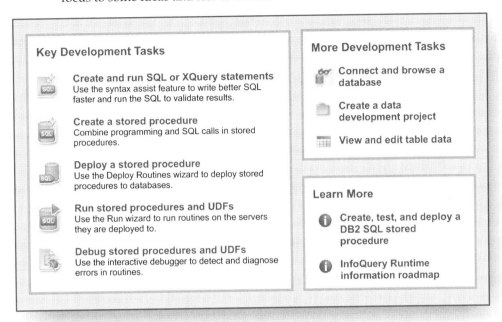

The main tasks in the window are shown on the left with extra details and larger icons, and related tasks and supporting links are shown on the right with no details and smaller icons. This window gives users access to several tasks, but it puts the most emphasis on the key tasks.

# Easy to find

Introduction and grouping are other ways to reinforce main points and relationships in the interface. The following dialog shows a list of fields with no particular organization:

Original

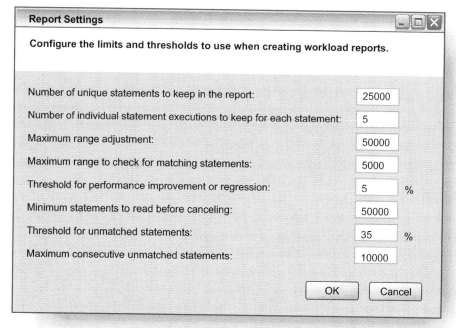

The original dialog treats all of the fields in the same manner and indicates no relationships between the fields.

Organization

Revision

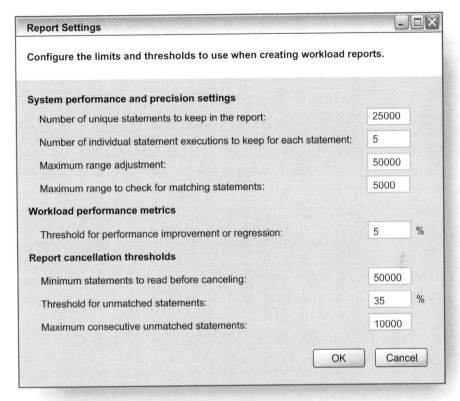

In the revision, the fields are grouped to show three main types of settings. The groupings and headings emphasize and introduce the three main outcomes that these settings can control. The emphasis that is provided by grouping the fields according to type gives users a sense of how the fields are related and makes the setting types easier to remember.

## Secondary points

When you give unwarranted emphasis to secondary information, you make that information seem more important than it is. Though a minor point might belong in the text, elaboration of the point probably does not.

Easy to find

**Original**

To record, replay, and compare a test case on two devices:

1. Clone the source device to the target device. You can use a tool such as InfoClone, or you can copy the device settings from the source to the target manually.

2. Open the InfoTest console at: https://example/infotest/console/index.jsp. When the console opens, a process is started that keeps track of your user ID and the privileges that you have. If you do not type your password correctly, you will get an error.

3. From the console, start recording by clicking **Record test case**, and then clicking **Start**.

    Instead of recording a test case, you can also import one from a different console. The import/export process involves four steps due to compatibility between devices. You would first save the test case in one console, and then export it to a second server. From the second server, you need to accept the import and then you can import the test case.

4. Run the test case on the source device. Your actions are recorded along with the responses of the device. The recording is saved as a .rec file that you can access from any connected computer or device if you have an authorized user ID. However, the .rec file is not easy to read outside of the console.

5. Click **Replay** and replay the test case on the target device. The test case is replayed, and the responses of the device are recorded in a .rec file that you can access from any connected computer or device. However, the .rec file is not easy to read outside of the console.

    If you want to export this replayed test case, you can click **Export** and follow a four-step process.

6. After the test case replays, click **Compare** to compare the responses from the two devices. You can see which device performs best or troubleshoot any commands that did not replay correctly.

The information about the import and export process is a minor point and probably not relevant to most users. Details about importing should be in a task about importing, not in a task for comparing a test case. Users don't expect details about importing to be in a task about comparing and so will not know to look there.

The original task topic also contains too much detail about errors that might occur and processes and files that users don't care about. Users probably are interested to know that their actions are recorded, but they don't need to know the file name of the recording and how to access it from outside the process.

**Revision**

> To record, replay, and compare a test case on two devices:
>
> 1. Clone the source device to the target device. You can use a tool such as InfoClone, or you can copy the device settings from the source to the target manually.
> 2. Open the InfoTest console at: https://example/infotest/console/index.jsp.
> 3. From the console, start recording by clicking **Record test case**, and then clicking **Start**.
> 4. Run the test case on the source device. Your actions are recorded along with the responses of the device.
> 5. Click **Replay** and replay the test case on the target device. The test case is replayed, and the responses of the device are recorded.
> 6. After the test case replays, click **Compare** to compare the responses from the two devices. You can see which device performs best or troubleshoot any commands that did not replay correctly.

The revision includes a couple of minor points but does not go into detail. Information about the consequences of typing a password incorrectly is removed as well as all mention of importing and exporting files. Information is removed from two steps about the files that are saved because those details are not pertinent to the task. Consequently, the revision is much more streamlined and on task.

For topics, web pages, and even user interface windows, be sure that secondary points do not appear in the first paragraph or at the top.

**Original**

> **What's new in InfoProd V8**
>
> InfoProd is developed in parallel with InfoDB and is in use throughout the world across job functions, industries, and every InfoDB database platform and economy of scale.
>
> Beginning with Version 6 the InfoProd features are all available for mobile devices, and now InfoProd V8 has new data visualization, solution-building, web-enablement, and solution-sharing capabilities.
>
> Watch the video to see the new features.

Easy to find

The first paragraph is touting the product's past accolades, which seems inconsistent with the main point of a page about what's new. Even the second paragraph starts with old information.

**Revision**

> **What's new in InfoProd V8**
>
> InfoProd V8 has new data visualization, solution-building, web-enablement, and solution-sharing capabilities.
>
> Watch the video to see the new features.

In the revision, the new features are mentioned first. Users do not have to read through the marketing text to pick them out.

Using notes is a tempting way to insert new information into previously written topics, but notes generally interrupt the organization and hide the main point. If you use notes in this way, you'll soon find that your topics have a lot of notes and no flow and that users can't follow your main points. In the following passage, a note label is used to call attention to information that does not warrant special treatment:

**Original**

> You can cancel registrations if you no longer need to copy data or if the registration is erroneous. You can cancel only the registrations that you created.
>
> **Note:** Canceling registrations ensures that the InfoDBase control tables are properly updated and prevents database errors.

The original passage gives unnecessary emphasis to the statement about the benefits of canceling a registration. That statement is not more important than the other two statements, yet the note label gives users the impression that it must be more important.

**Revision**

> You can cancel registrations if you no longer need to copy data or if the registration is erroneous. Canceling registrations ensures that the InfoDBase control tables are properly updated and prevents database errors. You can cancel only the registrations that you created.

In the revision, the information is added to the original paragraph and none of the sentences are rewritten. In addition, moving the sentence about which registrations users can cancel to the end of the paragraph improves the flow of the paragraph. The reasons and benefits of canceling registrations are covered first, followed by the restriction about which registrations can be canceled.

Too many notes dilute the effectiveness of the note label to visually emphasize a point. Take the time to fit the new information into the existing structure, or reorganize the information to accommodate the new information.

For more information about note labels and when to use them, see *The IBM Style Guide,* Chapter 3, "Notes and notices."

Easy to find

# Organization checklist

Organization is how pieces of information fit together to form a coherent arrangement of parts that makes sense to users.

Organization deals with much more than how you arrange your topics. Well-organized information never leaves users feeling lost. Contextual information is provided where users expect it, and information is arranged so that it is easy to navigate. The order of tasks is clear, links and topic order suggest postrequisite tasks and lead users in the right direction, and the overall structure is always evident to users.

You can use the following checklist in two ways:

- As a reminder of what to look for, to ensure a thorough review
- As an evaluation tool, to determine the quality of the information

| Guidelines for organization | Items to look for | Quality rating<br>1 = very dissatisfied,<br>5 = very satisfied |
|---|---|---|
| Put information where users expect it | - Elements are delivered where users expect to find them.<br>- Contextual information is not in noncontextual topics.<br>- Contextual information is provided in context.<br>- Contextual information is provided in the appropriate type of embedded assistance according to a pattern.<br>- Noncontextual information is separated into discrete topics.<br>- Topics do not contain mixed types of information.<br>- Topics that should stay together are organized into topic sets. | 1 2 3 4 5 |
| Arrange elements to facilitate users' goals | - Information has a clear structure.<br>- Topics are not arranged by arbitrary categories.<br>- The topic organization scheme makes sense.<br>- Topic maps are organized sequentially.<br>- Content of task topics is organized in the order users need it.<br>- Similar information is organized consistently across products and product sets.<br>- Contextual information follows a consistent embedded assistance pattern. | 1 2 3 4 5 |

| Guidelines for organization | Items to look for | Quality rating<br>1 = very dissatisfied,<br>5 = very satisfied |
|---|---|---|
| Reveal how elements fit together | • Relationships between topics are clear.<br>• Parent topics, transitions, and navigation elements provide sufficient context to help users to understand relationships. | 1 2 3 4 5 |
| Emphasize main points; subordinate secondary points | • Content is organized to emphasize main points.<br>• Secondary points are not given too much emphasis.<br>• Notes do not give extra emphasis where emphasis is not warranted. | 1 2 3 4 5 |

To evaluate your information for all quality characteristics, you can add your findings from this checklist to Appendix A, "Quality checklist," on page 545.

CHAPTER **10**

# Retrievability

*You've got to be very careful if you don't know where you're going because you might not get there.*

—Yogi Berra

Retrievability is the characteristic that enables users to find the information they are looking for—not just to get back to something they read before, but also to discover it for the first time. Users want to find what they need quickly and easily. When information has high retrievability, users can navigate easily from one piece to the next without getting lost in a maze.

Making sure that your tasks are well documented without making those tasks easy to find is like telling a friend that she can borrow your car but hiding the key. No matter what the context, when users need information, you need to supply the keys that will get them to their destination.

Imagine your users looking for a specific piece of information that they need to complete a task:

- Will they find the information by searching the web, even if they use jargon or synonyms instead of the terms used in your topics?
- Does the user interface provide ways to get information; for example, is there a link to more information from hover help?
- If there is an index, does it help users find what they need even if they use different terms than what the topics use?

# Easy to find

To ensure that your information is retrievable, anticipate your users' needs and make sure that the answer to each of these questions is yes.

By providing information about a particular control in embedded assistance rather than in a help topic, you ensure that help is right where users need it. By ensuring that your system administration information contains the right keywords, you know that users will find that topic about, for example, the PerfMon filter just by searching the Internet.

Many elements contribute to retrievability: obvious ones like links, tables of contents (navigation panes are a type of table of contents), tag clouds, and headings, and not so obvious ones like icons, highlighting, and revision markers. All of these elements are *entry points*—signposts that orient and direct users so that they can find the information that they want. Good retrievability requires effective use of appropriate entry points.

To develop highly retrievable information, follow these guidelines:

- **Optimize for searching and browsing**
- **Guide users through the information**
- **Link appropriately**
- **Provide helpful entry points**

# Optimize for searching and browsing

Users rely on searching above all other methods to find information. Whether your information is installed with the product, is available on the web, or both, prepare your content for the search engines that can point users to what they are searching for. After they find a website or page, most users look to the site's organization and navigation aids to guide them to what they want.

Ensure that a search engine can find the key topics, tasks, and terms in your information. The algorithms that search engines use to respond to search queries are proprietary and constantly evolving. However, by strategically applying keywords to your content and validating them, you can improve the chances that users find your information at the top of the search results.

Users who browse content usually have different goals than users who search. When users search, they are looking for something specific, whether or not they know the keywords that will ensure success. When users browse, they've already arrived at the website that interests them, and they browse to see what's there. So instead of looking for something specific, browsing users look to discover something of interest. Whether users are searching or browsing, good topic titles are important signposts to point them in the right direction.

To ensure that users who search and users who browse can find the information that they need, follow these guidelines:

- Use clear, descriptive titles
- Use keywords effectively
- Optimize the table of contents for scanning

## Use clear, descriptive titles

A topic title is the most prominent retrievability aid in your information, whether users see it in a list of search results, a table of contents, or a link from another topic. If titles don't clearly describe the topic's content, users might not take the link to find pertinent information. Or worse, they might take the link and find that it really doesn't meet their needs because the topic didn't contain what the title seemed to promise.

The organization of your content should enable users to browse the content that they are interested in, but the titles of your topics must support browsing by clearly describing the topic's subject matter.

A topic title must be specific, giving a brief, clear description of the topic. Avoid abstract headings such as *Considerations*, *Overview*, or *Important concepts*. The form that a title takes can convey the type of information that the topic contains. For example:

- Task topic titles should always have a verbal form, for example, *installing*, *configuring*, and *deploying*.
- Concept topic titles should not include the word *concepts*, as in *Integration framework concepts*, but should instead use concrete words that summarize the concept being discussed. If the topic is a conceptual overview or introduction, place that information at the end of the topic, as in *Integration framework overview*.
- Reference topic titles should typically consist of a noun or noun phrase that indicates the name of the item and a common noun, for example *installp command*, *MapTypeId API*, and *Maximo to Primavera data mappings*.
- Troubleshooting topic titles should state the problem that the topic addresses, for example, *Cannot connect to server*, *Installation program does not start*, and *Check engine light is on*.

Define and adhere to a convention for titles of concept, reference, task, and troubleshooting topics that differentiates each type for users.

Titles should contain the distinguishing information at the start and less important or repetitive information at the end, a practice that is sometimes referred to as *front-loading*. When the first words of the title contain the important information or keywords, users can spot the information easily when they browse a table of contents. In some contexts, such as on a mobile device or navigation pane, a title can be truncated, so don't waste those first words.

For example, if all of your task topic titles start with "How to", you are wasting valuable space with words that will be repeated many times. It's better to start with the verb that defines the goal of the task or the important keyword for the concept.

# Retrievability

The following table shows several types of problems with topic titles:

| Original | Revision | Explanation |
|---|---|---|
| Understanding the Functional Subsystem Interface | Functional Subsystem Interface overview | The word *understanding* in a topic title is usually an indicator of conceptual information and should be avoided in all titles. This topic is an introduction to a product component, so adding *overview* at the end of the title provides a better description of the content. |
| Considerations for UNIX installation | Installing InfoFinder on UNIX | The content of the original topic was improperly organized. For the revision, the content was moved to the relevant task topic that describes a UNIX installation, where the information is needed. A topic that contains only a list of considerations rarely deserves to exist as its own topic. Anytime that you find yourself wanting to use *considerations* in a topic title, reorganize your content instead to avoid such a grab-bag topic. |
| Restrictions | Restrictions for running InfoFinder in a cloud environment | One-word titles are often problematic, especially in a list of search results. While a topic about restrictions can be useful for users who are doing planning tasks, a complete title ensures that the right users will read the topic. |
| Adding objects | Adding business objects to the database | The original title is another example of an incomplete title. The revision includes an important prepositional phrase and a modifier for the somewhat abstract term *objects*. |
| Using InfoFinder | Finding information for your device | *Using* is another word that should not appear at the start of a topic title—*using* is not a real task or user goal. Be sure that task topic titles reflect the real tasks that users want or need to do with the product, rather than just "using" the product and its features. |
| Working with web services to create custom applications | Creating custom applications by using web services | Just as with *using*, *working with* does not reflect a real task. The original title actually had the real task hidden at the end of the title. The revision places the task at the start, followed by the method. The second half of the revised title also implies that another method is available to create custom applications. If that's not the case, then the second half isn't needed at all. |

**383**

# Easy to find

Questions are problematic as topic titles because you cannot place keywords in the most important part of the title. Unless the question style is important to the style of the information (in FAQs, for example), consider using noun phrases to better support scanning.

**Original**

> When do you want your database dumps performed?
>
> When would you want to set up default segments?
>
> When you want to remove virtual volumes and terminal library is unavailable

In the original titles, the first several words of each of the titles are ignored by search engines. Also, titles that are too long can be truncated in Google search results, so important information might not appear in the third title. The keywords appear towards the end of each title, meaning that users work harder to find what they are looking for and possibly even miss important information.

**Revision**

> Timing of database dumps
>
> Configuring default OMVS segments
>
> Removing virtual volumes when virtual library is unavailable

The revised titles are no longer questions, the style of each title gives a clue as to what type of content is in the topic, and keywords are closer to the start of each title.

## Use keywords effectively

Keywords are the important terms in your content that identify subjects. Keywords are also terms that users search for to find your content. Where and how you apply keywords in your content can greatly affect users' ability to find what they need.

Using keywords in technical information is similar to applying keywords to web sites in that you need to know your target audience and what they are looking for. However, businesses that want to optimize their websites for

search engines are looking to draw the audience—and more business—to their products or solutions. Users searching for technical information are a subset of this audience and are most likely looking for a specific answer. Be sure that your users can find the answers that your technical information provides.
The keywords for a topic are the words that express the subject matter of the topic. For example, in topics about installing a software product, the keywords would include:

- Name of the product, including shortened forms that users might use
- Version of the product
- Operating systems that the product is installed on
- The word *installation* and variations

You optimize your topics for searching and browsing by putting the keywords in the right places and with appropriate frequency. If you have an index or use index tagging, you also need to add the keywords to the index.

A title should accurately reflect the content of the topic by using one or more keywords that describe the topic. The description (or short description element in DITA) and body of the topic should also contain the keywords that appear in the title, and possibly other keywords. When you use clear, concise language, avoid jargon, and focus on the meaning, many of your keywords will fall naturally into place.

In addition to using keywords based on the subject matter, you must also identify keywords based on what you know of your target audience. Research which terms related to your product are the most searched-for terms. Talk to product architects and marketing and sales representatives to help refine your list. Then, choose the terms that you believe your users are most likely to search for and the terms that describe the product.

## Keyword location

Keywords typically belong in titles, descriptions, and body text. Eye-tracking research by Nielsen Norman Group shows that users read websites in an F-shaped pattern and spend more time reading the left side and top of a page. This research indicates that the upper left is the magic area for scanning. When users are looking for something quickly, they typically do not read beyond the first few words of a title or first paragraph. Be sure to place keywords in this magic area. This rule is especially important for mobile devices, where space is limited.

As shown in "Use clear, descriptive titles" on page 381, by front-loading titles, you can ensure that keywords appear at the start rather than toward the end of titles.

Descriptions should also contain keywords because:

- Search engines use descriptions to rank a page.
- Descriptions typically appear in search results.
- Users who scan content read titles and some initial content.

The description or first paragraph after the title should contain keywords and preferably content that differs from but supports the title.

**Original**

> **Connecting to the VMware vCenter server**
>
> This topic describes how to connect the VMware vCenter server to select your data store and target host.

In the description for this topic, the first five words are not keywords—they do not provide value for either search engines or people who are scanning. Self-referential information is rarely useful. The next several words repeat the title, which increases keyword density but otherwise provides no valuable information. It's not until users read the end of the sentence that they find something useful.

**Revision**

> **Connecting to the VMware vCenter server**
>
> Before you can select your data store and target host, you need to connect to the Vmware vCenter server.

In the revision, unnecessary words are removed, and additional keywords are closer to the upper left for scanning. The first part of the description has the reason for doing the task (and reading further) at the start, so users find what they need quickly.

## Keyword density and proximity

*Keyword density* refers to how frequently the keyword in a title appears in the body of web page or topic. Most search engine optimization (SEO) guidelines recommend a keyword density of 1 to 4 percent. For a topic of 250 words, you'd need from three to ten instances of your keyword to meet that criterion. Any fewer, and search engines might ignore your content; more than 4% puts you at risk of keyword stuffing and having your content ignored by search engines. *Keyword stuffing* refers to the practice of packing a web page with multiple instances of keywords to try to improve search rankings. Most search engines ignore content that appears to be stuffed with keywords.

*Keyword proximity* refers to how closely the words that make up a multiple-word keyword appear together. For example, *Linux server distributions* is a multiple-word keyword. The three words in this keyword do not need to be in the same order or even the same sentence in a topic for them to be associated by search engines. However, the closer the words are to each other, the better the search results are for users who search for this term or variations, such as *server distributions Linux*. Topics that refer to Linux distributions and Linux servers will appear in the search results, but topics with all three words in close proximity are more likely to be at the top of the results.

**Original**

> **Cron task setup overview**
>
> You can reschedule cron tasks and change parameter values without stopping and restarting the server. You can create cron tasks and cron task instances, remove cron tasks or cron task instances, and change cron task parameters.

The original topic is a parent to several task topics about setting up cron tasks. Although the description paragraph is the only content in this topic, two sentences have six instances of the keyword, which is nearly 17% of the topic!

**Revision**

> **Cron task setup**
>
> *Cron tasks* are background jobs that run automatically and on a fixed schedule. You can automate tasks such as backing up the database and synchronizing files.

In the revised topic, the keyword *cron task* appears only once in the description and once in the title. The new description provides a better overview of cron tasks and why you might want to use them. The word *overview* has also been dropped from the title because it's unnecessary. Because the term *cron tasks* is defined in the first sentence, italics highlight the term as an entry point.

Ideally, as you write and edit content, you are focused on quality of the content, so keywords will fall into place. But beware of keyword stuffing. Many tools, such as Google Analytics and SEOmoz, can check keyword density and proximity by scanning existing content; such tools can also help you determine the most-searched-for terms. One tool, Acrolinx, assists with search engine optimization in the authoring environment.

## Indexes

Indexing can refer to the processing of web pages by a search engine, but here it refers to manually creating locators to important subjects in your information. A well-constructed index can be a valuable tool for users looking for information. According to the American Society for Indexing, people who use indexes tend to have more success with finding the information that they are looking for than people who use full-text search.

Additionally, if you publish the index on its own web page, the links in the index help boost search rankings.

Meta tags and indexes work together to improve retrievability. Meta tags are HTML tags that provide information about the HTML page to browsers and other web services. A meta tag follows the title tag and looks like this:

```
<meta name="description" content="A good description tells
        users and search engines what a topic
        contains.">
<meta name="keywords" content="indexes, SEO, search engine
        optimization">
<meta http-equiv="content-type" content="text/
            html;charset=UTF-8">
```

The text in the meta tag provides important information about the contents of the page. Although Google, Bing, and Yahoo ignore the keywords meta tag, internal search engines, such as for a company intranet, might use this metadata. These keywords can also appear in the page info of some web browsers. In DITA, if you include your index terms in the prolog element, they appear in the keywords meta tag when the content is transformed to HTML.

Indexes are also important in PDF documents or other documents that can be printed. Many users value a good index, so don't let them down by not indexing your important content.

Chapter 10 in *The IBM Style Guide* has detailed guidelines for what and how to index.

## Optimize the table of contents for scanning

A table of contents can be something as familiar as the list of chapters in a book. However, you probably interact more often with other forms of tables of contents, such as the navigator in an application, a bookmarks list, or breadcrumbs on a website. The entries that you include in any table of contents need to show the appropriate detail to guide users to the right information.

The table of contents shows users at a glance how the information is organized. The top level of the contents hierarchy might correspond to the tasks that the information addresses, such as installing, customizing, and troubleshooting a product. Or the hierarchy might start with related products, as in the following table of contents:

**Figure 10.1** Table of contents showing hierarchy based on products on Microsoft Developer Network. Courtesy of Microsoft.

> ▷ Office and SharePoint development
> ▷ Office client development
> ▲ Office 2010
>     ▷ Access 2010
>     ▷ Excel 2010
>         ▷ Excel 2010 XLL SDK Documentation
>             ▷ Developing Excel 2010 XLLs
>                 ▲ Excel Programming Concepts
>                     Excel Commands, Functions, and States
>                     Excel Worksheet and Expression Evaluation
>                     Worksheet References
>                     Data Types Used by Excel

A table of contents helps guide users who like to browse for information. Users prefer browsing over searching when they are not necessarily looking for something specific, for example, when they are reading programming concepts for Microsoft Excel.

Or users might browse to see what sort of things they can do, such as creating applications from samples.

Figure 10.2 Table of contents showing hierarchy based on tasks from an IBM information center

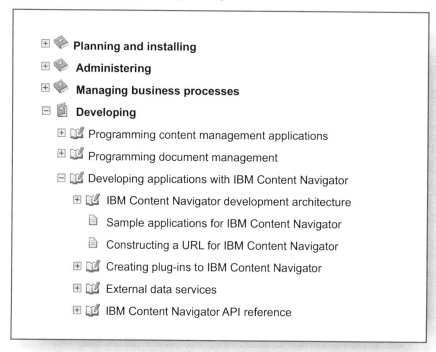

By using expandable sections in an online table of contents, you can include every topic in your information in the hierarchy. If you adhere to the organization guidelines, users can quickly grasp the overall organization and understand what each branch contains.

Entries in the table of contents do not necessarily need to match the topic title. By removing some words or changing the order of words in titles, you can make the contents easier to scan.

Instead of repeating words in titles, you can rely on titles higher up in the hierarchy to provide a sufficient level of detail.

Original

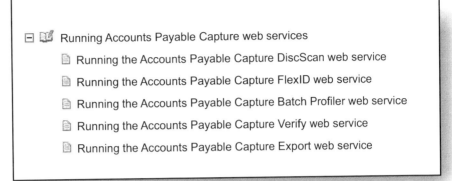

In the original table of contents, all but one of the subentries repeat the same words at the start of the title, while the distinguishing information is at the end. The repeated information is also in the top-level heading. Instead of repeating information unnecessarily and affecting scanning, you can remove the redundant information.

Revision

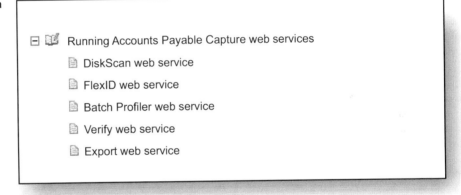

In the revision, the redundant information is removed, and the subtopics rely on the parent topic to provide context and identify the task. Users who want to run only the Batch Profiler web service can more quickly find the relevant topic. When they open the topic, the title will be complete as in the original table of contents; only the entry in the contents is shortened.

For documents that don't appear with a separate navigation pane, provide major headings in the table of contents. However, don't show more than two or three levels of headings in the table of contents unless you can use expandable sections. Too many levels impede rather than help a user's ability to find a

Easy to find

particular topic. The following reference information is a document that is displayed in a single frame on one web page:

**Original**

**XML Tools Reference**

Info Dbase XML tools overview
XML Editor
    Mixed schema elements
        Differences from earlier release
        Examples
        Content element
    Native data
        Differences from earlier release
        Example
        Content element
XML Tester
    Mixed schema elements
        Differences from earlier release
        Examples
        Content element
    Native data
        Differences from earlier release
        Example

The original table of contents provides links to the lowest-level headings in the document. Such links are not necessary in reference information when each section contains similar information. The level of detail also makes the contents too long—with this much detail, users must do quite a bit of scrolling to see the entire table of contents.

**Revision**

> **XML Tools Reference**
>
> Info Dbase XML tools overview
> XML Editor
>     Mixed schema elements
>     Native data
> XML Developer
>     Mixed schema elements
>     Native data
> XML Tester
>     Mixed schema elements
>     Native data

The revision removes the third-level headings and results in a much shorter table of contents. Users can easily find the heading for a specific tool and see the hierarchy of topics.

# Guide users through the information

To guide users through your information, you need to define the paths that they should take to get from one piece of information to the next. Whether from the user interface to a help topic or from one topic to the next, provide paths to guide users through the natural progression of information.

Navigation aids help users see where they are in relation to available information and the user interface. More importantly, navigation aids help users find their way around content. For example, a breadcrumb trail on a website shows the hierarchy of the content and ensures that users can get back to where they started or to any stop along the way. Other navigation aids include tables of contents and links that take users through the progressive disclosure path.

**Figure 10.3** Breadcrumb trail at the top of a task topic

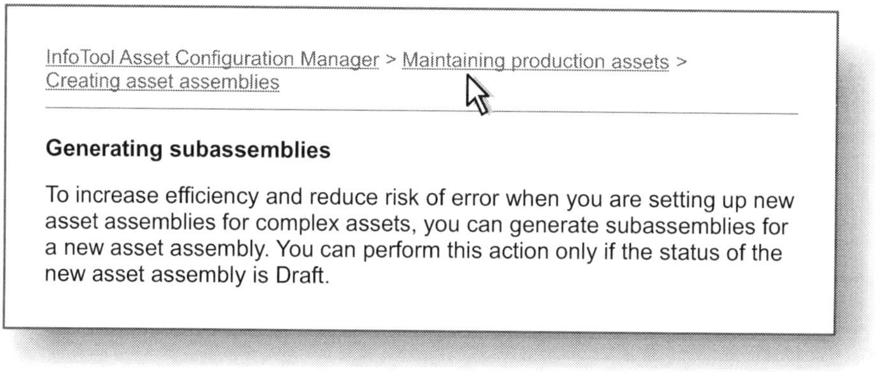

InfoTool Asset Configuration Manager > Maintaining production assets > Creating asset assemblies

**Generating subassemblies**

To increase efficiency and reduce risk of error when you are setting up new asset assemblies for complex assets, you can generate subassemblies for a new asset assembly. You can perform this action only if the status of the new asset assembly is Draft.

In a mobile environment, breadcrumbs often act as the only table of contents. Mobile users rely more on the device or application controls for going back to a previous topic or stepping through a sequence.

However, when topics are nested to too many levels, links can spill into multiple lines and take up too much of the available space. Also, because breadcrumbs are often truncated with ellipses, users might see only the first and last links in the trail. Because there's no hover help capability on mobile devices, the truncated information is essentially not there.

Good navigation is ultimately about guiding users from one place to the next when they need information and not just from one part of a website to another. For example, some users might need more information than is provided in the embedded assistance for a section of the interface, so you should provide a path to more detailed information.

**Figure 10.4** A link to more information from static text in the user interface

**Properties**

The values that you enter for the document properties can be used to find the document later. <u>Learn more</u>.

Class:

Document Title:

The paths that you provide through the information should be informed by your pattern for progressively disclosing information, as described in the completeness guideline "Apply a pattern for disclosing information." Paths to more information can simplify the way people navigate and find information. The following figure shows a path from an entry field to hover help and a path from inline text to a concept topic.

**Figure 10.5** Navigation paths from embedded assistance to more information

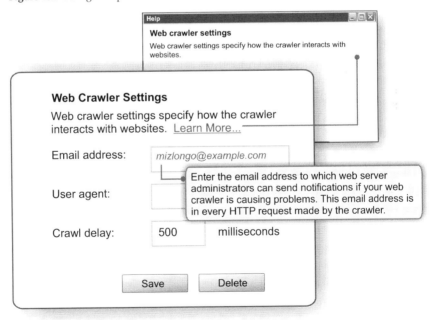

Another example of a path to *more information* is a link from hover help to more detailed help for users who might need more context. An example of a path to *related information* is the kind of linking commonly used in Wikipedia. However, such links can be more of a distraction than help.

In the following excerpt, each sentence has multiple links in it, and often a link goes to a topic that is very broad, so its relation to the original topic is questionable. Do people who want to learn about information architecture really want to follow a link to read a broad topic about science? Perhaps, if they are simply browsing and killing time on a rainy Sunday afternoon. But that's not how people typically use technical information.

**Figure 10.6** Links to related information, captured from Wikipedia October 2012

> **Information architecture (IA)** is the art and science of organizing and labelling websites, intranets, online communities and software to support usability.[1] It is an emerging discipline and community of practice focused on bringing together principles of design and architecture to the digital landscape.[2] [page needed] Typically it involves a model or concept of information which is used and applied to activities that require explicit details of complex information systems. These activities include library systems and database development.

While the multiple links in Wikipedia articles help to drive traffic to the site, users of technical information typically are not able to indulge themselves in a trip down the rabbit hole. The paths that you provide through the information should anticipate users' needs for more information so that they don't waste time going in the wrong direction or hunting for something that might take too much effort to get to.

Contrast the linking strategy in the previous Wikipedia example with the linking strategy of a topic that includes clearly relevant links intended to keep users on course through a complex task.

By defining a path for users to take through the information, you provide a guided journey instead of a scavenger hunt.

# Retrievability

In the following interface, not enough information is provided in the user interface to help users accomplish tasks:

Original

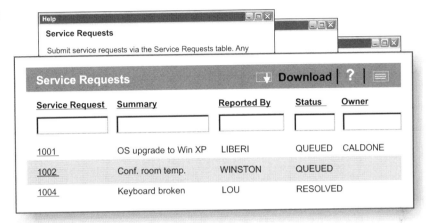

To find more information if they get stuck, users must go to the Help menu (by clicking the question mark icon), open the help system, and search for the information they need.

Revision

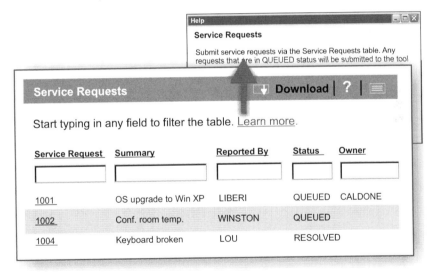

397

In the revision, the embedded assistance has a link to a topic with more detail about the screen.

Be sure to provide paths between levels of information. For example:

- Use hover help on labels that need more detail.
- From hover help, provide a link to more information if necessary to explain a concept or the task that the control is used for.
- In topics, link appropriately between related topics, as described in "Link appropriately" on page 399. For example, you might have a high-level task in which each step is documented in a separate task topic. Linking each task in sequence provides the path that users need to move from one step to the next.

> **Parent topic:** Preparing to create clusters
> **Previous topic:** Creating build files for clusters
> **Next topic:** Building the RMI registry file

Paths to more information need to be apparent, and they should not need explanation. For example, in the user interface, hover help is apparent when the pointer is moved over the label. But you don't need to add instructions that tell users to hover over a label to get more help.

Clear paths through the information ensure that all users easily find exactly the information that they need.

# Link appropriately

Links are essential in technical information. (The term *links* refers to hyperlinks in online information and cross-references in printed information.) However, when links are misused, they can be a distraction or, worse, a frustration.

Links provide:

- Access to more information than what's currently shown
- Information about related topics
- Navigation within the current topic or document
- Navigation within a website or domain such as a knowledge base

Decisions about when to provide a link and when to repeat information affect completeness as well as retrievability. The guideline "Repeat information only when users will benefit from it" deals with this issue from a completeness perspective.

Inappropriate linking can result in these problems:

**Broken links**

A "Page not found" message is always frustrating but especially so when users are trying to get help for a task.

**Links that take users on a wild goose chase**

Improper organization of linked content often means users must hopscotch from topic to topic in order to find the relevant bit they are looking for.

**Links with little to no relevance to the topic at hand**

If you include links to every topic that simply mentions the same subject, users can lose confidence in finding what they need.

**Links that don't describe the linked information**

Improper link text doesn't tell users what they'll find when they take the link. This problem can arise from inappropriate topic titles or link text such as "Click Here" that is vague or misleading.

To ensure that your links are appropriate:

- ❏ **Link to essential information**
- ❏ **Avoid redundant links**
- ❏ **Use effective wording for links**

# Easy to find

For additional advice about linking, see Chapter 7 in *DITA Best Practices* and the topic "Linking strategies" in Chapter 5, "References," in *The IBM Style Guide*.

## Link to essential information

When you organize content well and put the right information in the right place, you will find that links to related information are not needed as often as with poorly organized content. Necessary links are the ones that take users to what they need, not to what you think there's a chance they might be interested in.

Never add a link because you think the linked information *might* be useful—be *sure* that it will be useful. When you link from a topic to related topics, ensure that the linked information supports the current topic. When you link to topics that might be about the same subject but aren't completely relevant to the task at hand, users might waste time in taking the link and straying from the path that leads them to the information they need. To determine what information to link to, you must understand your audience's goals, as described in the completeness guideline, "Cover all subjects that support users' goals, and only those subjects."

Gone are the days when consistency was paramount, where if you had a Help button in one dialog, you needed one in every dialog. For example, you don't need to provide a help topic for a login window, so there's no need to link to help.

It's especially important to provide only essential links for mobile apps because screen space is minimal—and you can safely assume that users will be accessing information from mobile devices. Consider these users first and define a linking strategy around their needs and environment. Provide links that are accurate and targeted and that take users on a logical path to more information. By providing fewer, more precise and useful links, you'll also improve the desktop information experience.

Effective linking also supports search engine optimization. Too many links going in too many directions but with no links from outside sources can ensure that your topics are ignored by search engines.

**Original**

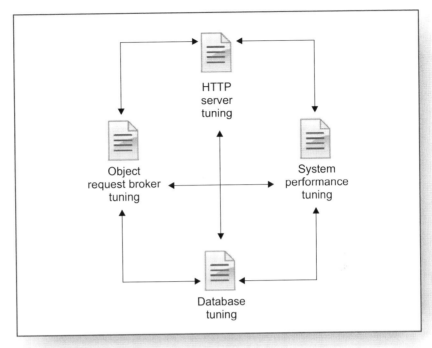

In the original linking model, every topic in the set links to every other topic, and there are no links from external sources. Never link child topics of the same parent to each other. The fact that the topics all use the same keyword (tuning) makes it less likely that search engines will find the topics relevant—they are essentially competing with each other.

**Revision**

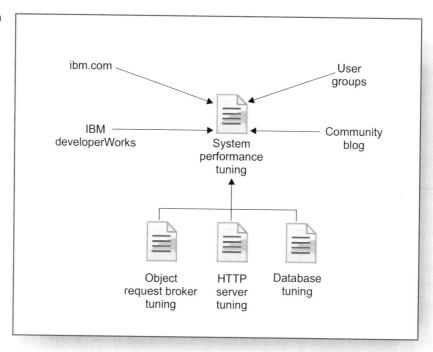

The revised linking strategy eliminates the peer-to-peer linking of the child topics. The only outgoing links from the child topics are back to the parent topic. The external links to the parent topic ensure that it has relevance to search engine algorithms—it's not a self-contained dead end.

Unnecessary links can get in the way of the information that users are looking for, as shown in the following hover help:

**Original**

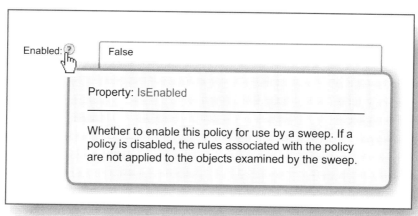

Prominently positioned at the top of the hover help is a link to the API that the field value interacts with. Users who want to understand the purpose of the **Enabled** field are most likely not interested in its API and will hopefully move to the paragraph after the link instead of taking the link.

**Revision**

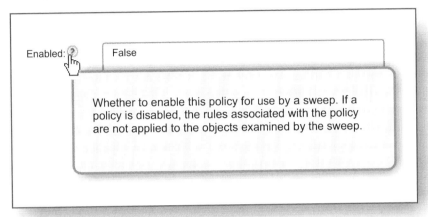

In the revision, the property name and link are gone, and the help starts at the top of the hover help window, ensuring that users who need help will get it immediately.

Opinions and research on whether to use in-sentence (inline) links vary, with some people firmly on the side of "never" and others who believe such linking is so standard that users don't notice it. The location of links should be based on your assessment of where they are needed, what your audience is focused on (seeking, doing, or learning), and to some extent, what your publishing environment supports. Ultimately, you should not add links simply because there is more information out there somewhere or to make one user or technical expert happy.

In particular, avoid unnecessary in-sentence linking because it can distract users, sending them someplace they don't need to go. If you have too many inline links, users begin to ignore them as noise and might miss an important path to more information.

# Easy to find

**Original**

> 3. Select the implementation type, as described in the topic <u>Generating models from workspace applications</u>.
>
> 4. Specify a location for the model.
>
> 5. Select the elements that you want to model. (See <u>Common monitoring templates</u> for more information.)

The original topic interrupts the procedure with links to other topics. Although the link in step 3 is introduced in a way that makes the link appear critical, it actually contains some conceptual information about implementation types, which users might or might not need when they are selecting the type. Following an appropriate pattern for progressive disclosure would put that information in the context of the task, perhaps as hover help. The second link is truly nonessential, as implied by the parentheses.

**Revision**

> 3. Select the implementation type.
>
> 4. Specify a location for the model.
>
> 5. Select the elements that you want to model.

Because the revised topic no longer links to nonessential information, users can complete the task without wondering whether to take a path to other information. The relevant information from the first link was moved to embedded assistance so that users who need more information about implementation types can get it in context. Neither linked topic in the original is closely related to the procedure at hand, so neither belongs as a related link at the end of the topic.

When you consider whether to include in-sentence links, make sure that the linked information is useful and worth the detour. You might find that the information should be moved to a place where it is most helpful to users when they need it.

Users of printed information rarely interrupt their reading to refer to remote information, necessary or not. They resist even a branch that they are told to take, and read on until they find that they really can't proceed without taking

# Retrievability

that branch. Often, they waste time by trying to go ahead. For example, many users will ignore a reference to a figure that is only as far away as the facing page. But when the figure appears on the same page, right after the reference to it, almost all users will look at the figure immediately.

The link in the following passage is to a topic that explains the syntax for a command:

**Original**

> **Administrator**
>   Authorizes other users and can register the data. The administrator is the owner of the user ID that runs the ABCONTROL command during installation. See "ABCONTROL command" on page 129 for more information about the command.
>
> **User**
>   Enters data into a subset of tables, as authorized by the administrator.

In the original passage, the user is reading about the difference between an administrator and a user. The user doesn't need to know the syntax of an installation command. The original passage provides the cross-reference in case a user might be immediately interested in that information, which is unlikely in this context.

**Revision**

> **Administrator**
>   Authorizes other users and can register the data. The administrator is the owner of the user ID that runs the ABCONTROL command during installation.
>
> **User**
>   Enters data into a subset of tables, as authorized by the administrator.

The revised passage saves most users the trouble of branching to unnecessary information. Any users who are especially interested in the ABCONTROL command can look it up through the index or search function.

## Avoid redundant links

Links that are repeated in the same topic are redundant. Two links that appear different but that go to the same topic can confuse or frustrate users. For example, a topic might include a link that goes to the top of a page and other links that go to certain parts of that page.

Easy to find

**Original**

### Presentation options for maps

The map view shows the location of assigned jobs and can provide routing directions. You can change the appearance of the map and the units of measure.

To change options for the map view, go to the Maps > Options page.

| Option | Comments |
| --- | --- |
| Colors | You can change the colors of location flags, route lines, and text. |
| Map grid | By default, the grid is turned off. |
| Units of measure | You can specify the units of measure that are shown on the map and in routing directions. |

**Related tasks**

Configuring the map view for routing directions

The topic contains four links to the same window where users can specify their preferences. The link near the top of the topic might be all that most users need—they just need to know where to go to set their preferences. Because the repeated links in each row of the table go to the same page, the link text can mislead users to believe they are going to other information. The repetition is not helpful.

**Revision**

> **Presentation options for maps**
>
> The map view shows the location of assigned jobs and can provide routing directions. You can change the appearance of the map and the units of measure.
>
> To change options for the map view, go to the <u>Maps > Options</u> page.
>
> | Option | Comments |
> |---|---|
> | Colors | You can change the colors of location flags, route lines, and text. |
> | Map grid | By default, the grid is turned off. |
> | Units of measure | You can specify the units of measure that are shown on the map and in routing directions. |
>
> **Related tasks**
>
> <u>Configuring the map view for routing directions</u>

The revised topic removes the redundant links and potential for confusion. By presenting the link once near the top of the topic:

- Redundant links are eliminated.
- Users who are simply looking for where to make such changes get the information first and don't have to read the rest of the topic.
- The possibility of opening multiple instances of the same page no longer exists.

Alternatively, the link to the preferences page can go at the bottom of the topic if you organize your content to provide links to related information after the body of the topic.

# Easy to find

If your information delivery system provides a breadcrumb trail, an automatically generated parent link might be redundant.

**Original**

> Maximo Asset Manager > Installing Maximo Asset Management > Installing on WebSphere Application Server > Preparing for installation
>
> **IBM Maximo Asset Management, Version 7.5**
>
> ### Software installation images
>
> You access the IBM Maximo Asset Management product software from IBM Passport Advantage or from the product DVD if you requested a product DVD.
>
> The installation images that you download from Passport Advantage can comprise multiple downloadable files. Download all files in the package to a single directory and extract the files for execution.
>
> **Related information**
> IBM Maximo Asset Management 7.5 Download Document
> Release notes for Maximo Asset Management 7.5
> Installing Maximo Asset Management
>
> **Parent topic:** Preparing for installation

The original topic has a breadcrumb trail, a parent link, and three related links. One of the related links and the parent link both appear in the breadcrumb trail, so they are redundant.

# Retrievability

**Revision**

> Maximo Asset Manager > Installing Maximo Asset Management > Installing on WebSphere Application Server > Preparing for installation
>
> **IBM Maximo Asset Management, Version 7.5**
>
> **Software installation images**
>
> You access the IBM Maximo Asset Management product software from IBM Passport Advantage or from the product DVD if you requested a product DVD.
>
> The installation images that you download from Passport Advantage can comprise multiple downloadable files. Download all files in the package to a single directory and extract the files for execution.
>
> **Related information**
> IBM Maximo Asset Management 7.5 Download Document
> Release notes for Maximo Asset Management 7.5

In the revised topic, the parent link has been omitted to provide a less-crowded list of links at the bottom of the topic. The redundant link to "Installing Maximo Asset Management" was also removed. Because this topic is quite short, the breadcrumb trail will always be visible at the top.

In longer topics that require scrolling, you might want to keep a link to the parent at the bottom. Be sure to consider the trade-off between too many links and helping users get back to where they came from. You also need to know how your content will be delivered. For example, if the delivery mechanism does not provide breadcrumbs, parent links will be an important path.

## Use effective wording for links

The text that you use for links can affect the relevance or weighting in search engines. Link text should clearly identify the linked information to both users and search engines.

Avoid creating links in text that say "Click here to read more about..." Links should keep the user's focus on the content, not on the links or the underlying structure of the topic. Instead of using unnecessary words such as "click here" or "go here," make each link a meaningful noun phrase or verb phrase, such as:

> Visual programming tutorial
> Installing InfoLight Server
> Download Version 3.0 beta code
> Application deployment process

There's no need to add extra words to tell users it's a link, and those extra words can cost precious space on a mobile device.

**Original**

> Recording and slides of Maximo 6.2 Upgrade are available here.
>
> You can also find more Upgrade information on the updated Maximo Wiki site here.
>
> Also, be sure to review this support tech note on key steps for upgrading reports here.
>
> Rebecca details performance considerations for your administration workstation here.

Each of the original links from a blog is on the word *here*, which is not only incorrect and colloquial, but can also negatively affect search engine ranking of the page.

**Revision**

> Download the recording and slides of Maximo 6.2 to 7.5 Upgrade.
>
> Find more upgrade information on the updated Maximo Wiki site.
>
> Also, be sure to review this support tech note on key steps for upgrading reports.
>
> On her blog, Rebecca describes performance considerations for your administration workstation.

The revised links are on text that includes important keywords, such as *upgrade*, *reports*, and *performance*. This change eliminates unnecessary words and provides effective links that are more compelling.

When you use lists of links with similar text, put only the words or phrases that are different inside the link, instead of the entire phrase. An even better solution is to move the redundant text to the top of the list.

The following links all start with the same words:

**Original**

> Software requirements for Android
> Software requirements for iOS
> Software requirements for Windows phone

In the original set of links, the words "Software requirements" are repetitive, and the operating systems are not immediately apparent.

**Revision**

> Android software requirements
> iOS software requirements
> Windows Phone software requirements

The revision moves the distinct information to the start of each heading so that users can quickly scan to find their operating system.

One type of unhelpful link often appears at the end of a paragraph or topic and points to another source of information: "See the readme file for more information." Such links give no indication of why the user should seek out the linked information. Users might have questions about what they are reading, but this cross-reference does not provide enough information for them to understand whether they might find the answers in the linked information.

A helpful cross-reference is something like: "For information about the XLINK language syntax, see the *XLINK Programming Guide*." An exception would be if you link to a topic with a clear title: "For more information, see XLINK programming language syntax". Alternatively, keep the link at the bottom of the topic and use only the topic title, which should clearly indicate that it is about XLINK language syntax.

In printed information, if the cross-reference is within the same book, make sure that it includes a page number. If the reference is in a different book, make sure that the text that you use to describe the related information helps the user to find that information quickly in the other book. Use the same wording that is used in the table of contents or index (or both) of the other book.

## Provide helpful entry points

Entry points are the titles, labels, lists, captions, and other elements that help identify different tidbits of information in a document or interface. As far back as 1997, Jakob Nielsen's research showed that users don't read websites—they scan. The same is true today, and users likely spend even less time per document or topic because of the overwhelming amount of available information.

User expectations change as technology changes. Users can access technical information from mobile devices, receive text messages when their favorite topics are updated, and build custom documents from available content.

Don't impede scanning by writing paragraphs that take up half the browser window or multiple mobile screens. Break up long blocks of text and provide entry points such as headings, labels, and lists. Aim for two to four short sentences per paragraph. Keep the content focused on the primary point that you are conveying, especially for mobile. As Nielsen writes, "Short is too long for mobile."

The following memo shows the difference you can make by restructuring the information to add entry points in the form of a list:

**Original**

> **New comparative checking feature**
>
> Shortly, we will enable a new feature for the Qchecker tool known as comparative checking. Comparative checking is generally transparent, but you might see a dialog box open briefly when you close a file or close an editor. In FrameMaker, you might see that the document is marked as changed and be prompted to save even though no change was made by the comparative checking feature. Be sure to save your file.
>
> **You do not need to take any action.**

In the original text, the most prominent information is in the heading and the bold highlighted sentence that follows the paragraph. Users' eyes will naturally jump to the highlighted text, and its message will convey to many users that they don't need to read anything else. However, the most important point is in the paragraph: the fact that a dialog will open might take users by surprise, and the Tools team wants to ensure that users know about this new behavior to prevent false problem reports. Additionally, FrameMaker users are advised to take an action, so the highlighted sentence is actually misleading.

Easy to find

**Revision**

> **New comparative checking feature**
>
> We have enabled a new feature for the Qchecker called comparative checking.
>
> Comparative checking is generally transparent, but you might see the following behavior:
>
> > **All editors that use the Qchecker plug-in:** A dialog will open briefly when you close a file or close your editor. **Allow the checking to compete. You do not need to take any action.**
> >
> > **Framemaker only**: The document will be marked as changed, and you will be prompted to save even though no change was made by the comparative checking feature. **Be sure to save your file**.

The revision restructures the information to focus the attention of users of all authoring tools and users of only Framemaker. Where the original made it appear that nobody needed to take action, the revision emphasizes the action that Framemaker users need to be aware of. By using a list and highlighting the most important information, the revised text provides the appropriate entry points and ensures that users see the right message.

Consider the most likely sources of links into your information. A user might link from other pages within your information set, from other pages controlled by your organization, from a product interface, or from a search engine. In each case, try to anticipate what information users will be looking for, and then use highlighting, color, icons, labels, figures, or headings to make that information stand out.

Although highlighting is a good way to make key terms stand out, excessive highlighting does more harm than good. Too many highlighted phrases compete for attention and defeat the purpose. And too many different kinds of highlighting (bold, color, italics, underlining) confuse users instead of directing them to key terms and concepts.

**Original**

> InfoDBase provides four predefined relationship categories. *Hierarchical* relationship types connect objects that have a hierarchical relationship. *Peer-to-peer* relationship types connect objects that have a peer relationship. *Support* relationship types connect precedence objects to data resources (for example, you can connect a **News** object to a **Spreadsheet** object). *Precedence* relationship types connect precedence objects to data resources (for example, you can connect a **Precedence** object to a **File** object). Objects that are connected with this category of relationship are displayed in the Show Lineage Tree window.

The original text shows a paragraph in which several items are highlighted, using different types of highlighting. The emphasis makes the passage difficult to scan because it has too many entry points.

**Revision**

> InfoDBase provides four predefined relationship categories:
>
> **Hierarchical**
> > Connects objects that have a hierarchical relationship.
>
> **Peer-to-peer**
> > Connects objects that have a peer relationship.
>
> **Support**
> > Connects supporting objects to another object. For example, you can connect a News object to a Spreadsheet object.
>
> **Precedence**
> > Connects precedence objects to data resources. For example, you can connect a Precedence object to a File object. Objects that are connected with this category of relationship are displayed in the Show Lineage Tree Window.

The revised passage uses a list and one type of highlighting to emphasize the terms being defined and make their definitions easy to see. The last sentence now clearly belongs with the last term. The object names are no longer highlighted because the use of initial capitals on the names helps to distinguish them from normal words. When you are deciding what types of highlighting to use, consider the effect that highlighting will have on retrievability. The style guideline "Apply consistent highlighting" discusses practical and consistent highlighting.

Tables can improve the retrievability of some kinds of information, but tables must also have good entry points for the data that they contain.

**Original**

| Option | Default | Comments |
|---|---|---|
| CalStackSize | 10 | CalStackSize=15<br><br>Sets the number of displayable items in the combination box for the Inspector Call Stack |
| CreationMode | ControlArray | CreationMode=AskUser—User confirmation<br><br>CreationMode=ControlArray—No warning<br><br>CreationMode=IndControl—Group created from Toolbox |
| UndoStackSize | 10 | UndoStackSize=#<br><br>For example, a value of 5 limits Edit > Undo and Edit > Redo to five actions each |

The last column of the original table is a catchall for different kinds of information.

**Revision**

| Option | Default | Example | Explanation of example |
|---|---|---|---|
| CalStackSize | 10 | CalStackSize=15 | Sets the number of displayable items in the combination box for the Inspector Call Stack. |
| CreationMode | Control Array | CreationMode=AskUser | Requests user confirmation. |
| | | CreationMode=ControlArray | No warning needed. |
| | | CreationMode=IndControl | Creates a group from the Toolbox. |
| UndoStackSize | 10 | UndoStackSize=5 | Limits **Edit > Undo** and **Edit > Redo** to five actions each. |

The revision puts the examples in one column and the explanation of the examples in another column so that both are easier to see and understand.

Users also need entry points in the user interface when they are interacting with a product. Labels, section headings, and other embedded assistance are all entry points in an interface.

You might be constrained as to where you can place embedded assistance in the user interface. Ideally, the user assistance mechanisms in your product support various levels of embedded assistance. The labels on the interface are one level of assistance that is always available, but input hints, programmatic assistance, and inline text can also be helpful.

Where you place inline text is also important. When inline text precedes a field, users often don't read it—they go straight to the action areas of the interface.

**Original**

### Deployment Environment

Specify information that describes your deployment environment. If you specify a default middleware password, it is used for all local and remote middleware system users. Required Maximo system users are the adminstrator, registration user, and system integration user.

**Database Information**

Database type: DB2    Host name: foundation48

☐ Automatically create and configure the database

User name: Administrator    Password:

**Application Server Information**

Server type: WebSphere    Host name:

☐ Automate configuration of the WebSphere deployment

**Default passwords**

☐ Use a default password for middleware password fields

Default middleware password:    Confirm password:

☐ Use a default password for the required Maximo users

Default Maximo password:    Confirm password:

# Easy to find

The original window includes inline text only at the top of the window, even though most of the information is about specific items in the window. While users might read the information, they are also just as likely to go directly to the entry fields. And if they need more information about a particular field, they won't think to go to the top of the window.

**First revision**

> **Deployment Environment**
>
> ---
>
> Database Information
>
> Database type: [ DB2 ▾ ]   Host name: [ foundation48 ]
>
> ☐ Automatically create and configure the database
>
> User name: [ Administrator ]   Password: [ ]
>
> ---
>
> Application Server Information
>
> Server type: [ WebSphere ▾ ]   Host name: [ foundation48 ]
>
> ☐ Automate configuration of the WebSphere deployment
>
> ---
>
> Default passwords
>
> ☐ Use a default password for middleware password fields
>
> If you specify a default middleware password, it is used for all local and remote middleware system users.
>
> Default middleware password: [ ]   Confirm password: [ ]
>
> ☐ Use a default password for the required Maximo users
>
> Required Maximo system users are the adminstrator, registration user, and system integration user.
>
> Default Maximo password: [ ]   Confirm password: [ ]

In the first revision, inline text specific to two check boxes is moved to directly under each check box, closer to where users might need it.

**Second revision**

### Deployment Environment

#### Database Information

Database type: [DB2 ▼]    Host name: [foundation48]

☐ Automatically create and configure the database

    User name: [Administrator]    Password: [          ]

#### Application Server Information

Server type: [WebSphere ▼]    Host name: [          ]

☐ Automate configuration of the WebSphere deployment

#### Default passwords

☐ Use a default password for middleware password

Default middleware password: [     ]    Confirm

> If you specify a default middleware password, it is used for all local and remote middleware system users.

☐ Use a default password for the required Maximo users

Default Maximo password: [     ]    Confirm

> Required Maximo system users are the administrator, registration user, and system integration user.

---

Inline text in user interfaces should be used judiciously, for example, to show a sample of what can be entered in a field. Because the information about the check boxes is useful only to some users who might need additional information, the second revision moves the inline text into hover help. The entry points become the mechanism for showing the hover help rather than taking up real estate in the interface.

Helpful entry points depend on good organization of information—from the information embedded in the interface to how tables are segmented. By following the organization guideline "Arrange elements to facilitate navigation" on page 345, you can ensure that your information has useful entry points.

# Retrievability checklist

Retrievability is the characteristic of information that ensures that users can find the information they need, no matter where it is or how it is presented. If information is not retrievable, it doesn't matter how accurate or complete it is: users must find the information to be able to use it.

Use the guidelines in this chapter to ensure that your technical information is retrievable. Refer to the examples in this chapter for practical applications of these guidelines.

You can use the following checklist in two ways:

- As a reminder of what to look for, to ensure a thorough review
- As an evaluation tool, to determine the quality of the information

| Guidelines for retrievability | Items to look for | Quality rating<br>*1=very dissatisfied, 5=very satisfied* |
|---|---|---|
| Optimize for searching and browsing | • Keywords appear in the appropriate locations and with appropriate frequency.<br>• Titles effectively describe the content.<br>• Titles and descriptions have keywords at the start.<br>• Where applicable, an index is included.<br>• The table of contents is optimized for scanning. | 1 2 3 4 5 |
| Guide users through the information | • The UI and the information contain effective paths to more information.<br>• Navigation aids such as breadcrumbs and links are effective for the type of delivery. | 1 2 3 4 5 |
| Link appropriately | • There are no broken links.<br>• Links are provided only when necessary.<br>• There are no redundant links.<br>• Links do not clutter a UI or topic.<br>• Wording of links describes the target information. | 1 2 3 4 5 |
| Provide helpful entry points | • Headings, labels, tables, lists, and other elements balance the information.<br>• Embedded assistance appears where users need it.<br>• Highlighting does not distract from the main point.<br>• Information in tables is segmented appropriately. | 1 2 3 4 5 |

To evaluate your information for all quality characteristics, add your findings from this checklist to Appendix A, "Quality checklist," on page 545.

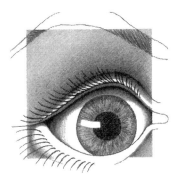

CHAPTER **11**

# Visual effectiveness

*Confusion and clutter are failures of design, not attributes of information.*
—Edward R. Tufte

Visual effectiveness is a measure of how the appearance of information and the visual elements within it affect how easily users can use, understand, and find the information they need. Users must deal with a huge volume of information, and they must do so on increasingly smaller devices. Visual effectiveness plays a key role in helping users quickly navigate and process all of this information.

Technical communicators generally focus on words, so obvious visual elements such as illustrations and layout of user interfaces might seem beyond their purview. Some writers shy away from this aspect of quality, thinking that they are "not visual," or "no good at drawing."

In fact, visual effectiveness is as integral to information quality as any of the other quality characteristics. There are many things that you can do as a writer, some of them quite simple, to ensure that your information is both visually pleasing and highly usable.

# Easy to find

As the sample guidelines in the following table show, many style guidelines also affect visual effectiveness:

| Visual element | Guideline | Example |
| --- | --- | --- |
| In hover help, links that users can click to get more information | Use the label "Learn more" in bold blue font. | Use the minus sign (-) to precede terms to exclude. **Learn more**. |
| Code that users must type exactly as it is written | Use a monospace font. | **Find this** / **Sample query in search box** — Documents about the philosopher `Calvin` and not the cartoon character `Calvin` / `Calvin -Hobbes` |
| An invalid field entry in a user interface | Use a red symbol next to the invalid field entry. | ⊗ Crawl delay: [0.pp] seconds between pages |

Typically writers make decisions about when, where, and how to use the following visual aspects and elements, all of which affect visual effectiveness:

- Embedded assistance in user interfaces, including organizational groupings, labels, static text, hover help, tooltips, and links
- Textual and nontextual elements in illustrations
- Screen captures of windows or user interface elements
- Highlighting, shading, and color
- Navigational graphics and headings
- Tables, lists, and paragraphs
- Line spacing, fonts, and font sizes (to the degree that these elements aren't controlled by writing tools and templates)
- Syntax diagrams of programmatic elements

Visual effectiveness *does* include illustrations and user interface designs—and the importance of collaborating with visual designers and usability engineers in the development of quality technical information can't be overstated. But visual effectiveness is more than that. The visual effectiveness of technical information influences, to a greater or lesser degree, all of the other quality characteristics, and it can help enhance them in the following ways:

- Provide a big picture of **tasks**, emphasize the sequence of tasks and steps, and show the interrelationships between tasks
- **Accurately** depict facts and relationships
- **Complete** a description or explanation
- **Clarify** information that might otherwise be confusing or too complex when expressed in words alone
- Make abstract information, especially concepts, more **concrete** and succinct
- Set a visual tone and reinforce a consistent **style**
- Make **organization** and structure more obvious
- Support navigation and make information easy to **retrieve**

Everything that users can see in the interface or on the page affects the visual effectiveness of the information, which in turn affects the overall quality and ultimate success of that information. Even though other professionals might be responsible for elements such as illustrations, icons, and user interfaces, technical communicators play a vital role in creating a usable experience in the product.

Users are motivated when they perceive a reasonable chance of success at their task. The effective use of visual elements can help users be confident and can make their search for information more interesting, efficient, and maybe even fun. Your goal for visual effectiveness should be to attract and encourage users to access your information, to motivate them to continue reading and interacting, and, ultimately, to help them accomplish their goals.

To make information visually effective, follow these guidelines:

- ❑ **Apply visual design practices to textual elements**
- ❑ **Use graphics that are meaningful and appropriate**
- ❑ **Apply a consistent visual style**
- ❑ **Use visual elements to help users find what they need**
- ❑ **Ensure that visual elements are accessible to all users**

Easy to find

# Apply visual design practices to textual elements

Elements such as headings, tables, labels, color and shading, and the layout of pages and interfaces are entry points to information that visually convey hierarchy, importance, and focus. But even the text itself and the *white space* (the areas of a page, user interface, or image that aren't used) around it are visual elements. When you make choices about when and where to use such elements, you're also making decisions that impact visual effectiveness.

The following page contains both task and reference information (the example is in Greek so that you can focus on the visual aspects of the page):

**Original**

Της μοστ χομμον στοραγε σηορταγε προβλεμσ ινωολωε ΚΘΗ. ΜΧΑΤ υσεσ λαργε αμουντσ οφ ΚΘΗ φορ ιτσ χοντρολ βλοχκσ ανδ βυφφερσ χονταινινγ δατα τηατ ισ βεινγ σεντ αρουνδ τηε νετωορκ. Τηισ προβλεμ χαν βε δυε το α ωαριετψ οφ χαυσεσ. Φορ εξαμπλε, ιφ ονε αππλιχατιον ισ φλοοδινγ ανοτηερ ωιτη δατα, ανδ τηε σεχονδ αππλιχατιον ισ υναβλε το ρεχειωε τηε δατα ανδ προχεσσ ιτ ατ αν αδεθυατε ρατε, τηεν τηε βυφφερσ βυιλδ υπ ιν ΜΧΑΤ στοραγε. Τηισ σιτυατιον χαν βε αωοιδεδ βψ υσινγ σεσσιον παχινγ.

Ιφ ΜΧΑΤ ισ υναβλε το γετ ενουγη στοραγε το ισσυε α μεσσαγε, τηεν τηε νορμαλ στοραγε σηορταγε μεσσαγε ισ αχχομπανιεδ βψ αν ΙΣΤ999Ε μεσσαγε.

ΦΓΗΜΠ ανδ ΘΦΡΔΤ: Της Πριμαρψ ρετυρν χοδε (ΦΓΗΜΠ) ανδ σεχονδαρψ ρετυρν χοδε (ΘΦΡΔΤ) αρε βοτη γιωεν το ΔΔΦ ωηεν αν ΑΠΠΙΧΧΜΔ μαχρο ηασ χομπλετεδ. Ον οχχασιον, τηεσε χοδεσ μαψ ινδιχατε τηατ τηερε ισ α στοραγε σηορταγε ιν ΜΧΑΤ. Φορ εξαμπλε, ιφ ΦΓΗΜΠ ισ ξ 0037 ανδ ΘΦΡΔΤ ισ ξ 0000, τηεν τηισ ινδιχατεσ α στοραγε σηορταγε ωηιλε ΜΧΑΤ ωασ ρεχειωινγ δατα ορ σενδινγ α παχινγ ρεσπονσε.

Α ΦΓΗΜΠ οφ ξ 0093 ωιτη ΘΦΡΔΤ οφ ξ 0000 ινδιχατεσ τηερε ισ α τεμποραρψ στοραγε σηορταγε ωηιλε σενδινγ δατα. Υσυαλλψ τηισ ρετυρν χοδε μεανσ τηατ τηε σενδ ρεθυεστ ηασ τεμποραριλψ δεπλετεδ τηε βυφφερ ποολ το συχη αν εξτεντ τηατ τηε ποολ μυστ βε εξπανδεδ. Της εξπανσιον ηαδ νοτ οχχυρρεδ βεφορε τηε χομπλετιον οφ τηε ΑΠΠΙΧΧΜΔ μαχρο.

ΣΝΑ σενσε χοδε: Σομε ΣΝΑ σενσε χοδεσ ινδιχατε τηερε μαψ βε α στοραγε σηορταγε αλσο. Φορ εξαμπλε, α υσερ μιγητ βε αχχεσσινγ δατα φρομ τηε ρεμοτε δαταβασε, ανδ ρεχειωε 037Χ0000. Χηεχκινγ τηισ σενσε χοδε ιν τηε μανυαλ ινδιχατεσ τηατ τηερε ισ α περμανεντ ινσυφφιχιεντ ρεσουρχε χονδιτιον. Τηισ ρεσουρχε χουλδ βε στοραγε. Οτηερ σενσε χοδεσ, συχη ασ 300Α000 ινδιχατε α στοραγε τψπε προβλεμ, βυτ νοτ νεχεσσαριλψ α στοραγε σηορταγε.

Ερρορσ: Τηερε αρε α σεριεσ οφ ΠΘΛ ερρορσ ωηιχη ινδιχατε στοραγε σηορταγε προβλεμσ. Φορ εξαμπλε, ΕΡΡΟΠ373 ανδ ΕΡΡΟΠ30Α. Τηεσε αρε υνχομμον ιν ΜΧΑΤ.

Ηανγσ: Δεπενδινγ ον τηε στοραγε σηορταγε ανδ τηε προχεσσινγ τηατ ισ οχχυρρινγ, τηε στοραγε σηορταγε χουλδ μανιφεστ ιτσελφ ιν α ηανγ σιτυατιον. Φορ εξαμπλε, ιφ α ωιρτυαλ ρουτε βεχομεσ βλοχκεδ δυε το στοραγε σηορταγεσ, τηεν αλλ τηε σεσσιονσ τηατ ηαδ βεεν υσινγ τηατ σεσσιον ηανγσ υντιλ τηε στοραγε σηορταγε ισ ρελιεωεδ ανδ τηε ωιρτυαλ ρουτε βεχομεσ οπεν αγαιν. Ωηεν ανψ οφ τηεσε ινδιχατιονσ οφ α στοραγε προβλεμ ισ ρεχειωεδ, τηε φολλοωινγ στεπσ χαν βε υσεδ το φινδ ουτ μορε ινφορματιον αβουτ τηε σηορταγε:

Δετερμινε τηε αρεα οφ στοραγε σηορταγε. Τηισ ισ ψμπορταντ, ασ τηε ΜΧΑΤ δισπλαψ χομμανδ ηασ ινφορματιον αβουτ τηε ΜΧΑΤ ΚΘΗ υσαγε. Τηισ δοεσ νοτ ηελπ διαγνοσισ ιφ τηε στοραγε σηορταγε ισ ιν ΜΧΑΤ πριωατε. Ηοωεωερ, περηαπσ ΚΘΗ υσαγε σηουλδ προβαβλψ βε χηεχκεδ ανψωαψ.

Δισπλαψ βυφφερ υσαγε: Ιφ τηε στοραγε σηορταγε ηασ οχχυρρεδ ιν ονε οφ τηε ΚΘΗ συβποολσ, τηεν α Π ΣΕΤ,ΔΧςΡΦ χομμανδ χαν βε υσεδ το δετερμινε τηε αμουντ οφ ΚΘΗ στοραγε βεινγ υσεδ βψ ΜΧΑΤ. Υσυαλλψ τηε βυφφερ υσε δισπλαψ γιωεσ α γοοδ δεα οφ ωηιχη ΜΧΑΤ ποολ ισ χαυσινγ τηε στοραγε σηορταγε.

Μονιτορ βυφφερ υσαγε: Ωηεν λοοκινγ ατ α βυφφερ σηορταγε, ιτ ισ οφτεν ηελπφυλ το κνοω ωηετηερ τηε ονσετ οφ τηε προβλεμ ωασ γραδυαλ ορ ιμμεδιατε. Ιφ ρεγυλαρ βυφφερ υσαγε δισπλαψσ αρε δονε, τηεν γραδυαλ ινχρεασεσ ιν βυφφερ υσε χαν βε σεεν. Τηεσε γραδυαλ ινχρεασεσ μαψ τακε δαψσ το μανιφεστ τηεμσελωεσ ιντο στοραγε σηορταγε προβλεμσ. Ιν φαχτ, ιφ ΜΧΑΤ ισ τακεν δοων ρεγυλαρλψ,

This page is so visually dense and monotonous that users can't quickly determine whether the information meets their needs. Other than the spacing between paragraphs, there are no signposts, such as subheadings, lists, or tables, to help users understand what they're looking at. Users must read the entire passage to determine whether the content is what they're looking for. Imagine trying to look at a page like this on a mobile phone.

Before attempting to rewrite the information, you can make it more approachable and visually inviting simply by separating it into distinct information types. This activity is inextricably related to the organization guideline "Separate noncontextual information into discrete topics by type" and the retrievability guideline "Provide helpful entry points," but it is just as relevant to visual effectiveness.

Easy to find

**Revision**

The revised page provides structure and organization. The headings, table structure, numbered steps, and associated white space facilitate scanning and enliven the information. The white space provides places for eyes to pause and helps delineate groupings. Users who are looking for task steps can more easily see where those begin. Even a cursory glance at the page gives users a clear idea of how to approach the information.

In a user interface, something as simple as the alignment of field labels and entry fields and the use of white space can have a dramatic impact on retrievability.

## Visual effectiveness

**Original**

[First name, Last name, Email, Password, Retype password, Address, City, State, Postal code, Phone fields with left-aligned labels and variable-width entry fields]

In this web form, the field labels are left-aligned, which facilitates scanning. But the entry fields that begin and end in varying positions are distracting. Also, the form appears cramped because there's so little white space between the field label rows.

**First revision**

[First name, Last name, Email, Password, Retype password, Address, City, State, Postal code, Phone fields with right-aligned labels]

# Easy to find

In the first revision, all of the entry fields begin in the same position, so the overall effect isn't as jumpy. However, the field labels are right-aligned, which makes scanning difficult, and the field label rows are still cramped.

**Second revision**

**Account owner**

First name: _____

Last name: _____

**Account information**

Email: _____

Password: _____

Retype password: _____

**Contact information**

Address: _____

City: _____

State: ____

Postal code: ____

Phone: ___ ___ ___

In the second revision:

- The field labels are left-aligned to facilitate scanning.
- The entry fields are the right size for the information.
- Bold group labels separate the groups of fields into more approachable chunks of information.
- Additional white space between the field labels and the entry fields reinforces the organization and makes the form more inviting.

When blocks of text are broken up by other visual elements, including white space, users are able to pause, evaluate, and absorb the information before moving on.

## Attributes of visually effective text

The jury still seems to be out on the ongoing, decades-long debate over the readability of *serif* versus *sans serif* typefaces. In his article, "Serif vs. Sans-Serif Fonts for HD Screens," Jakob Nielsen notes:

> There's no strong usability guideline in favor of using one or the other, so you can make the choice based on other considerations—such as branding or the mood communicated by a particular typographical style.

Nielsen's observation about the mood that is conveyed by fonts is an important one. Just as you probably wouldn't use a cartoon in a technical illustration because of the casual, unprofessional tone it would convey, you probably wouldn't use Comic Sans font for your textual elements, as the hover text examples in the following table show:

| Font | Use for textual elements | Example |
| --- | --- | --- |
| Comic Sans | No | Use the minus sign (-) to precede terms to exclude. Learn more. |
| Times New Roman | Probably not | Use the minus sign (-) to precede terms to exclude. Learn more. |
| Arial | Yes | Use the minus sign (-) to precede terms to exclude. Learn more. |

Sans serif fonts such as Arial have a more modern appearance than serif fonts such as Times New Roman. As a result, sans serif fonts seem to be preferred for web pages and user interfaces, as evidenced by their preference in Apple iOS, Microsoft Windows, and other interface design guidelines.

Follow these tips for visually effective text:

- Use a font size of 10 point or larger so that most users won't have to manually increase the font size and to ensure that the information is readable to all users.

  A font size of 8 point is acceptable for tables, illustrations, screen captures, sample code, and syntax diagrams because these elements are read as small blocks of text.

- Be judicious in the number and type of font variations that you use:
  - Limit the use of font variations such as bold, italics, and all uppercase letters.
  - Don't use vertical text in illustrations or in text elements such as table headers.
  - Don't use reverse fonts (white text on a dark background) for large blocks of text because they're more difficult to read than dark text on a white background.
  - Don't use blinking or moving text.
- Leave enough white space around the text and objects in illustrations. Cramped text makes information unapproachable and impedes translation.

# Use graphics that are meaningful and appropriate

Effective illustrations clarify and enliven technical information and make it much more concrete and interesting. Because most users process visual elements more quickly and efficiently than they process text, your information will be more successful if you find ways to make it visually interesting and dynamic.

In his 2012 book, *The Power of Infographics*, Mark Smiciklas summarizes:

> Fifty percent of the human brain is dedicated to visual functions, and images are processed faster than text. The brain processes pictures all at once, but processes text in a linear fashion, meaning it takes much longer to obtain information from text. Furthermore, it is estimated that 65% of the population are visual learners (as opposed to auditory or kinesthetic), so the visual nature of information graphics caters to a large portion of the population.

To make your information more meaningful and dynamic, follow these guidelines:

❑ **Illustrate significant tasks and concepts**
❑ **Make information interactive**
❑ **Use screen captures judiciously**

## Illustrate significant tasks and concepts

Illustrations in technical information are often considered secondary in importance to text, or simply decorative, when in fact they're integral. It's usually easier to convey task flows, complex technical concepts, relationships, and processes more succinctly through both illustrations and text than it is through text alone. Meaningful illustrations help make information more concrete, clear, and interesting for users. Even if you don't consider yourself a "visual person," it's fairly easy to create simple but effective graphics that can make a big difference for users.

Illustrations give users the big picture at a glance, even before they read the accompanying text. In some cases, they might be able to skip reading the text altogether.

### Task flows and processes

You can use illustrations to show the overall task flow, as in the following flow chart, which describes the preinstallation steps for a software product. It's much easier and more effective to show the various paths through a complicated task by using a flow chart than it is to explain all of those paths by using words alone.

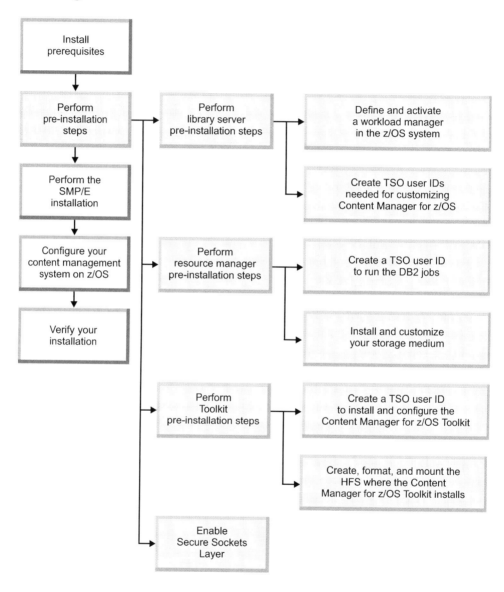

# Visual effectiveness

Similarly, you can show the order of tasks and subtasks, as in the following illustration for installation and configuration tasks:

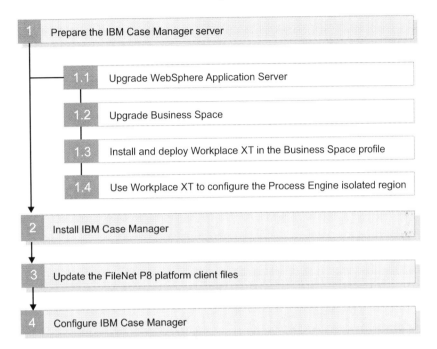

These flow illustrations quickly convey the number, order, and hierarchy of installation tasks and are fairly easy to create without the assistance of a visual designer.

Illustrations can also be used to make the information in scenarios more concrete.

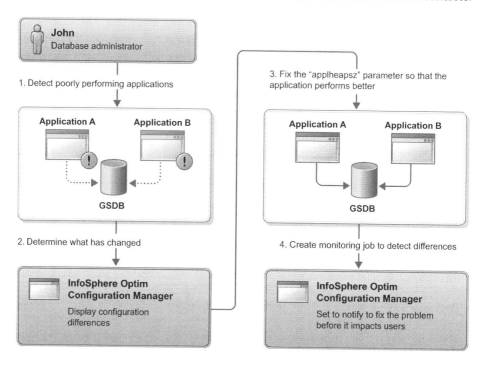

The blue boxes show what John does with the product. All the elements of the task flow stand out without being distracting and clearly show the order of events.

# Visual effectiveness

The following topic describes how data can be archived and retrieved by using a software product:

**Original**

> ### InfoSphere Optim Data Growth Solution
>
> InfoSphere Optim Data Growth Solution helps DBAs manage data growth, archival, and retention. Oftentimes, both active and inactive data is stored on the production system because inactive data might be required for audit and compliance or for other reporting applications. However, by archiving inactive data, DBAs can help their production applications perform better. InfoSphere Optim Data Growth Solution safely moves sets of relational data from one or more data sources to an archive, which can be stored in a variety of environments. The Data Growth Solution can archive data on the basis of both DBMS and application-managed relationships, whether within a single data source or across heterogeneous data sources.
>
> If needed, the archived data can be easily retrieved to an application environment when additional business processing is required. Users and applications can still access this data by using traditional access methods, such as ODBC and JDBC, and so they can use the data in report writers, such as Cognos, Microsoft Excel, and enterprise applications.

Easy to find

**Revision**

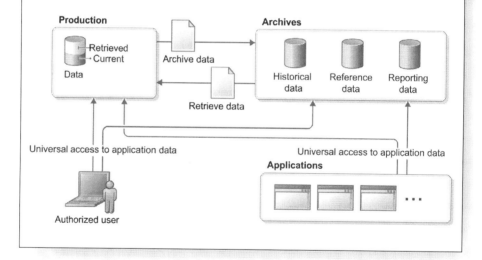

The illustration that's included in the revised topic makes the conceptual information more concrete because it depicts the physical workflow for the archiving capability.

You can also make such illustrations interactive by changing them into image maps or interactive illustrations, as described in "Make information interactive" on page 441.

## Spatial relationships

Illustrations are also a good way to convey conceptual information that's difficult to explain solely by using text. The following illustration maps out database storage elements and concepts, including the physical order of elements and their spatial relationships, in a way that would be difficult for users to conceptualize without the supporting image:

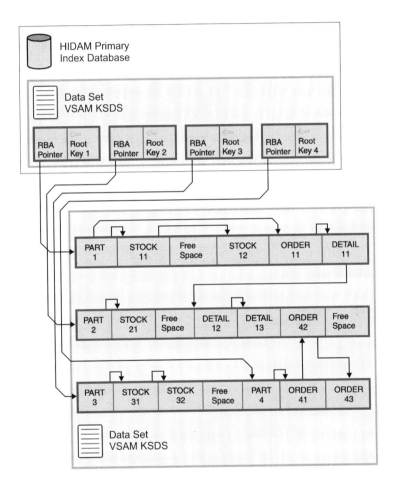

This illustration isn't a replacement for text. You still need to explain the information that it conveys in the surrounding text so that visually impaired users also have access to it. But some proportion of users will be able to skip reading and instead get the same information by skimming this illustration.

# Easy to find

## Big picture

Illustrations can help complete users' understanding by conveying a big picture of products and their relationships. The following table from a topic describes a group of products that can be used to manage the phases of the data lifecycle:

**Original**

Table 1 Key products that help IT staff manage the various phases of the data lifecycle

| Data lifecycle phase | Products |
| --- | --- |
| Design | InfoSphere Data Architect |
| Develop | IBM Data Studio, InfoSphere Optim pureQuery Runtime |
| Test | InfoSphere Optim Test Data Management |
| Administer | IBM Data Studio, InfoSphere Optim Configuration Manager |
| Back up and Recovery | DB2 Advanced Recovery Solution, InfoSphere Optim High Performance Unload, DB2 Merge Backup, DB2 Recovery Expert |
| Monitor | InfoSphere Optim Performance Manager |
| Tune | InfoSphere Optim Query Workload Tuner |
| Archive and Retire | InfoSphere Optim Data Growth Solution |

Although the table contains all of the necessary data, it doesn't effectively articulate the concept of a lifecycle.

# Visual effectiveness

**Revision**

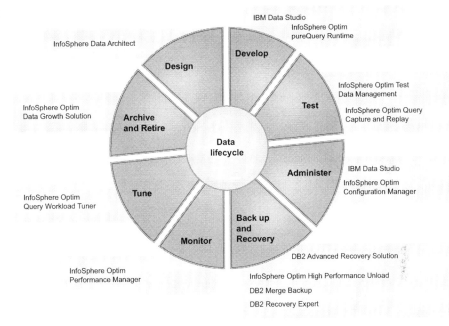

The illustration conveys the same information as the table, but the wheel image better conveys the idea of a lifecycle. As with the flow chart examples, the design is simple.

## Practical ways to get started

If you defer creating illustrations until the end of your development cycle, it's unlikely that you'll ever get around to doing them. It's easier and more productive to write the text and create the visual elements in tandem. In fact, if you don't work this way, information and design are often disjointed.

Write your content and develop a supporting illustration at the same time to fine-tune your message, decide where to place emphasis, and cull out what's not needed. In the end, you'll find that text and illustrations are better than if you develop each separately.

If the idea of creating illustrations is new or daunting to you and you don't know how to begin, here are some ideas:

### Know the guidelines

Become familiar with your organization's design guidelines. Your illustrations will be much more effective if you're aware of the visual possibilities for enhancing and clarifying your information (similar to understanding your options for adding embedded assistance within the constraints of your interface design guidelines).

**Brainstorm ideas**

Just as you wouldn't write text without input and feedback from product team members, developing the elements of an illustration isn't something that you should tackle alone. Use your meetings with product team members as an opportunity to brainstorm about visual solutions. Sketch out ideas during the discussion:

- What new and complicated concepts are being introduced to users?
- What does each new concept "look like" to users?
- What are the main steps to any complicated task that you want to map out? What supporting information will users need along the way to succeed?
- How do the new tasks and concepts relate to other concepts or tasks that users are likely already aware of, and how can you build on that knowledge?

The materials that you develop might very well help other team members to better understand a concept and might also be reusable in other contexts (for design documents, education materials, marketing purposes, and so on).

**Standardize visual elements**

Look for examples of the types of illustrations you'd like to create and use them as the starting point for your illustration. In the best case, your organization has visual guidelines and a library of elements to choose from. Regardless, ensure that your illustrations for an information set follow the same visual style. For example, if you create a series of illustrations to describe database concepts, use the same object to represent a database in all of the illustrations.

**Iterate**

Take the time to refine your illustrations. Your text isn't perfect the first time you write it, and illustrations won't be either. The text and illustrations will continue to evolve together until you get the content balanced and to the right level of detail and focus.

## Make information interactive

The use of mobile devices greatly increases users' demand for technical information that's more dynamic than static text and static illustrations. These devices offer rich media capabilities that aren't available in print such as customizable interfaces and the ability to view and interact with images, video, and sound.

Interactive and dynamic visuals help address the changing user expectations described in Chapter 1. Especially in mobile contexts, users would much rather see and hear focused content than have to read it to try to find the relevant portions.

You can use dynamic visuals such as interactive illustrations and videos to provide tutorial, overview, and process information or to capture expert knowledge, user interactions, and task flows. When you plan your information and decide how and where to deliver it, think about how you might visually enliven it and move beyond static text and illustrations. Your challenge as a writer is to find ways to provide interactive information that addresses users' universal plea: don't tell me how it works; *show* me how to use it!

### Interactive illustrations

Image maps are one way to add interactivity to illustrations. You can activate areas of static illustrations by adding hot spots that link users to associated topics. The hot spots can have hover text that describes the content of the associated topics. Image maps are a kind of interactive "visual launchpad" to associated information. Users can interact with the image and learn more as needed. Hover text over hot spots might be all the information that users need.

The following illustration maps out the tasks for installing a software product. The numbered tasks in this illustration correspond to a series of task topics:

# Easy to find

**Original**

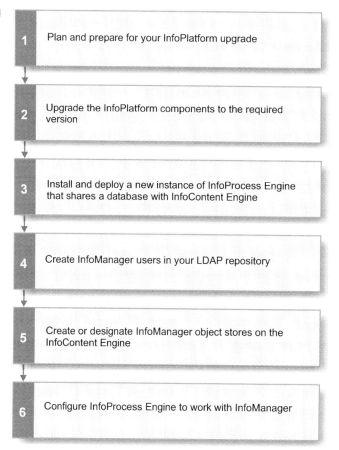

By adding hot spots to each of the numbered tasks in the illustration, you can help users find the related task information. Hover text shows a link preview for the associated task topics.

# Visual effectiveness

**Revision**

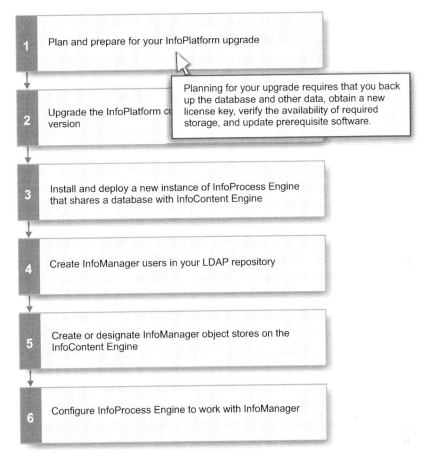

You can add both interactivity and animation to illustrations. The following interactive illustration provides an overview of the tasks involved in using a software product. Users can hover over hot spots to learn more about elements of the graphic and see simple animations. The hover help can link to more information outside of the illustration.

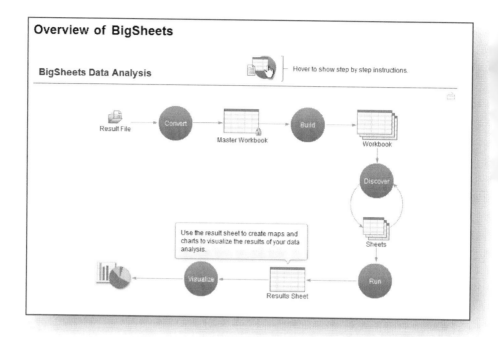

Interactive illustrations are a powerful way to involve users in their own learning. Users can explore freely and decide what additional information they want.

## Videos

*Video* refers to any kind of audiovisual recording and can take many forms. Types of videos include recordings of interviews with product experts and computer animations or demonstrations that are captured in video format.

Videos can be a useful format for presenting product documentation. Like illustrations, videos enable you to convey information more succinctly and rapidly than you can by using text alone. For example, you might be able to make a point in a short, animated *viewlet* (a video that is created by using a screen recording tool such as Camtasia) that would otherwise take 12 pages of text to describe.

Some types of content lend themselves better to video than others:

- Installation, configuration, or complex tasks
- Tutorials and step-by-step scenarios
- Product overviews, tours, and introductions to user interfaces (especially ones with new conventions)
- Processes (such as how an engine works)
- General tasks (such as replacing the memory card in a computer)
- Interviews with experts

Information that's not helpful as text is also not helpful in video. For example, video tours of interfaces can be effective if they show users how to accomplish a task or goal in the user interface and are based on scenarios. However, a video tour that does no more than show each field and control ("Click the **Close** button.") is ineffective.

In his article, "User Expertise Stagnates at Low Levels," Jakob Nielsen summarizes the results of his user testing of even well-designed and highly usable interfaces:

> People don't read manuals. People don't go exploring all over the user interface in terms of neat features ... Users would gain substantial benefits if only they bothered to spend a few moments looking around the interface.

Nielsen's observations underscore how valuable a video tour that is based on a scenario can be. Users want or need to be shown how to do something or to see how something works.

Suppose that your product has a powerful capability, but few users are able to use it fully because of its complexity. You could interview the product expert to capture her expertise and use that information to write concept and task topics. Conveyed in text alone, however, her expertise takes the form of lengthy topics that still don't provide users with the information that they need to use the product's powerful capability. Will users even have the patience to read through lengthy topics? Alternatively, you could convey the information in a video that includes audio of the product expert narrating as she completes the tasks.

Developers of an IBM product information website noticed during discussions with users that even after years of the website's availability, users remained unaware of how to search effectively. They were also largely unaware of other features of the website, such as the ability to subscribe and be notified of changes.

# Easy to find

**Original**

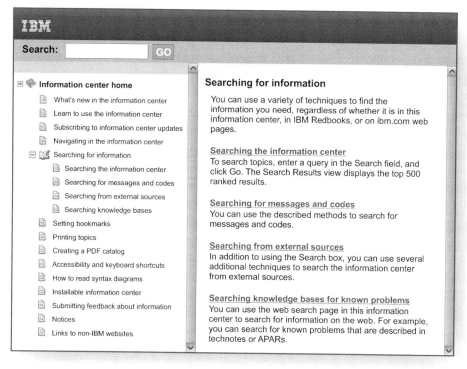

Although this information was provided on the website as topics with titles such as "Searching for information," users were unaware of the topics or didn't want to invest time in reading them.

# Visual effectiveness

**Revision**

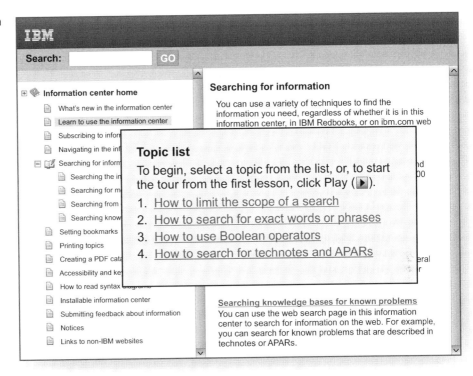

The website developers created a series of viewlets that step users through the interactions that are required to do tasks such as effectively search for information. The script for the viewlets is based on the same information that's in the topics, but that information is much more dynamic and interesting in viewlet form, and users are more likely to explore it in that form.

The publicly available Apple "Tour the Mac interface" video tutorials are another good example of how short videos can quickly convey how-to information that's focused on specific tasks (such as "Setting up user accounts") and areas of a user interface.

**Design tips for video**

Video must have the same quality and production value as your traditional information and must be stylistically consistent, visually pleasing, and accessible to visually impaired users. A quick perusal of YouTube product expert videos shows a wide variety of styles—some of such poor quality that text is illegible or the sound is difficult to hear.

**447**

Follow these tips when you create videos:

- Set and follow standards for your video content to ensure a consistent user experience and enable users to focus on the content. If templates don't already exist, create them. Templates can specify attributes such as screen resolution, design elements and fonts, opening and closing sequences, and output formats.
- Keep it short. Jakob Nielsen advises that web videos should be less than a minute long, but videos about complex technical concepts or features might be as long as 4 - 5 minutes.
- Determine how you'll deliver your videos and establish conventions for their presentation and access. Videos might be started from or viewed within a user interface, from links in hover help or other user assistance, within an online information library, or from a web page.
- Design videos for the appropriate screen resolution.
- Produce your videos with high resolution in a format that can be played on most platforms.
- Resist the urge to use all the gimmicks at your disposal. Even though video editing software offers a seemingly endless array of window backgrounds, transitional effects, and animations, simplicity is more important with technical information.
- To help users focus on the information rather than the animation techniques:
  - Select a screen background that complements your animation but doesn't overpower it, and use the same background throughout the video.
  - If your video uses dynamic transitional effects such as fades, choose one or two types and use them judiciously and consistently.
  - Use consistent navigation and button placement throughout the video.
  - Use consistent placement of standard elements such as headings and body text. This way, the content won't appear to jump as the video progresses.
- Develop and repeatedly test your information in its intended format and delivery vehicle. Reversing a poor first impression is difficult after the content is published.

## Use screen captures judiciously

If a product interface is well designed and any needed assistance is embedded within that interface, you shouldn't need to show many screen captures of the interface. However, when used appropriately, screen captures can help focus users' attention and complete their understanding. If you use screen captures, be aware of the maintenance and translation overhead involved in their use.

## Visual effectiveness

### Appropriate uses for screen captures

Users will probably look at the user interface first. If they get stuck or can't complete their task, they'll look at the supporting information. In some cases, screen captures might help compensate for complex or poorly designed products. An intuitive, easy-to-use user interface is always your goal, but sometimes you can't overcome design complexities.

Screen captures can also be used in task topics or tutorials to help confirm for users that they're taking the appropriate action. You might also use a screen capture to highlight a dramatically enhanced user interface in an overview topic.

The following annotated screen capture is of a product interface element called a *time slider*. During usability testing, users struggled to understand how the time slider worked and also overlooked some of its features. There wasn't time to simplify the product design or create a video tour of the interface, so a "Learn about the time controls" link was added to the time slider. The link opens a help topic that describes the time slider's features and includes the following image:

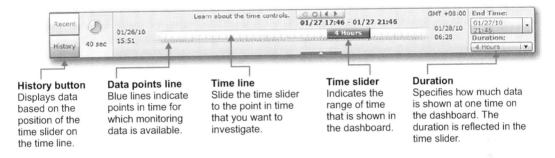

**History button**
Displays data based on the position of the time slider on the time line.

**Data points line**
Blue lines indicate points in time for which monitoring data is available.

**Time line**
Slide the time slider to the point in time that you want to investigate.

**Time slider**
Indicates the range of time that is shown in the dashboard.

**Duration**
Specifies how much data is shown at one time on the dashboard. The duration is reflected in the time slider.

If this help topic needed to be translated, translators would also need to translate all the text in the graphic. In such cases, it might be preferable to use only the labels in the figure and describe the time slider elements in a list in the text.

The following list describes some of the reasons why you might decide that showing the windows is integral to users' understanding of the material:

**Design problems**
A user interface might be poorly designed and nonintuitive, and there's no possibility of improving the interface or embedding information in the interface, as is preferable. To compensate, you need to write supporting information that explains areas of the interface that are confusing (as in the previous example).

**449**

**Reassurance**

Your users need reassurance that they're taking the correct actions. Users might need this feedback because they're novices and feel tentative, or because the task is complex and the repercussions of making mistakes, such as losing data or overwriting an existing installation, are significant. In this case, you might include screen captures at significant milestones in a task rather than including every screen that's involved. For example, in a step-by-step tutorial it might be appropriate to show, along with the text, the toolbar buttons that users should click as they progress through the task steps or the window that results at the end of a complex task.

**Usability testing results**

Your usability testing indicates that users are more successful or speedier at completing their tasks if the screen captures are included, or calls to product support suggest that these images would help users solve problems without additional assistance.

**Complex interactions**

Your user interface requires free-form interaction, such as dragging and dropping objects onto a canvas and connecting them or drawing on a canvas. Because these interactions are free-form and there aren't specific fields or user interface controls to interact with, it's difficult to explain what to do and how to accomplish a task. For example, using a screen capture is a much more effective way to show how to connect two objects on a canvas than using text alone:

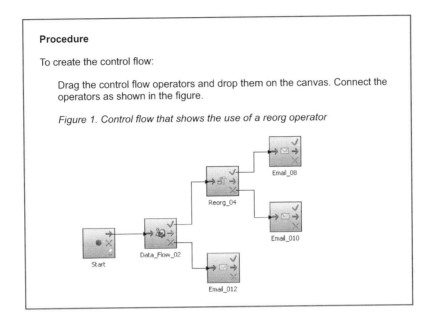

Usability testing is a good way to determine whether users will benefit from seeing a window image in the information about a task. Try testing the information without the images. Are users just as successful? Although they might prefer to have the images for the reassurance, if they can succeed without the images, the images probably can be eliminated.

The following topic is an example of when including a screen capture might make sense. The image in the example topic isn't really a screen capture of an interface; it is an abstracted graphical representation of an interface. This type of representation is used to call out and describe areas of the interface so that the documentation can later use the names of those areas in other parts of the information.

# Easy to find

When you start InfoSpace Portal, it opens to your list of tasks. The following figure shows the main areas of the InfoSpace Portal interface. It includes areas for your work, dashboards, and user preferences. The right part of the interface contains a tabbed area for working with processes and collaborating with others.

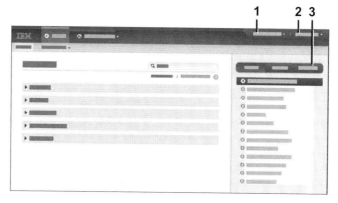

Table 1. Main areas of the InfoSpace Portal interface

| Area | Description |
|---|---|
| 1 | Link to the User Preferences page. This page is available only if your system is configured to use InfoSpace. |
| 2 | Link to the Other Spaces page. This page is available only if your system is configured to use Business Space. Click this link to navigate to other business spaces. |
| 3 | The tabs that are shown here depend on whether you are working with your list of tasks in the My Tasks tab, or working on an individual task to complete the work. |
| | and so on... |

This image was included in the topic because there's no mechanism for adding hover text for these fields in the interface. This information will be translated into dozens of languages, so this solution minimizes both translation and maintenance impact. Showing the interface in a visually abstract way is useful for maintenance because changing and maintaining one image is easier than recapturing multiple screens in different languages.

The following topic was written to guide users through a nonintuitive user interface that couldn't be changed. Because the order of elements in the interface is different from the order of task steps, users might inadvertently click **Run** before reviewing the generated commands that they're running. The callouts in the screen capture show the correct sequence of task steps. Only the section of the interface that is relevant to the task step is included in the screen capture.

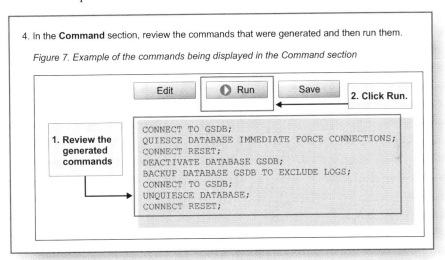

In this case, including an annotated screen capture in a topic was a quick solution to the problem of a nonintuitive user interface.

## Unnecessary screen captures

Even if you need to explain an interface in the documentation, you don't necessarily need to include images of what users can already see on a screen as they work with a product. Imagine looking at an image of a mobile user interface window from within the mobile user interface window itself.

Easy to find

The following window shows a user interface pane on the left and its associated help topic:

Original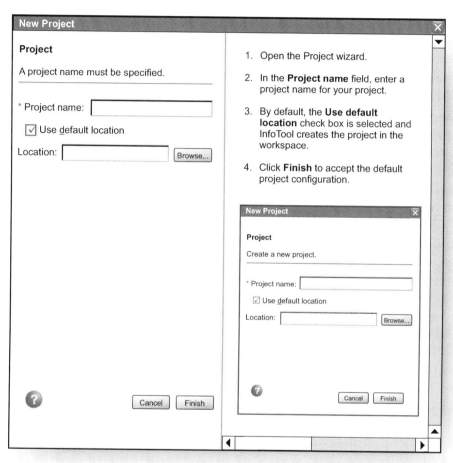

The help topic, which opens in a pane on the right when users click the **Help** icon, unnecessarily describes how to use the Project wizard and includes a screen capture of a pane that users can already see. The screen capture of the New Project window in the help topic doesn't match the New Project window in the product because the screen capture was taken before the design was finalized. This mismatch highlights key drawbacks to including screen captures in information: maintenance and currency. More importantly, such mismatches affect accuracy and can affect users' confidence in the information.

# Visual effectiveness

**Revision**

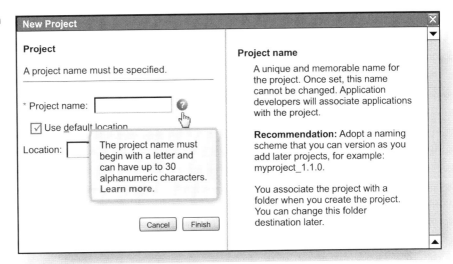

The revised help topic focuses on reference content that can't be embedded in the interface and no longer includes the unnecessary screen capture.

# Easy to find

The following help topic includes a screen capture of a product menu that's displayed after users enter a command to open that menu:

**Original**

> ### Starting the IVP from the Application Menu
>
> You can start the IVP dialog from the Application menu:
>
> 1. Open an ISPF application dialog.
>
> 2. Start the Application menu by running the following TSO EXEC command:
>
>    ```
>    EXEC 'qqq.SPGDEXEC(ELAPPL)' 'HLQ(qqq)'
>    ```
>
>    The Application menu opens.
>
>    ```
>                    Application Menu
>
>    Command ===> _____
>
>    Select the application and press Enter.
>
>    1  Single Point of Control (SPOC)
>    2  Manage resources
>    3  Installation Verification Program (IVP)
>    ```
>
> 3. In the Application menu, select Option 3 to start the IVP.

In this case, users who enter the command can already see the Application Menu. Showing it in the topic to confirm what users should be seeing is unnecessary. This menu will also be translated into other languages, which means that the image must be recaptured for each language. And if the menu items change in future releases of the product, this screen capture will also need to be updated and translated again.

# Visual effectiveness

**Revision**

> **Starting the IVP from the Application Menu**
>
> You can start the IVP dialog from the Application menu:
>
> 1. Open an ISPF application dialog.
>
> 2. Start the Application menu by running the following TSO EXEC command:
>
>    `EXEC 'qqq.SPGDEXEC(ELAPPL)' 'HLQ(qqq)'`
>
> 3. In the Application menu, select Option 3 to start the IVP.

In the revision, the screen capture has been removed and the topic adequately conveys the complete task information without it.

The following screen capture from a product user's guide is included in a topic that describes the elements of the user interface. Because novice users quickly grasp how links such as breadcrumbs work the first time they click one, an image like this (not to mention the topic itself) is superfluous and adds nothing that users actually need:

| Breadcrumb | **Home > Projects** |
|---|---|
| | Breadcrumbs give the user a view of where they are in the navigation hierarchy and provide access to higher levels of the navigation hierarchy from the current location. Clicking any part of a breadcrumb takes you to that page. |

Other examples of unnecessary screen captures include:

- Toolbars (which should always have tooltips)
- "Accept the license" screens
- Error, warning, or information message dialogs
- Login dialogs
- Wizard windows
- Help buttons or Help welcome pages

Heavy use of screen captures in information is sometimes a symptom of a broader issue: namely, documenting the interface rather than the tasks that users want to do.

## Drawbacks of using screen captures

Before you include screen captures, consider the drawbacks of using them in information:

### Currency

It can be difficult to keep screen captures up to date with the current level of the product. Product development cycles are rapid, and designs change continually. If the image in your information doesn't match the window that users see, users might be confused. Keeping screen captures current also adds maintenance work to different releases of the product.

### Translation

If the product and the information are translated, the window will need to be recaptured from the translated product. But translators might not have that capability or might choose not to include screen captures of the translated interface. If they don't include the screen captures, and the surrounding information is written to rely on the image, the translated information will seem incomplete.

### Accessibility

As discussed in "Ensure that visual elements are accessible to all users" on page 478, visually impaired users cannot see screen captures. If you include screen captures, ensure that the same level of detail is otherwise available to those users.

## Design tips for screen captures

When you create screen captures:

- Capture all of the windows or window fragments at the same screen resolution so that the text and graphical elements within the windows display at a consistent size.

  If some product windows are much larger than others, resize the larger windows, if possible, before capturing them.

- Include the pointer in the screen capture only when its presence is significant.

- Add callouts and text to your screen captures if you need to draw users' attention to specific areas of the interface and describe them.

Follow these specifications for screen captures:

- Specify a maximum screen area of 1024 × 768 pixels for desktop applications or 1200 × 1024 for web applications. If you use a higher resolution than these, your screen captures might be larger than necessary, and users might have to scroll to see the whole image.
- Use consistent settings to ensure that all of your captures have a consistent appearance:
    - Set your display resolution to use "Small Fonts" of 96 dpi to ensure that the fonts in your screen captures display at the size and resolution that are expected by most users (roughly equivalent to 8-point type).
    - Set the interface appearance to its default or standard color and font scheme. Don't use customized settings such as brightly colored window title bars.
    - If possible, do not capture window decorations as part of screen captures because these differ by operating system, desktop interface (Windows, Macintosh, and so on), and web browser. Capturing screens without the window decorations enables you to reuse the same screen capture in information for multiple platforms in cases where the content in an application window is the same.

# Apply a consistent visual style

Whether you're working on user interfaces, conceptual illustrations, or videos, you must follow relevant design guidelines, just as you must follow style guidelines for written content. If the guidelines that you need for your work don't exist, work with your team to establish visual style guidelines as a complement to your style guidelines for textual elements. When you follow established guidelines for the content that you create, you're not only ensuring consistency for users, but you'll be in the best possible position to offer relevant design feedback to product developers and usability engineers.

The following illustration has the sort of problems that tend to occur when a visual designer isn't consulted or when visual design guidelines aren't well defined or adhered to:

**Original**

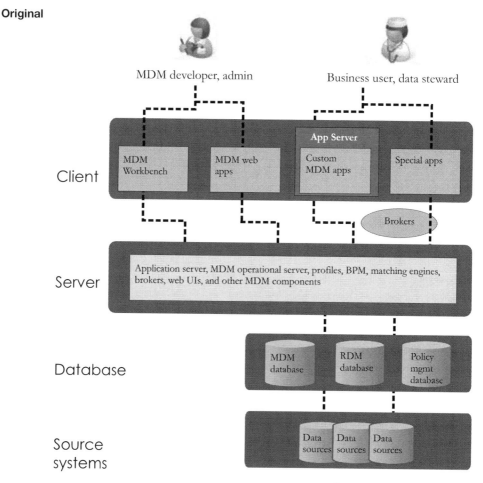

# Visual effectiveness

This illustration has several visual inconsistencies:

- Truncated text is used, such as "mgmt" for "management" and "admin" for "administrator." (These abbreviations are a problem particularly for nonnative speakers of English.)
- Multiple fonts and sizes are used, and font weights (bold) aren't used to help convey organization or groupings.
- The objects have different dimensions: both 2D and 3D elements are used.
- The two 3D icons at the top are the primary focus because of their detail and bright color, but that emphasis is misplaced. Those elements shouldn't receive any more focus than the other elements in the image.
- The heavy, busy lines and line weights detract from the content.
- The blue and magenta color combination is distracting and will be difficult to read in black and white.
- The text in the "Policy mgmt database" element is cramped.

**Revision**

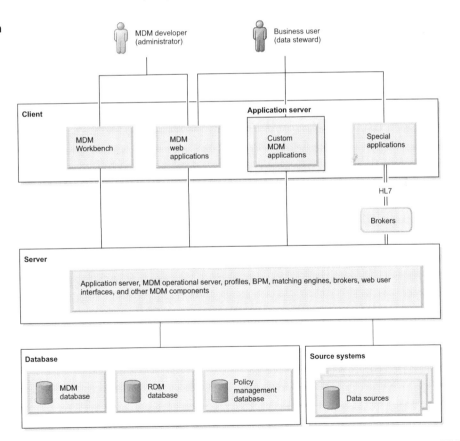

461

The revised illustration is improved to follow a consistent visual style:

- All truncated text has been changed to full words.
- Bold font and boxes around groups of elements help to convey the four major groupings.
- Line weights are standardized so that the heavy black lines no longer dominate the image and draw users' eyes away from the more meaningful content.
- The distracting, nongender-neutral icons are changed to generic, gender-neutral icons.
- A simpler color palette puts the focus on the elements, not the colors themselves.
- The lines through the Brokers element are corrected to represent flow through rather than flow over.

The following draft illustration maps out the elements, relationships, and interactivity for an interactive illustration that was created to help users understand the relationships between a complex array of products and components. Areas of the illustration are activated when users point to them.

Original

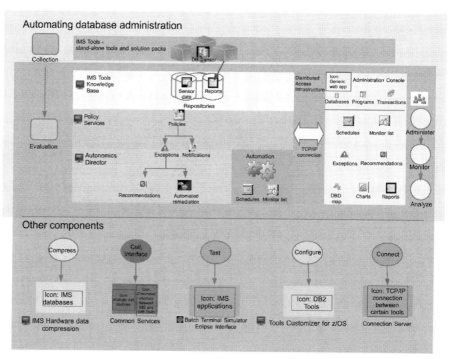

## Visual effectiveness

This illustration includes all the products and components and maps out the basic structure, hierarchy, and relationships, but it has the same types of visual inconsistencies and problems as the previous example in this guideline.

**First revision**

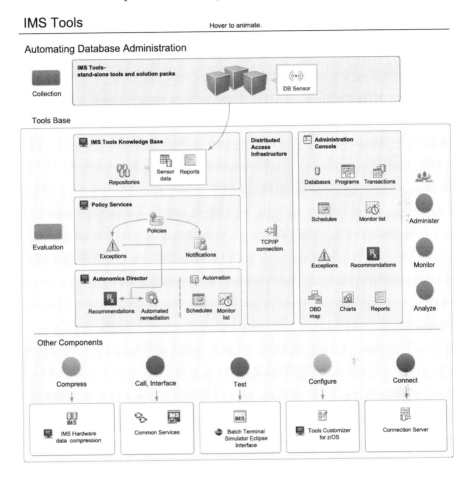

In the first revision, the layout of the illustration is improved and better emphasizes the hierarchy and groupings of elements (images and text). The text is also somewhat more standardized in size and weight, and the color scheme

# Easy to find

is more subtle and attractive. However, this illustration has the following visual inconsistencies:

- The number of disparate, detailed elements, the scale differences between the objects, and the variety of colors make it difficult to know where to focus first. The bright colors compete for users' attention.
- The brightly colored circles along the bottom give too much emphasis to "Other Components," which are secondary in importance to the main Tools Base components.
- Font sizes are inconsistent and in some cases as small as 6 points, which is difficult to read. For illustrations, a minimum font size of 8 points is recommended.

**Second revision**

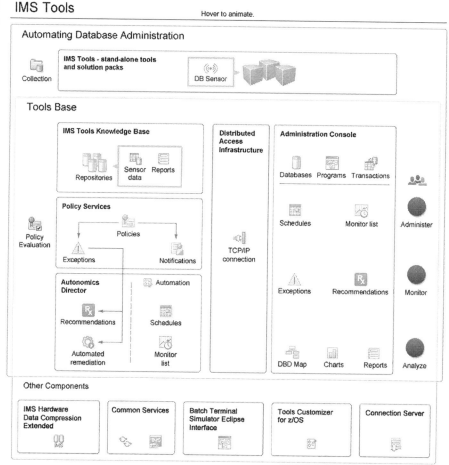

The revised illustration is improved to follow a consistent visual style:

- A monochromatic color scheme is applied consistently to almost all of the objects. Elements become colorful only when users point to them. Instead of being distracting or misplacing focus, color is used to reinforce the interactivity and the changes in the illustration.
- The scale of objects, line styles and weights, fonts, and highlighting of fonts are standardized and consistent. For example, it's now clear that the "Other Components" section is not the focal point.

If you work for an organization that does not have defined guidelines, it's worth taking the time to develop them before any product or information development begins. If you've been a writer for some time, you know how difficult it is to change something after it's already been externalized. It's much easier to design things correctly from the start. Think of guidelines as the frame of a house that you build your information around, not as something you try to apply later, after the structure is built.

Depending on the type of product that you're working on, the ways that you deliver information, and the tools that you plan to use, your visual and user interface design guidelines could define, for example:

- User interfaces and elements, including layout, color palettes and themes, icons, buttons, and highlighting styles
- Elements for illustrations such as color palettes, fonts and font sizes, line styles and weights, and a standardized library of objects and symbols
- Conditions for the proper use of product screen captures in information and stylistic specifications for screen captures (such as image size and styles for callouts)
- Standards and templates for videos
- Contexts and stylistic specifications for the use of cueing graphics

Talk to your product engineers, usability engineers, and visual designers about the following topics:

- What standards or guidelines are they following as they design the product?
- Are they following any external standards such as those for Eclipse, Apple iOS, and Microsoft Windows?
- What guidelines are they following for illustrations? Is there a library of reusable graphical elements that should be drawn from and built on?

Consistency and predictability enable users to focus on a task without having to puzzle over possibly trivial visual differences in things like imagery and interactions: users' experience is similar across different products and interfaces. When you use a consistent visual system, users can learn it quickly and then focus their attention on achieving their goals. The visual elements become so familiar to users that the elements almost merge with the background, supporting the information and the users' journey through it.

# Use visual elements to help users find what they need

You can help users find what they need more quickly by using visual elements systematically and consistently. White space and textual elements such as tables and lists, and visual elements such as callouts and cueing graphics, can aid retrievability and enhance the clarity of your information.

Use visual elements to call attention to important information such as errors or to highlight parts of code or illustrations. For example, a red symbol next to a field that contains an error shows users exactly where to look. In a long code example where only a small section has changed, you might highlight the changes by using bold font so that the changes are easy for users to spot. In an illustration, if one object or portion of the illustration is particularly significant, emphasize it with a slightly heavier outline, shading or color, or a highlighted label.

## White space

The Greek example earlier in this chapter shows how critical white space is in information, whether in an interface, illustration, or on a page; what you leave out is as integral to the success of visual effectiveness as what you include. White space isn't necessarily white, but the term refers to the areas of a page, user interface, or image that aren't used. *White space* offers "breathing room," a place for the eyes to rest, and it can make a user interface, page, or illustration appear less dense and intimidating. It also aids retrievability because distinct chunks of information are easier to identify.

White space is often considered expendable or isn't even considered at all. After all, it contains no information, so why not fill it up? But too much content without breathing room can overwhelm users, who might avoid the information altogether—just as most people would quickly close a topic that looks like the Greek example. White space also provides important buffer space for expansion or contraction of text during translation.

Apple's iOS user interface design guidelines mention the importance of negative space (another name for white space):

> Use plenty of negative space. Negative space makes important content and functionality more noticeable and easier to understand. Negative space can also impart a sense of calm and tranquility, and it can make an app look more focused and efficient.

The idea of leaving empty space on a small mobile user interface might sound counterproductive, but too much information crammed into a small screen is even more overwhelming than on a larger screen.

# Easy to find

Textual elements such as tables and lists are one way to build white space into text and help users quickly find what they need.

## Tables

A table is a visual aid that helps users scan information. Tables in the same book, user interface, or website that are inconsistently styled are distracting and possibly confusing. Avoid drawing attention to the style rather than the content by ensuring that tables are consistent.

The following image demonstrates how a large block of text becomes much more appealing and retrievable when it's broken into chunks of information in a table format:

The following reference topic includes a definition list of extensions that are available for AdminProduct (only one list item is shown here). The list describes the extensions, their primary use, and what tools provide them.

## Visual effectiveness

**Original**

### AdminProduct extensions

Extensions from other products provide enhancements to AdminProduct.

#### Viewing maps with the InfoMap Viewer

Provides a graphical visualization of a database structure map, the macro source, and the XML document that is generated based on the descriptor in the PROD.LIB library. This extension is supported by InfoMap Utilities 2.0 and InfoProd Solution Pack.

This extension helps you with the following tasks:

- View the structure of the database and graphically analyze the relationships between segments.
- Save the graphical view of the database structure in an HTML file and view it locally or share it with your team.
- Examine the control statements that define the database characteristics.

This list will be updated over the life of the product as the number of extensions grows, which means it will get longer and more difficult to scan.

**Revision**

### AdminProduct extensions

Extensions from other products provide enhancements to AdminProduct.

*Table 1. AdminProduct extensions*

| Extension | Description | Tasks | Provided by |
|---|---|---|---|
| InfoMap Viewer | Provides a graphical visualization of a database structure map, the macro source, and the XML document that is generated based on the descriptor in the PROD.LIB library | • View the structure of the database and graphically analyze the relationships between segments.<br>• Save the graphical view of the database structure in an HTML file and view it locally or share it with your team.<br>• Examine the control statements that define the database characteristics. | InfoMap Utilities 2.0 and InfoProd Solution Pack |

# Easy to find

It's much easier to quickly retrieve the four key chunks of information when they're presented in a table format instead of in a list. If users want to know what tools contain InfoMap Viewer, they don't need to read a paragraph to unearth that information.

Be sure to apply consistent styles, colors, formatting, and interaction to the tables in documentation and in user interfaces so that users will focus on the content and not on the style.

The following application uses two styles of tables in different parts of its interface:

**Original**

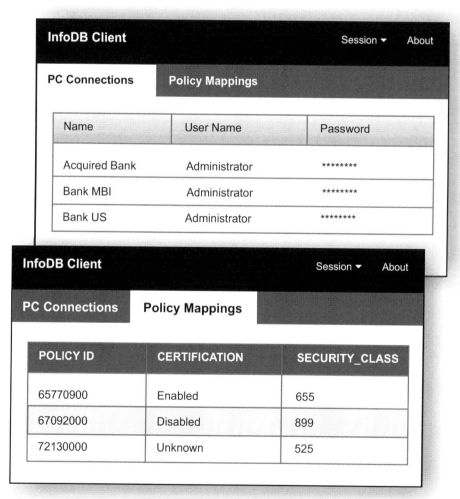

## Visual effectiveness

The first table uses a gray column heading and headline-style case for the column names. The second table uses a blue column heading, all uppercase letters for the column names, and alternate colors for rows. The inconsistency might lead users to think that they can interact with the tables in different ways. For example, some might think that the yellow highlighted row is selected or clickable.

**Revision**

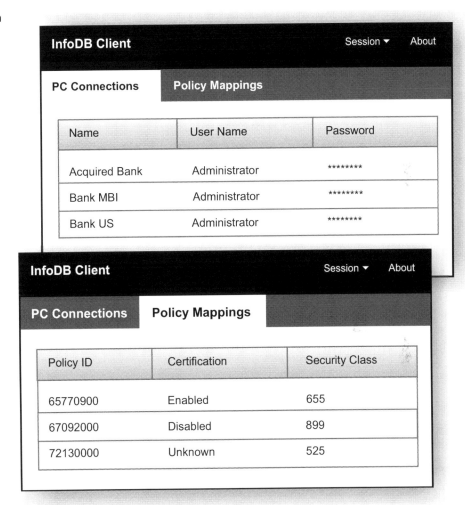

# Easy to find

The following table describes the fonts that are available with a product and the variations that they are available in, such as italic:

**Original**

Table 1. Fonts available with InfoGraphics Tool

| Name | Plain | Bold | Italic | Underline |
|---|---|---|---|---|
| Courier | ✓ | ✓ | ✗ | ✓ |
| Arial | ✓ | ✗ | ✗ | ✗ |
| Serif | ✗ | ✗ | ✓ | ✗ |
| Sans Serif | ✓ | ✓ | ✗ | ✓ |
| Tahoma | ✗ | ✓ | ✓ | ✓ |

Users might assume that the check mark indicates that a font is available in a variation and that an X indicates that a font is not available in a variation. However, because those symbols aren't explained, users have no way of knowing what these symbols mean. The horizontal and vertical rules break the information into cells that, combined with the clutter of the symbols, make the table difficult to scan quickly.

**Revision**

Table 1. Fonts available with InfoGraphics Tool

| Name | Plain | Bold | Italic | Underline |
|---|---|---|---|---|
| Courier | ✓ | ✓ |  | ✓ |
| Arial | ✓ |  |  |  |
| Serif |  |  | ✓ |  |
| Sans Serif | ✓ | ✓ |  | ✓ |
| Tahoma |  | ✓ | ✓ | ✓ |

In the revision, only the check mark is used, which eliminates any confusion about which fonts are available. Other stylistic changes include adding more white space between rows, removing the vertical rules to enable easier horizontal scanning, left-aligning the text and headers, and removing the bounding frame around the table.

Follow these general guidelines for tables:

- Select a common style for table formatting in an information collection or user interface. For example, if some tables don't have vertical rules, don't use vertical rules in any tables in the collection. Also decide whether to use table captions or titles in documentation. If you do use table captions, decide how they should be punctuated.
- Top-align and left-align the text in all table body cells. This alignment helps clarify the row and column boundaries.
- Don't use vertical rules to delineate cells because they impede horizontal scanning. The left-aligned column of text is sufficient to delineate the cells.
- Avoid putting too much information into a single table. If a table requires horizontal scrolling when it's viewed online, it might be too wide. In these cases, break the information into separate tables.
- Use column widths that are proportional to the amount of information that the cells contain.
- Don't scale the text in tables to smaller than 80%. Use a minimum font size of 8 point for text in tables.
- If you include captions, use italic font for them.

For more information about tables, see *The IBM Style Guide*.

## Callouts

Callouts are an effective way to help users quickly map textual information to information in an area of an illustration, screen capture, or code example.

# Easy to find

The following topic describes an installation process and includes numbered steps that map to the numbered callouts in the illustration:

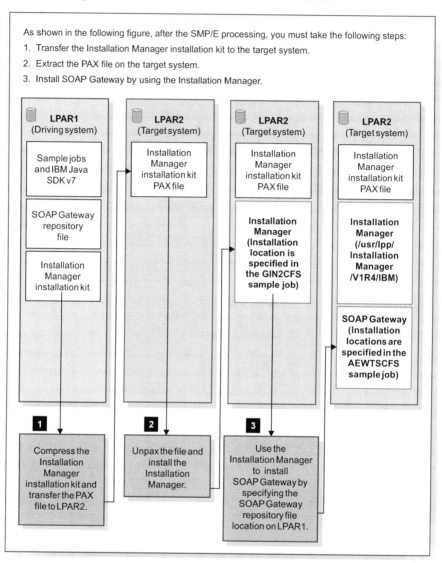

Users can quickly identify the areas of the illustration that correspond to the step numbers.

Similarly, when screen captures are included in information to help users navigate through a complex interface or learn about a product's features, callouts can be used to draw users' attention to key areas of the screen captures:

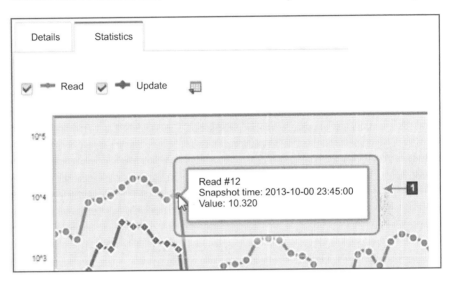

If your information is translated and the English version of the interface is acceptable in other languages (that is, won't need to be translated and recaptured), you can add numbered callouts to your screen captures. Then, add the textual description that corresponds to the numbered callouts to the body of your information; the callouts in the image won't need to be translated, and the description will be translated with the rest of the textual information.

Adhere to your team's visual guidelines for callouts and text (font, font size, and line style).

See the concreteness guideline "Make code examples and samples easy to use" for information about using numbered callouts in code examples to highlight specific areas of the code.

## Cueing graphics

*Cueing graphics* are small graphical cues that help users quickly identify types of information, for example, a light bulb icon might be used as a cueing graphic to identify tips in topics. Cueing graphics are also effective in helping users quickly assess whether information pertains to their scenario or task. These graphics can be particularly helpful when most of the information is common to several environments, scenarios, or user subgroups, but some of the information applies to only a subset of the larger group.

Easy to find

---

**Installing SOAP Gateway**

For z/OS, use the standard SMP/E installation process to install IMS Enterprise Suite SOAP Gateway. For Windows systems, use the IBM Installation Manager to install.

| z/OS | For z/OS, order a CBPDO through ShopzSeries for IMS Enterprise Suite.

| Windows | For distributed systems, download SOAP Gateway from the IMS Enterprise Suite download site.

---

Be careful not to overuse such graphics because they can clutter the information and defeat the purpose of their use.

**Original**

**Administering and configuring IBM HTTP Server**

Learn how to administer and configure IBM HTTP Server, including: Secure Sockets Layer (SSL), Key management, Lightweight Directory Access Protocol (LDAP), and System Authorization for z/OS systems.

| z/OS | Performing required z/OS system configurations
Before starting the IBM HTTP Server, you must configure the z/OS system.

| Distributed operating systems | z/OS | Starting and stopping IBM HTTP Server
You can start or stop IBM HTTP Server using the WebSphere Application Server administrative console or using other methods depending on your platform.

| Distributed operating systems | z/OS | Configuring IBM HTTP Server
To configure the IBM HTTP Server, edit the `httpd.conf` configuration file.

| Windows | AIX | Serving static content faster with the Fast Response Cache Accelerator
The Fast Response Cache Accelerator (FRCA) can improve the performance of IBM HTTP Server when serving static content such as text and image files.

| Distributed operating systems | z/OS | Enabling IBM HTTP Server for FastCGI applications
FastCGI applications use TCP or UNIX sockets to communicate with the web server. This scalable architecture enables applications to run on the same platform as the web server.

| Distributed operating systems | z/OS | Configuring IBM HTTP Server for SMF recording
Use System Management Facilities (SMF) to record operational statistics for IBM HTTP Server.

| z/OS | Classifying HTTP requests for WLM (z/OS operating systems)
Classify HTTP requests for workload management (WLM) by first enabling WLM support in IBM HTTP Server. Then, map HTTP requests to one or more transaction classes.

Visual effectiveness

If you're using cueing graphics to help distinguish three or more situations with alternatives, split the information into separate topics so that users don't have to wade through information that doesn't apply to them.

Revision

**Administering and configuring IBM HTTP Server on z/OS**

Learn how to administer and configure IBM HTTP Server, including: Secure Sockets Layer (SSL), Key management, Lightweight Directory Access Protocol (LDAP), and System Authorization for z/OS systems.

Performing r̲
Before starti̲
z/OS system

Starting and
You can star̲
Application S
methods dep

Configuring
To configure
configuration

Enabling IBM
FastCGI app̲
communicate
enables app̲
server.

Configuring
Use System
operational s

Classifying
Classify HTT
first enabling
HTTP reque̲

**Administering and configuring IBM HTTP Server on distributed operating systems**

Learn how to administer and configure IBM HTTP Server for FRCA, FastCGI applications, and SMF recording.

Starting and stopping IBM HTTP Server
You can start or stop IBM HTTP Server using the WebSphere Application Server administrative console or using other methods depending on your platform.

Configuring IBM HTTP Server
To configure the IBM HTTP Server, edit the `httpd.conf` configuration file.

**Windows** **AIX** Serving static content faster with the Fast Response Cache Accelerator
The Fast Response Cache Accelerator (FRCA) can improve the performance of IBM HTTP Server when serving static content such as text and image files.

Enabling IBM HTTP Server for FastCGI applications
FastCGI applications use TCP or UNIX sockets to communicate with the web server. This scalable architecture enables applications to run on the same platform as the web server.

Configuring IBM HTTP Server for SMF recording
Use System Management Facilities (SMF) to record operational statistics for IBM HTTP Server.

In the revision, the topics are broken out into two sets: one for z/OS systems and one for distributed systems. In the latter topic, because only the differences are identified by cueing graphics, the differences are much easier to identify.

# Ensure that visual elements are accessible to all users

The adage "a picture is worth a thousand words" doesn't take into account users who are visually impaired. When visual elements aren't adequately supported by text, visually impaired users cannot get the same information that sighted users have access to. Visual elements and techniques that enhance the information for sighted users can sometimes impede visually impaired users' understanding of the same information.

Your information might be accessed and viewed in multiple ways, for example:

- As a PDF book or EPUB book on a laptop
- As an HTML topic in a web browser on a mobile device
- In an app that's used in both a laptop and a mobile device
- In color or not (depending on a user's settings)

To comply with Section 508 of the US Rehabilitation Act, which addresses the needs of people with disabilities, you should ensure that all of your information is accessible to visually impaired users in at least one format.

## Alternative text for graphics

Visual elements such as illustrations and screen captures can't be translated into Braille or audibly read by screen readers, so these elements aren't accessible to visually impaired users. If your text doesn't sufficiently cover the same information that's conveyed in a graphic, you must provide an explanation of the graphic in an alternative text element to ensure that visually impaired users have access to equally complete information.

The following table describes the alternative text to use for different types of graphics:

Table 11.1 Alternative text to use for different types of graphics

| Purpose of graphic | Alternative text to use | Assistive technology response |
|---|---|---|
| Included only for visual effect or is otherwise unimportant | Use a null alternative text element (alt="") | Ignores the image |
| Provides relevant or important information | Use a description of the image (alt="*description*") | Reads the image |

# Visual effectiveness

The following illustration on a web page represents a particular arrangement of XML data in a database:

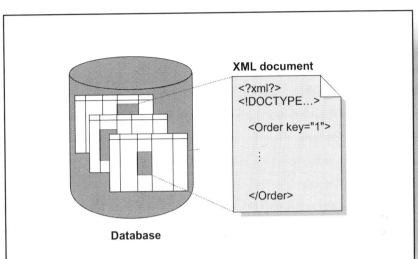

The alternate text describes the graphic vaguely.

**Original**

XML database.

This alternative text doesn't provide visually impaired users with all the information that's depicted in the graphic.

**Revision**

On the left a database contains three tables, which hold data from the XML document that is outlined on the right. The three tables comprise an XML collection.

The revised alternative text describes the elements of the illustration in detail.

The following illustration in a concept topic describes a typical scenario for use of a product:

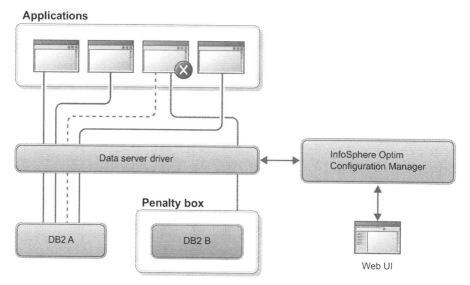

The alternative text doesn't adequately describe the illustration.

**Original**

Penalty box.

**Revision**

A diagram that shows a typical scenario of using InfoSphere Optim Configuration Manager to reroute a misbehaving application to a member of a DB2 for z/OS data sharing group known as the penalty box.

The revised alternative text more thoroughly summarizes the components of the illustration and the concepts it conveys.

See *The IBM Style Guide* for more information about writing for diverse audiences, including how to write alternative text for graphics.

## Color and contrast

Don't rely on color differences alone to convey meaning in your content. Color-blind users might not be able to see the differences. For example, a red asterisk that indicates a required field in a user interface is still identifiable to color-blind users, but red shading of a field without the asterisk is not.

Some color-blind people can see no color; most color-blind people have difficulty distinguishing red from green. Test your interfaces and other visual elements to make sure that red and green aren't used as the only way to distinguish between two states or values.

Also ensure that any visual elements that use color, whether icons or illustrations, are also legible in black and white.

- Ensure that the values of the colors, that is, the darkness or lightness of the colors, are different.
- Test and verify that the colors you choose, if converted to grays, still contrast with one another well enough for the information to be clearly visible.
- If you use colored text or a colored background, ensure that the contrast between the text and background is strong enough so that users can read the text.

In the following interface, red text is used against a green background on the **Log Out** button and the drop-down field:

Original

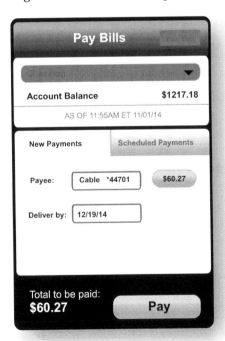

Easy to find

Color-blind users and many fully sighted users likely would not be able to read the text on those two elements. In this case, using white or black text is preferable.

Revision

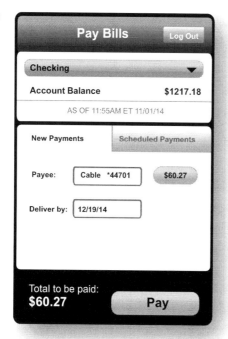

Tables

Complex tables can pose a problem for visually impaired users. Keep your tables simple and avoid using complex layouts such as cells that span two or more rows or columns. If you're using a markup language such as HTML or DITA, you can use table row and cell coding techniques that help to define the table structure for visually impaired users.

To make tables accessible, follow these guidelines:

- Avoid using tables to achieve specific layout effects on a page, for example, using a one-row table to present a code example. Screen readers convey to users the grammatical or document structure that is being used so that users have an idea of how the information is organized. A table that doesn't present related, tabular information can be confusing.
- Identify row and column headers for data tables. Row and column headers enable screen readers to provide information about the relationship of data cells in a table to some visually impaired users.

# Visual effectiveness checklist

Visual effectiveness is a measure of how the appearance of information and the visual elements within it affect how easily users can use, understand, and find the information they need.

Visual effectiveness cannot be separated from the other quality characteristics. By itself, it can do little to improve the quality of your information. But if you use it as a vehicle to aid in improving each of the other quality characteristics, you can enliven your information and make it more effective. Information that is visually effective helps the structure of the information disappear and helps focus users' attention on the information itself. You can help users find and understand information quickly by using visual elements systematically and consistently.

You can use the following checklist in two ways:

- As a reminder of what to look for, to ensure a thorough review
- As an evaluation tool, to determine the quality of the information

| Guidelines for visual effectiveness | Items to look for | Quality rating 1=very dissatisfied, 5=very satisfied |
|---|---|---|
| Apply visual design practices to textual elements | • Visual elements help users identify the information that they're looking for.<br>• Document structures and elements are used consistently, and they help users quickly identify types of information.<br>• Typeface and size are appropriate for the information delivery.<br>• The number and variation of fonts and font styles is limited. | 1 2 3 4 5 |
| Use graphics that are meaningful and appropriate | • Graphics make the text more meaningful.<br>• Significant concepts and tasks are illustrated.<br>• Graphics are interactive and facilitate users' exploration.<br>• Videos are high quality and work in all delivery formats.<br>• Screen captures are used judiciously and show only what users need to see. | 1 2 3 4 5 |
| Apply a consistent visual style | • Visual elements are standardized and consistent.<br>• Layout helps convey hierarchies and groupings.<br>• Adequate space is left for translation of text. | 1 2 3 4 5 |

## Easy to find

| Guidelines for visual effectiveness | Items to look for | Quality rating<br>1=very dissatisfied,<br>5=very satisfied |
|---|---|---|
| Use visual elements to help users find what they need | • Text is broken up by white space or varied structures.<br>• Document structures such as tables are used appropriately and are well designed.<br>• Cueing graphics and callouts are used judiciously.<br>• Visual elements help guide users to correct paths. | 1 2 3 4 5 |
| Ensure that visual elements are accessible to all users | • Graphics support, rather than replace, text.<br>• Alternative text adequately describes the graphics.<br>• Color is not used alone to convey meaning.<br>• Document structures such as tables are simple and accessible. | 1 2 3 4 5 |

To evaluate your information for all quality characteristics, you can add your findings from this checklist to Appendix A, "Quality checklist," on page 545.

# PART 5

# Putting it all together

Developing quality technical information involves all of the quality characteristics. The chapters in this part of the book look at the interplay of characteristics. One of the chapters also looks at the roles of other people besides writers in the development cycle: technical reviewers, users, usability engineers, testers, editors, and visual designers.

| | |
|---|---:|
| **Chapter 12. Applying more than one quality characteristic** | **487** |
| Applying quality characteristics to progressively disclosed information | 488 |
| Applying quality characteristics to information for an international audience | 494 |
| Applying quality characteristics to topic-based information | 501 |
|     Applying quality characteristics to task information | 501 |
|     Applying quality characteristics to conceptual information | 506 |
|     Applying quality characteristics to reference information | 509 |

# Chapter 13. Reviewing, testing, and evaluating technical information — 515

- Reviewing technical information — 516
    - Reading and using the information — 516
    - Finding problems — 516
    - Reporting problems — 517
- Testing information for usability — 518
    - Prototyping — 519
    - Testing in a usability laboratory — 519
    - Testing outside a usability laboratory — 522
- Testing technical information — 524
    - Tools for testing — 525
    - Test cases — 525
    - Testing the user interface — 526
- Editing and evaluating technical information — 527
    - Preparing to edit — 528
    - Getting an overview — 528
- Reading and editing the information — 531
    - Looking for specific information — 533
    - Summarizing your findings — 533
    - Conferring with the writer — 535
- Reviewing the visual elements — 536
    - Preparing to review — 537
    - Getting an overview — 538
    - Reviewing individual visual elements — 540
    - Summarizing your findings — 541
    - Conferring with the editor or writer or both — 541

# CHAPTER 12

# Applying more than one quality characteristic

This book presents each quality characteristic in a separate chapter, but in practice you will apply the guidelines for more than one quality characteristic to the information that you create. With practice, you can discern the effects of each quality characteristic and how to make improvements.

Good embedded assistance, for example, must be task oriented, accurate, well organized, clear, and retrievable. Good embedded assistance primarily contributes to task orientation, but it also contributes to organization by providing information where users expect it and to retrievability by providing paths to more detailed information where necessary.

You can best improve technical information by applying the guidelines for more than one quality characteristic. This chapter gives examples of the quality characteristics to apply in the following situations:

- Applying a pattern for progressive disclosure of information
- Writing for an international audience
- Writing task, concept, and reference topics

# Applying quality characteristics to progressively disclosed information

When you apply a pattern for progressively disclosing information, you naturally apply the organization guideline "Separate contextual information into the appropriate type of embedded assistance." But you also need to pay attention to other characteristics when you develop embedded assistance and topics.

Many software products will have different types of users that you will create information for. For example, you'll write for system administrators who install and maintain the product, and end users, who might have little time or interest in reading documentation.

For example, the following set of topics supports a product for deploying a mobile app that integrates with a web client application. System administrators must build and deploy the app to a server before the app can be downloaded to mobile devices. The app communicates with the web client application, which is used by the supervisor to assign and change work. Users of the mobile app are workers who get the details of their work assignments from the app.

The topics cover different methods for deploying the mobile apps.

**Original**

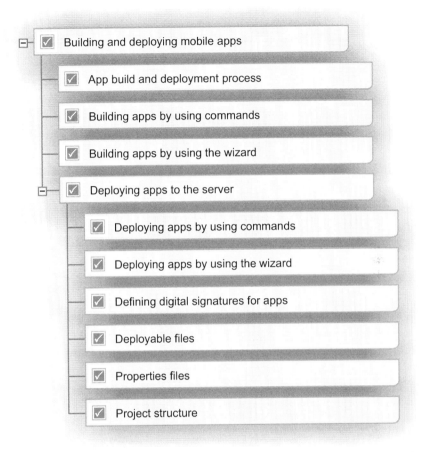

The topics about building and deploying apps include concepts, tasks, and reference topics. The process appears rather complicated because there are so many topics and because the documentation covers two methods of building and deploying apps: one from the command line and one from a wizard.

Putting it all together

**First revision**

The build task consists of running one command or one step in a wizard. By including the build step in the deploying topics, you can remove the topics "Building apps by using commands" and "Building apps by using the wizard." You can also remove the concept topic "App build and deployment process" because it describes the process of building and deploying apps by using commands. The redundant content is unnecessary, and other relevant information should be moved to either the task topic or the reference topics so that the information is where users expect or need it.

The first revision addresses problems with completeness and organization. However, the Build and Deploy Apps wizard has no embedded assistance, so the topic about deploying apps with the wizard provides information that's not otherwise available and must remain in the topic set.

**Second revision**

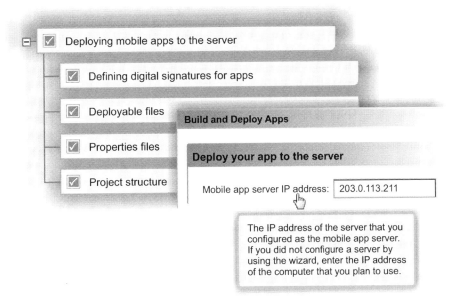

In this revision, hover help is added to the wizard and a label is clarified, so the task topic that described how to use the wizard is no longer necessary. The top-level topic, "Deploying mobile apps to the server," explains that there are two methods for doing the task.

The second revision addresses problems with completeness, retrievability, organization, and task orientation.

The information set for this product also includes topics about working with mobile apps.

**Original**

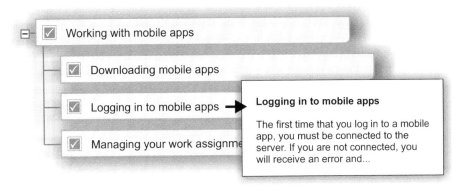

Putting it all together

Topics aren't needed to support mobile users. Someone who is using a mobile app is not going to go to an online manual or information center to read about how to use the app. In particular, a topic about logging in to the app is not helpful when users are trying to log in. The most noticeable problem with this set of topics is with task orientation: the information doesn't address the user's perspective. However, the biggest problem is with organization: the information is in the wrong place.

First revision

In the first revision, all of the topics about using mobile apps have been removed from the information set. Topics that describe how to interact with mobile apps are always unnecessary. A mobile app should have a good design and appropriate embedded assistance so that users don't need to look for help. If help is necessary, it should be in the app, for example, on a separate screen that's linked from the app's menu. In this revision, important information about logging in for the first time now appears on the login screen, closer to the task that it supports.

In addition to addressing task orientation and organization problems, the first revision corrects completeness and retrievability issues.

Second revision

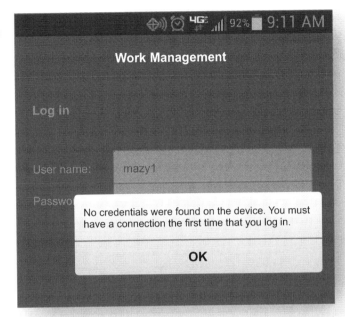

In the second revision, information about logging in is no longer on the login screen where all users see it every time they log in. If a user does attempt to log in the first time without a connection, that user gets an error message. By putting information where users need it, you don't bother users who don't need the information.

The revised embedded assistance and topics in each set of examples address problems that are related to the following quality characteristics:

| Quality characteristic | Relevant guidelines |
|---|---|
| Completeness | • Apply a pattern for disclosing information<br>• Repeat information only when users will benefit from it<br>• Cover all subjects that support users' goals and only those subjects<br>• Cover each subject in as much detail as users need<br>• Make user interfaces self-documenting |
| Organization | Put information where users expect it:<br>• Separate contextual information from other types of information<br>• Organize contextual information into the appropriate type of embedded assistance |
| Retrievability | • Optimize for searching and browsing<br>• Guide users through the information |
| Task orientation | Present information from the user's point of view |

# Applying quality characteristics to information for an international audience

When your information is on the web, users anywhere in the world have access to it. Although all the quality characteristics apply to information for an international audience, some characteristics deserve particular focus.

To make your information easy to translate and easy to apply to other cultures without offense or confusion, focus mainly on the quality characteristics that contribute to ease of understanding:

- Clarity
- Concreteness
- Style

To reduce the time and cost of translation, many companies rely on a combination of human and machine translation. In such situations, the clarity of the information is even more important because the more costly human translators must do the work when machine translation tools cannot. In addition, reducing the amount of information has the side benefit of fewer words to translate.

If information is hard to understand in its original language, a translator almost certainly cannot make it easy to understand. Very few translators have ready access to development groups or the time to get answers to the questions that the information might raise. For example, if the translators perceive an ambiguity, they might preserve the ambiguity or select the meaning that they consider more likely, which might not be the right choice.

Suppose that the marketing plan for a product that you work on is to sell the first release in the United States and to sell subsequent releases in more and more places throughout the world. This information for the product will be available on the web. What preparations would you make so that you could easily adapt the information to an international audience? Suppose that the following task information is being developed for this product, with much input from marketing and development:

**Original**

> Congratulations on buying your global positioning system (GPS) device, InfoGPS! With InfoGPS you can always make a determination about where you are, and you can discover and locate the best route to wherever you want to go.
>
> InfoGPS was programmed at the factory with the date and time and the satellites that are appropriate there. However, chances are that you're not at our factory! Therefore, you may need to reset the device for where you are, though the date is probably not incorrect in the continental United States. So first check the time to make sure it is correct, given the different time zones in the United States. InfoGPS does the rest for you. About five minutes are spent looking for the satellites it expects to find. If they are not found, the Autoinitialize search facility is started, which (since the latest satellite search technology was used) will take a maximum of five minutes to search for whatever satellites are available where you are. The next time you turn on InfoGPS, it will find its position in less than a minute unless you have moved more than 100 miles from the last location where it was used. For example, if you go from Chicago to Cincinnati, you must initialize InfoGPS again, but if you go from Chicago to Kenosha, you don't.
>
> Rather than wait for the Autoinitialize search facility, you can try initializing InfoGPS yourself. To initialize the device, press the Menu key. Then press the down arrow key until you see **Initialize GPS** highlighted. Then press the right arrow key to display the map screen. Use the arrow keys on the map screen to move the crosshairs to your approximate location. There are In and Out keys that you can use to resize the map and so find your location more easily. You can check the latitude and longitude of the cursor position in the box at the top. Once you are satisfied that the position of the crosshairs matches your position, press the Enter key.

The quality characteristics that primarily affect ease of use (task orientation, accuracy, and completeness) apply to technical information throughout the world. Regardless of where the InfoGPS product will be sold, the tasks and the steps must be clear. This passage looks as though it has no steps because the steps are buried in a paragraph. Extraneous pieces of information (such as mentions of the factory, time zones, and search technology) mask the choices that a user has. In addition, the three paragraphs have no real entry points. The following guidelines for task orientation, completeness, and retrievability apply here:

- Present information from the users' point of view.
- Provide clear, step-by-step instructions.
- Cover each subject in only as much detail as users need.
- Provide helpful entry points.

# Putting it all together

The tone of the passage has a marketing flavor. Marketing language tends to be idiomatic and therefore hard to translate and hard for nonnative English speakers to understand. The passage has many other clarity and style problems that could confuse translators and nonnative speakers, such as:

- Nominalizations such as *make a determination about*
- Repetitive expressions such as discover and locate (which are also not as direct as *find*)
- Negative expression (*not incorrect*) instead of a positive expression
- Vague referent *they* instead of *these satellites*
- Missing *that* and *that are* (making the text harder to parse and translate)
- Ambiguous use of *since* instead of *because* and *once* instead of *after*
- Long noun strings (such as *satellite search technology* and *AutoInitialize search facility*)
- Use of *may* instead of *can* or *might*
- Misplaced modifier (*in the box at the top* should not modify *cursor position*)

In addition, these paragraphs do not provide visual cues such as lists that could help users skim the information.

The following revision remedies many of these problems:

*Applying more than one quality characteristic*

**First revision**

> When you first turn on InfoGPS, your global positioning system (GPS) device, it knows the current date but maybe not the local time and your position. Reset the time if it is incorrect. The satellites that InfoGPS automatically looks for are not necessarily the ones that are available where you are. If InfoGPS doesn't find these satellites after five minutes of searching, it starts looking for any satellites. This Autoinitialization takes a maximum of five minutes until InfoGPS finds the local satellites and shows your position. The next time that you turn on InfoGPS, it will find your position in less than a minute unless you have moved more than 100 miles from the last location where you used it. For example, if you go from Chicago to Cincinnati, you must initialize InfoGPS again, but if you go from Chicago to Kenosha, you don't.
>
> Rather than wait for the Autoinitialization, you can initialize InfoGPS yourself.
>
> 1. Press the Menu key.
> 2. Press the down arrow key until you see **Initialize GPS** highlighted.
> 3. Press the right arrow key to display the map screen.
> 4. Use the arrow keys to move the crosshairs on the map screen to your approximate location.
>
>    **Tip:** You can use the In and Out keys to resize the map and find your location more easily. In the box at the top, you can check the latitude and longitude of the cursor position.
>
> 5. When you are satisfied that the position of the crosshairs matches your position, press the Enter key.

The first revision fixed many of the task-orientation problems in the second paragraph and provides entry points with the addition of the ordered list. The tone of the first paragraph is more appropriate than in the original passage, and many of the clarity and style problems have been fixed throughout.

However, the first paragraph is still hard to understand in terms of what to do and when. Imperative statements are mixed with descriptive statements.

If you analyze the task of establishing one's position, you find that a user must do this task more than once. The original passage suggested that the user could do this task at the outset and then the device would be set forever, but the information about going farther than 100 miles implies that the user would need to redo this positioning. Therefore, the task must be presented in a way that suits not only first-time users, but also users who have gone beyond the 100-mile radius where the old positioning applied. In fact, the latter situation is probably more common for users who need this information.

## Putting it all together

Terms that are used in a product interface can cause problems in using the product and also problems in understanding the information when it uses these terms. The name of the AutoInitialize feature might be difficult for nonnative speakers and for translators to understand if they do not recognize *Auto* as an abbreviation for *automatic*. In addition, some users might not understand the technical term *Initialize*.

You need to work with application developers and with usability engineers to make the terminology in the interface acceptable. Refining terminology is important even in the first release of a product because a change in the interface later could confuse current users. The date format, for example, should be in the international format *yyyy-mm-dd* and should allow users to change it to their preferred format. You might also consider whether to start out using *kilometers* rather than *miles* as the unit of measurement for distance.

The example is specific to the United States, but it seems unnecessary because the distance information is already specific. Besides, only people who are familiar with the area of the Midwestern United States that the example highlights might find the example helpful. Deleting the example is probably more appropriate than adapting it to every locale.

**Second revision**

### Orienting your InfoGPS

When you turn on InfoGPS, your global positioning system (GPS) device, check the date and time to make sure that they are correct.

The satellites that InfoGPS is programmed to look for might not be the ones that are available where you are. You can choose either of two methods to acquire signals from the local satellites:

- Let InfoGPS search automatically
- Set up your position yourself

Regardless of the method that you use to orient your InfoGPS, the next time that you turn it on, it will find your position in less than a minute unless you have moved more than 100 miles from the last location where you used it.

**Automatic searching for satellites**

When you let InfoGPS manage its own searching for local satellites, it spends a maximum of 10 minutes:

- Up to 5 minutes to check for the satellites that it expects to fid in the area that it was programmed for
- Up to 5 minutes to look for any local satellites

To activate automatic searching, simply turn on InfoGPS.

1. Press the Menu key.
2. Press the down arrow key until you see **Position GPS** highlighted.
3. Press the right arrow key to display the map screen.
4. Use the arrow keys to move the crosshairs on the map screen to your approximate location.

    **Tip:** You can use the In and Out keys to resize the map and find your location more easily. In the box at the top, you can check the latitude and longitude of the cursor position.

5. When you are satisfied that the position of the crosshairs matches your position, press the Enter key.

# Putting it all together

The second revision clarifies the relationship between automatic searching and manual positioning. The tasks can work anywhere in the world. This revision also replaces *initialize* in the interface.

In addition to the task orientation, completeness, and retrievability guidelines that affect the revisions of this passage, the following guidelines are most relevant to improving this information for nonnative users of English and for translators:

| Quality characteristic | Relevant guidelines |
| --- | --- |
| Clarity | - Eliminate wordiness<br>- Avoid ambiguity<br>- Use technical terms consistently and appropriately |
| Concreteness | Consider the skill level and needs of users |
| Style | - Use active and passive voice appropriately<br>- Convey the right tone |

# Applying quality characteristics to topic-based information

When you write topic-based information, guidelines that apply to each type of information might differ.

## Applying quality characteristics to task information

Task information provides instructions and information such as prerequisites, examples, and the rationale or context for the task.

Task topics should provide users with what they need to know to successfully complete a task. The following task topic might challenge some users.

Putting it all together

**Original**

### Migrating InfoDev Studio to version 8.1

To migrate InfoDev Studio to version 8.1, it must be installed into Eclipse version 4.2.2 as a new installation. It cannot be upgraded, nor installed on top of an earlier version of Eclipse.

**About this task**

InfoDevStudio cannot be directly upgraded to V8.1.0.0 from earlier versions in a single update installation. If you want to install InfoDev Studio v8.1.0.0 into the Eclipse instance where your current version of InfoDev Studio is installed, you must first uninstall your current version of InfoDev Studio, and then upgrade Eclipse to the supported version. When InfoDev Studio V7.1.0.0 is installed, you can then point to your earlier workspace and work with your existing projects.

**Procedure**

1. Optional: Uninstall your existing instance of InfoDev Studio (only if you install InfoDev Studio V8 into the same Eclipse as the one where you have your current version of InfoDev Studio).
    a. Open the Help menu and click **About Eclipse**.
    b. Click **Installation Details**.
    c. Select all of the following if they are installed: Dojo Tools, jQuery Tools, InfoDev Studio.
    d. Click **Uninstall**.
    e. On the Uninstall window, select the items you want to uninstall, and click **Finish**.
2. If you already have Eclipse version 4.2.4, skip to step 11.
3. If you have an earlier Eclipse version, open the Window menu and click **Preferences**.
4. In the left panel, click **Install/Update** to display more options.
5. Click **Available Software Sites**, and click **Add**.
6. In the Add Site window, enter a value in the **Name** field (for example, `Eclipse Update`).
7. Click **OK** to exit the Preferences window.
8. Open the Help menu and click **Check for Updates**.
9. In the Available Updates window, select the items that you want to install or update.
10. Click **Next**.
11. Verify the items that you want to install or update and click **Next**.
12. If required, accept any licenses and click **Finish**.
13. Install InfoDev Studio V6.0.0. For instructions on InfoDev Studio installation, see installing InfoDev Studio.

**Results**

InfoStudio is now installed. **Note:** If the update appears to hang, it might be because you are using a bad mirror. Add this line to your eclipse.ini file to solve the problem:

```
- Declipse.p2.mirrors=false
```

## Applying more than one quality characteristic

This topic has several problems:

- The title is repeated in the first paragraph.
- The procedure is too long with 13 main steps.
- Step 2 points to step 11 instead of step 13.
- Incorrect version numbers, probably a result of multiple updates to this topic, appear just before the procedure and in the last step.
- Step 13 includes a textual cross-reference but no link to required information.
- Conditional and optional steps aren't correctly identified.

You can also find a range of clarity and style issues, from a dangling modifier in the first sentence and terminology problems (*migrate, update, upgrade*) to inconsistent highlighting throughout the steps.

Putting it all together

**Revision**

## Upgrading InfoDev Studio to version 8.1

InfoDev Studio version 8.1 requires Eclipse version 4.2.2. If you have an earlier version of Eclipse, you must upgrade Eclipse before you install InfoDev Studio.

### Before you begin

To prevent potential problems with the upgrade process, add this line to your eclipse.ini file:

```
-Declipse.p2.mirrors=false
```

### About this task

To install InfoDev Studio version 8.1 into the Eclipse instance where your current version of InfoDev Studio is installed, you must first uninstall your current version of InfoDev Studio and then upgrade Eclipse to the supported version. After you install InfoDev Studio, you can configure Eclipse to point to your earlier workspace and work with your existing projects.

### Procedure

1. If you are installing the new version of InfoDev Studio in the same Eclipse instance as the previous version, uninstall InfoDev Studio from Eclipse:

    a. Select **Help > About Eclipse**.
    b. Click **Installation Details,** select Dojo Tools, jQuery Tools, and InfoDev Studio.
    c. Click **Uninstall** and follow the instructions to complete the uninstallation.

2. If your version of Eclipse is earlier than 4.2.2, upgrade Eclipse:

    a. Click **Install/Update > Available Software Sites > Add**.
    b. In the Add window, specify a name, for example, `Eclipse Update`, and click **OK**.
    c. Select **Help > Check for Updates**.
    d. In the Available Updates window, select the most current version of Eclipse, click **Next**, and follow the instructions to complete the installation.

3. Install InfoDev Studio.

### What to do next

Configure Eclipse to use your existing workspace and projects.

The revision is improved in the following ways:

**Task orientation**

The introductory paragraph provides a rationale for doing the task.

Steps are grouped for usability.

Step 1, which was erroneously identified as optional, is now clearly conditional.

**Organization and retrievability**

The information that was at the end of the topic and labeled as a note is now a prerequisite, presenting the information in the right sequence.

Grouping the steps helps to emphasize the main points of the procedure and subordinate the secondary steps.

The reference to installation instructions in the last step is replaced with a helpful link.

**Accuracy and completeness**

Product versions have all been corrected and the number of instances reduced to make it easier to maintain.

Unnecessary information, such as the instruction to accept the product license and the obvious result, is removed.

**Clarity and style**

The dangling modifier in the first sentence of the topic is gone.

Inconsistent use of the terms *upgrade, migrate,* and *update* has been corrected by using *upgrade* consistently.

Jargon (*hangs, on top of*) and unnecessary modifiers (*directly, existing*) have been addressed.

Wording of steps is consistent.

Highlighting is consistently applied.

The following guidelines apply to these revisions:

| Quality characteristic | Relevant guidelines |
|---|---|
| Accuracy | Maintain information currency |
| Completeness | Include only necessary information |
| Clarity | • Place modifiers appropriately<br>• Avoid empty words<br>• Use terms consistently |
| Organization | • Arrange information sequentially<br>• Emphasize main points; subordinate secondary points |
| Retrievability | Link to essential information |
| Style | Apply consistent highlighting and formatting |
| Task orientation | • Provide clear, step-by-step instructions:<br>  – Group steps for usability<br>  – Clearly identify steps that are optional or conditional<br>• Indicate a practical reason for information |

## Applying quality characteristics to conceptual information

Conceptual information provides the background that many users need before they can understand something or successfully work with it. Concepts can help users feel comfortable with new information and relate it to what they already know.

The following subjects, for example, could be treated as concepts:

- The process of how something works
- An overview of a product, parts of a product, or several products
- A comparison of related technologies
- A scenario that demonstrates how a product helps users achieve business goals

The following passage is about the process of managing the presentation of help topics in response to actions by users. This description is of interest to software developers of a help engine or to the people who provide service for the help engine.

# Applying more than one quality characteristic

**Original**

> A link is established for each function request to the help engine. Each link has its own "current" help. Help selection from one link does not affect the current help on another link. Any link may select any help; more than one link may select the same help.
>
> The default help, 0, is considered current for the link unless changed by request on that link. The default is used before any help is created or selected on the link, or if the link explicitly deletes its own current help.

This passage is hard to understand. The initial hurdle seems to be the passive voice, which obscures the agent. Who does what? When do they do it? The following revision deals with these problems:

**First revision**

> The help engine creates a link for each request that requires processing. Each link connects to a help topic. Connecting to a help topic does not change the connection between any other link and help topic. Any link can connect to any help topic; more than one link can connect to the same help topic.
>
> Initially, a link connects to the default help topic. The help engine can connect the link to a new help topic or delete the help topic that the link connects to. If the help engine deletes the help topic, the engine also resets the link to connect to the default help topic.

The first revision uses an active style and clarifies when the help processor is responsible for an action. Each sentence is easier to understand, but the meaning of the passage is still hard to grasp. The passage is still two dull paragraphs. To make the meaning obvious, you can apply some guidelines from retrievability and visual effectiveness, as in the second revision:

Second revision

**Process of linking help topics**

In response to actions by users who request help, the help engine takes the following actions:

| User action | Help engine action |
|---|---|
| None | Connects to the default help |
| Requests help | Opens the appropriate help topic |
| Requests different help | Opens a requested help topic |
| Exits help | Returns to the default help topic |

The second revision uses a heading, a table, and a list to show at a glance what the information is and how the parts relate to each other. The passage now can stand as a discrete topic. You might decide that the topic needs to define some terms (such as *link* and *default help topic*) or link to a concept topic that explains these terms. Examples would also help make the passage more concrete and therefore easier to understand.

The following guidelines apply to these revisions:

| Quality characteristic | Relevant guidelines |
|---|---|
| Clarity | • Write coherently<br>• Define each term that is new to the intended audience |
| Concreteness | • Use concreteness elements that are appropriate for the information type<br>• Use specific language<br>• Consider the skill level and needs of users |
| Organization | Emphasize main points; subordinate secondary points |
| Retrievability | • Link appropriately<br>• Provide helpful entry points |
| Style | • Use active and passive voice appropriately<br>• Apply consistent highlighting and formatting |
| Task orientation | • Present information from users' point of view<br>• Indicate a practical reason for information |
| Visual effectiveness | Use visual elements to help users find what they need |

## Applying quality characteristics to reference information

Reference information provides quick access to facts, often as lists or tables. For example, a reference topic might give command syntax, a table of part numbers, or a list of software requirements.

The following excerpt from a topic uses a list as the main organizer:

**Original**

> The keywords are:
>
> - The component identification keyword.
>
>   This is the first keyword in the string. A search of the database with this keyword alone would detect all reported problems for the product.
>
> - The type-of-failure keyword. The second keyword specifies the type of failure that occurred. Its values can be:
>
>   - ABEND
>   - ABENDUxxx
>   - DOC
>   - INCORROUT
>   - MSGx
>   - PERFM
>   - WAIT/LOOP
>
> - Symptom keywords
>
>   These can follow the keywords above and supply additional details about the failure. You select these keywords as you proceed through the type-of-failure keyword procedure that applies to your problem.
>
>   The suggested approach is to add symptom keywords to the search argument gradually so that you receive all problem descriptions that might match your problem. You can AND or OR additional keywords in various combinations to the keyword string to reduce the number of hits.
>
> - Dependency keywords
>
>   These are program- or device-dependent keywords that define the specific environment in which the problem occurred. When added to your set of keywords, they can help reduce the number of problem descriptions that you need to examine. See Appendix B, "Dependency keywords," on page 303 for a list.

# Putting it all together

The original passage highlights the keywords, but the rest of the information is hard to sort out. The brief descriptions of the keywords leave users wondering "So what?" Buried in some of the paragraphs is information about how to use a keyword. However, users must read carefully to find this information, and then they need to figure out what to do with it. The passage also appears to be part of a larger whole because it has no title.

You might decide that this passage communicates reference information about keyword strings and can stand on its own as a topic. You can therefore give it a noun as an appropriate title, add a definition, and organize the information to clarify the types of keywords and their general use. You can also link to the related reference information about dependency keywords.

**First revision**

> **Keyword strings**
>
> A *keyword string* is a set of terms that you use to describe a problem with the product when you report the problem to software support.
>
> | Type of keyword | Description or value | Use for this type of keyword |
> |---|---|---|
> | Component identification | A set of characters that represents the product or an orderable feature of the product | Find all reported problems with the product or with an orderable feature. |
> | Type of failure | - ABEND<br>- ABENDUxxx<br>- DOC<br>- INCORROUT<br>- MSGx<br>- PERFM<br>- WAIT/LOOP | Refine your search to a particular type of failure for the product or orderable feature. |
> | Symptom | Details about the failure | Refine your search gradually (by combining keywords in various ways) so that you receive all problem descriptions that might match your problem. |
> | Dependency | Program- or device-dependent keywords that define the environment in which the problem occurred | Help reduce the number of problem descriptions that you need to examine. |
>
> **Related reference**
>
> Dependency keywords

The first revision makes the meaning obvious by its structure. The revision also presents the information from the user's point of view rather than from the point of view of the product. The information in the revision not only looks easier to understand; it *is* easier to understand. By making the passage a discrete topic, you have improved the clarity, task orientation, retrievability, and visual effectiveness of the original passage.

Putting it all together

The completeness of the original passage is also improved because gaps that the pattern of information revealed are now filled in. The original passage did not describe the first keyword, other than to say that it comes first. The table now conveys this information about sequence implicitly through the order of the keywords from top to bottom. The original passage also did not give a use for the type-of-failure keyword. The table in the first revision includes both of these items.

**Second revision**

## Keyword strings

A *keyword string* is a set of terms that you use to describe a problem with the product when you report the problem to software support.

| Type of keyword | Description or value | Use for this type of keyword | Example of a keyword string |
|---|---|---|---|
| Component identification | A set of characters that represents the product or an orderable feature of the product | Find all reported problems with the product or with an orderable feature. | CKJPROD |
| Type of failure | - ABEND<br>- ABENDUxxx<br>- DOC<br>- INCORROUT<br>- MSGx<br>- PERFM<br>- WAIT/LOOP | Refine your search to a particular type of failure for the product or orderable feature. | CKJPROD WAIT |
| Symptom | Details about the failure | Refine your search gradually (by combining keywords in various ways) so that you receive all problem descriptions that might match your problem. | CKJPROD WAIT CYCLE |
| Dependency | Program- or device-dependent keywords that define the environment in which the problem occurred | Help reduce the number of problem descriptions that you need to examine. | CKJPROD WAIT CYCLE AS4 |

**Related reference**

Dependency keywords

Developing Quality Technical Information

The second revision adds a column for examples. These examples help users see what the rest of the information is about and relate the information to a keyword string of their own. You have made the topic more concrete by adding the examples.

The revisions reflect the following characteristics and guidelines:

| Quality characteristic | Relevant guidelines |
| --- | --- |
| Clarity | Define each term that is new to the intended audience |
| Completeness | Cover each subject in only as much detail as users need:<br>• Include enough information |
| Concreteness | Use concreteness elements that are appropriate for the information type |
| Organization | • Organize information into discrete topics by type<br>• Emphasize main points; subordinate secondary points |
| Retrievability | Provide helpful entry points |
| Task orientation | • Present information from users' point of view<br>• Indicate a practical reason for information |
| Visual effectiveness | Use visual elements to help users find what they need |

CHAPTER **13**

# Reviewing, testing, and evaluating technical information

Reviewing, testing, and evaluating technical information can take several forms, depending on who the reviewer is. This chapter deals with the role of the following kinds of reviewers:

- Technical reviewers
- Actual users or people similar to them
- Testers (who might also be writers, editors, information architects, visual designers, or usability engineers)
- Editors (who might also be information architects or writers)
- Visual designers

Every writer needs reviewers—other people to see and use the information. Reviewers bring their own skills and perspectives to the information.

Just as reviewers have special skills, they also have different areas of focus. Not every reviewer needs to look for all of the same items. For example, technical reviewers don't need to comment on style because that's a domain for editors. See Appendix B, "Who checks which characteristics?," on page 549 for a table that summarizes what the various kinds of reviewers should look for.

Not every project requires the same level of reviewing, testing, and evaluating. Some projects have few quality problems or low visibility, for example, and so do not warrant intensive efforts by many people. However, in general, the fewer resources spent on reviewing, testing, and evaluating the information for a project, the greater the risk of poor quality.

# Reviewing technical information

A technical reviewer is an expert in the subject that the technical information covers. In general, technical reviewers are best qualified to evaluate the accuracy, completeness, and concreteness of technical information. However, technical reviewers might want more technical precision in the information than a particular audience needs.

The technical reviewer's role is to:

- Read the information critically, or use it, or both.
- Find problems with the information.
- Report the problems in a way that helps you understand the problems and fix them.

To help reviewers provide useful comments, you can talk with them about their role before a review and discuss the following advice about reading technical information, finding problems, and reporting them. You can also provide that information in your cover note to the technical reviewers when you announce a review. You can give them examples of the kinds of problems to look for (based on the pertinent guidelines in Table 13.1) and of helpful kinds of review comments.

## Reading and using the information

Reviewers need to put themselves in the place of users when they review technical information. Reviewers should actively question what they read and continually ask themselves: "If I were the user, could I find, understand, and use this information?"

Reviewers tend to assume that because they understand something, a user will also. The people who create a product or technology often have a difficult time assuming the perspective of users and can have trouble maintaining objectivity.

Reviewers should review the information in the same context that users will see it and not just read through a printed set of topics. Reading the embedded assistance in the UI rather than in a properties file helps reviewers see what help users will get in the user interface.

## Finding problems

The following table shows the quality characteristics and guidelines that are most likely to benefit from scrutiny by technical reviewers.

Table 13.1 Quality characteristics for technical reviewers to focus on

| Quality characteristic | | Guidelines |
|---|---|---|
| Easy to use | Task orientation | • Tasks support user goals.<br>• Step-by-step instructions are clear. |
| | Accuracy | • Information is current.<br>• Information has been verified.<br>• Information about all subjects is consistent. |
| | Completeness | • Only needed subjects are covered.<br>• Each subject is covered in only as much detail as users need. |
| Easy to understand | Clarity | • Technical terms are defined and used consistently.<br>• Technical terms are appropriate for the audience. |
| | Concreteness | • Concreteness elements are focused, realistic, and up to date.<br>• Code examples and samples are easy to adapt.<br>• Language is specific. |
| Easy to find | Organization | • Emphasis and subordination are appropriate. |
| | Retrievability | • Keywords are used effectively.<br>• Links are helpful and appropriate. |
| | Visual effectiveness | • Graphics are meaningful and appropriate. |

## Reporting problems

When technical reviewers report problems, comments like "I don't like it," "It's wrong," or "This needs fixing" are not much help. Ask them to be specific about what's wrong and to suggest a solution, as in:

- "This section has too much detail. Users need only the first two paragraphs."
- "All message identifiers have eight characters. Change to eight throughout."
- "Aligning the columns in this code as I've shown is what most users expect."

Even problems of accuracy and completeness can have more than one fix that will work. Ask reviewers to give enough information so that you can understand the reasoning behind a comment and discern the best fix. Applying the guidelines from the earlier chapters in this book can help you advise reviewers how to be specific about typical problems and how to suggest solutions.

# Testing information for usability

One way that you can improve and validate the quality of your information is to observe users while they are using it. Through usability testing, you can find problems especially in task orientation and concreteness, as shown in the following table. Users can provide direct and indirect feedback on the quality characteristics.

Table 13.2  Quality characteristics for users to focus on

| Quality characteristic | | Guidelines |
|---|---|---|
| Easy to use | Task orientation | • Information is appropriate for the intended audience.<br>• Information is presented from the users' point of view.<br>• A practical reason for information is evident.<br>• Tasks support user goals.<br>• Step-by-step instructions are clear. |
| | Accuracy | • Information is current.<br>• Information about all subjects is consistent. |
| | Completeness | • User interfaces are self-documenting.<br>• Embedded assistance follows a predictable pattern.<br>• Only needed subjects are covered.<br>• Each subject is covered in only as much detail as users need.<br>• Information is repeated only when needed. |
| Easy to understand | Clarity | • The focus is on the meaning.<br>• Ideas are free from ambiguity.<br>• Elements flow from one to another.<br>• Technical terms are necessary and appropriate.<br>• New terms are defined. |
| | Concreteness | • Information is appropriate to the skill level and needs of users.<br>• Concreteness elements are focused, realistic, and up to date.<br>• Code examples and samples are easy to adapt.<br>• Examples and scenarios are presented in context.<br>• Language is specific. |

Table 13.2 Quality characteristics for users to focus on (continued)

| Quality characteristic | | Guidelines |
|---|---|---|
| Easy to find | Organization | • Emphasis and subordination are appropriate.<br>• Information is where users expect it.<br>• Elements are arranged to facilitate navigation.<br>• Users can see how the pieces fit together. |
| | Retrievability | • Information is easy to find by searching and browsing.<br>• Links are helpful and appropriate. |
| | Visual effectiveness | • Visual elements help users find what they need.<br>• Screen captures are used judiciously. |

Early testing is most valuable for pointing out problems while you can still make major changes. Flaws in a design, such as lack of information in the user interface, can be hard to remedy later in a project.

## Prototyping

A prototype is a mock-up of a design, perhaps on paper (sometimes called "low tech" or "low fidelity"). A prototype of a user interface, for example, can show how the supporting information in the interface works in various user situations. You can use prototypes to get feedback from potential users or from people similar to potential users. Prototyping avoids problems with installing actual software and can be carried out early in a project. Prototypes are also easy to produce and change while the design is still fluid.

## Testing in a usability laboratory

The usability laboratory offers the opportunity to:

- Observe a user and the user's work area through one or more cameras from another room
- Record the proceedings for later analysis
- Control the testing environment to limit interruptions

The usability laboratory makes it possible for more people to learn from the usability testing both while it is going on (through observation) and later (through the film or tape). A usability engineer can help you gather and analyze both qualitative and quantitative data. However, testing in a usability laboratory can also require a lot of time and resources.

Putting it all together

Techniques for testing information in a usability laboratory are roughly grouped into evaluation tests and validation tests:

- *Evaluation tests* are done early in the development process to determine whether a design will meet the usability criteria if fully implemented.
- *Validation tests* are done late in the development process to determine whether the implementation meets the usability criteria.

The type of testing that you choose depends on your goals, time, and resources, and where you are in the development process.

Both evaluation tests and validation tests are done with actual users or with people whose experience and skills make them representative of users. These representative users can come from inside or outside the company. If they are from inside the company, they should not be familiar with the information.

The following table gives an overview of the procedures and measurements for both types of tests.

Table 13.3  Techniques for testing information in a usability lab

| Type of test | Procedure | Measurements |
|---|---|---|
| Evaluation tests | Users do scenario-based tests on the information, maybe without the product. Tests are generally iterative (test, improve, retest, improve, retest). | • User comments<br>• Errors<br>• Ratings of satisfaction on quality characteristics<br>• Comparisons of the preceding measures for competitive information |
| Validation tests | Users do scenario-based tests on the product and information together. The test team observes users; records actions, errors, and comments; later does statistical analysis on the data and summarizes the results. | • Successful completion of a task<br>• Time to do a task<br>• Number of errors in retrieving information<br>• Ratings of satisfaction on quality characteristics<br>• Comparisons of the preceding measures for competitive information |

### Evaluation tests

Compared with validation tests, evaluation tests tend to be less formal in these ways:

- The test participants and the test administrator interact more, in terms of asking questions and giving directions.

- Fewer scenarios are used, with fewer users for each.
- The measures are more subjective.

Testing with fewer users does not mean that your results are necessarily less valid. After even a few users, you will probably see the same errors in the same places. Showing that a problem exists can be enough to make a usability test worthwhile.

Use evaluation tests to achieve the following kinds of goals:

- To evaluate the overall organization of the information or components (such as topic titles, contents list, and examples)
- To determine which of two or more alternative designs is more usable
- To evaluate new information technologies (such as new search or hover help mechanisms)
- To determine the best terms to use in the interface for new concepts and functions
- To compare a particular kind of information (such as tutorials) to that of the best-of-breed supplier of that type of information

## Validation tests

Compared with evaluation tests, the test participants and the test administrator interact less, more scenarios are used, and the measures are more objective. Validation tests provide a good way to determine how your information compares with the information of a best-of-breed product. You can, for example, compare satisfaction ratings, errors, time that was spent on a task, and successful completion of a task and thus gather benchmark measures.

You can use feedback from validation tests to:

- Pinpoint and fix major usability problems before you make the product and information available
- Assess the interaction and flow of information in the product and whether users can get to the needed information
- Test whether the information helps users complete product tasks
- Validate quality characteristics of the information such as task orientation, completeness, and retrievability
- Get a final benchmark against the information of a best-of-breed product

If you do not have enough time and resources to fix all of the problems, you can plan to resolve less serious usability problems through either of two means:

- A later release of the product
- Enhancements to the information that you make available on the web

Validation tests are generally more difficult to design and take longer to conduct, although they yield more practical results. A validation test is probably appropriate in these cases:

- If the impact of your test will be wide ranging (for example, the information that you are testing will be used by several products or will be a template for other information)
- If you have a fairly stable product to test with
- If you have the time and resources

If you have done evaluation tests during design and development of the information and changed the information to deal with the problems that you found, the validation tests should not produce surprising results.

## Testing outside a usability laboratory

Formal usability testing is often done outside a usability lab. Some testing outside the usability laboratory can be done internally, without real users. Tests that involve people such as technical reviewers, writers, and usability specialists can yield valuable information and yet not require a lot of time to prepare and run. In a walkthrough, for example, the people who are involved in developing the information try to use the information in a realistic way, by performing procedures and following instructions. This type of testing yields information that you can use to fix problems.

You have several possibilities for testing information outside a usability laboratory with actual users (perhaps during a beta program), as shown in the following table.

Table 13.4  Usability tests outside a usability laboratory

| Type of test | What is tested | Procedure | Measurement |
|---|---|---|---|
| Field observation | Product and information as they are used in a real-world situation | Writers visit users and observe them using the product and information. The contextual inquiry form of observation also involves asking questions to make sure you are observing accurately and making correct inferences. | Qualitative data about what users do |
| Survey or questionnaire | Product and information | Users answer questions about the quality of the information and product. | Comparison of answers |
| Remote usability test | Links to and content of online help and embedded assistance<br><br>Product information with or without the product | Users participate in tests in one location, while test administrators in another location use special software to view the users' screens and listen through a conference call. | Successful completion of a task<br>Time to do a task<br>Number of errors |
| Instrumented beta test | Links to and content of online help and embedded assistance | "Hooks" in the beta code enable recording which help topics or field-level help users access and in what circumstances. Application developers must provide the necessary hooks, and users must agree to this form of data collection. | Qualitative data about what users do |
| User edits or reviews | Information and the product in beta test | Users review the information and offer recommendations for improvement. | Periodic ratings of satisfaction on quality characteristics during the beta test |

Using any of these forms of feedback, you can change the information to fix problems and improve the quality. When you are uncertain about a problem or how to fix it, you can return to the users to get more information or ask another set of users.

## Testing technical information

Just as you cannot test quality into a product that has not been developed to meet quality standards, you cannot test quality into information that has been developed without regard for quality. However, appropriate testing can help you find and fix certain types of problems with information, as shown in the following table.

Table 13.5  Guidelines for testing technical information

| Quality characteristic | | | Guidelines |
|---|---|---|---|
| Easy to use | Task orientation | | • Tasks support user goals. |
| | | | • Step-by-step instructions are clear. |
| | Accuracy | | • Information has been verified. |
| | | | • Information is current. |
| | | | • Information about all subjects is consistent. |
| | | | • Tools have been used to check the accuracy of information. |
| | Completeness | | • User interfaces are self-documenting. |
| | | | • Embedded assistance follows a predictable pattern. |
| | | | • Each subject is covered in only as much detail as users need. |
| Easy to understand | Clarity | | • Language is unambiguous. |
| | Concreteness | | • Concreteness elements are appropriate for the information type. |
| | | | • Concreteness elements are focused, realistic, and up to date. |
| | | | • Code examples and samples are easy to use. |
| | | | • Examples and scenarios are presented in context. |
| | Style | | • Information has the right tone. |
| | | | • Terms are spelled and capitalized correctly. |
| | | | • Information is free from punctuation errors. |
| | | | • Templates are used correctly and applied consistently. |
| | | | • Markup language tagging is consistent. |

Table 13.5 Guidelines for testing technical information *(continued)*

| Quality characteristic | | Guidelines |
|---|---|---|
| Easy to find | Organization | • Tasks are organized sequentially. |
| | | • Elements are organized consistently. |
| | | • Contextual information is in the appropriate type of embedded assistance. |
| | Retrievability | • Content is optimized for searching and browsing. |
| | | • Paths guide users through the information. |
| | | • Links are appropriate. |
| | | • Entry points are helpful. |
| | Visual effectiveness | • Visual elements are accessible to all users. |

## Tools for testing

Writers should routinely test any kind of information by using a spelling checker, a grammar checker, and a link checker.

Be sure to look closely at the items that a spelling checker finds. Do not assume, for example, that a strange word or acronym has been found because it is not in the checker's dictionary but should be; the item might be misspelled. Also, be careful not to add unapproved abbreviations to your dictionary.

A grammar checker can be more time-consuming to use than a spelling checker, but is just as important if not more so. A grammar and style conventions checker is particularly important if you don't have an editor or a writing colleague to look at the information.

You can also use tools to test the following aspects of online information:

- The links are correct.
- Appropriate tags are used to facilitate the use of screen readers.
- Tags are properly nested and will work in various browsers.
- A document that is formatted for online viewing opens correctly and has correct links.

## Test cases

To test a product, testers write and run test cases, which have steps for exercising parts of the product. These test cases include ensuring that the product responds well when users make errors, such as when they enter the wrong format for a web address.

Writers can create test cases to test parts of the online information such as information that should show up by using the search capability. For example, a test case might include several strings to search for, to ensure that the expected information is available and that the results are presented appropriately.

Writers can also run product test cases to make sure that the product information covers the items that are in the test cases. Test cases can be a good source of information about how the product works. For example, some test cases test that the software responds appropriately when users enter inappropriate data. The information should help users enter appropriate data and help them recover if they do not.

## Testing the user interface

The writing team might have responsibility for at least some of the user interface, especially the embedded assistance. Testers (whether writers, editors, information architects, or usability engineers) can check the interface in the following ways:

- Check the appearance of interface controls and icons, particularly for consistency of color and size and for legibility
- Check the text on the screen and on interface controls, such as buttons.
- Test input methods, including the keyboard, mouse, and touch screen.
- Test navigation within and between windows, menus, or components.
- Test hover help and any links to help from the user interface.
- Test the expected behavior of interface controls and icons.
- Test whether appropriate messages inform users about problems or potential problems.

## Editing and evaluating technical information

Conventional editing produces comments about the weaknesses in information. By using the quality characteristics, however, you can categorize both strengths and weaknesses. You can also see which quality characteristics of the information most need work and prioritize revisions. When you incorporate the quality characteristics into your editing, you probably need to modify your approach to editing.

Editors look at all aspects of the quality characteristics as they edit technical information and user interfaces. To verify the accuracy of the information, editors often rely on technical reviewers who are more familiar with the technical details.

You have several choices for organizing your editing, particularly if more than one writer is working on the project:

- Edit the embedded assistance and any topics that it links to. By following the path that users might take, you can make sure that information is where they need it most.
- Edit a main task and all of its related task, concept, and reference information. With this approach, you can determine whether the information is complete and whether links are appropriate.
- In an agile environment, edit the information that was developed in the previous sprint.

You might combine some of these approaches depending on the schedule and the needs of the material.

When you edit technical information, you can apply the quality characteristics throughout the following steps:

- Prepare to edit.
- Get an overview.
- Read and edit the information.
- Look for specific information.
- Summarize your findings.
- Confer with the writer.

## Preparing to edit

Familiarize yourself with the general plan for the information before you begin editing it. The writer should provide this information through appropriate parts of the content plan and other materials. You should find out the following information before you begin:

- Audience, including job tasks and experience level, and environment
- User interface design guidelines
- Delivery medium or media, such as embedded assistance, EPUB or other softcopy format, help, wizard, multimedia
- Expected number of new panels, fields, and so on
- Expected number of topics and a hierarchical list of topics
- If a revision, planned changes in the content
- What style guidelines apply
- Plans for translation into other languages

For book-format information that isn't new, browse the most recent edition. You will probably see more clearly how the new edition is intended to differ from its predecessor.

The purpose of all this preparation is to develop an idea of what the finished information is meant to contain and look like. This preparation gives you a helpful set of expectations and enables you to determine what has and hasn't been realized in the information.

Before delivering draft information to you, the writer should use automated tools, such as a spelling checker and a grammar checker, and fix the problems that the tools finds.

## Getting an overview

When you begin editing information, survey it as a whole, as a user might.

Pay attention to the visual elements:

- Is the information presented in an attractive way?
- Are the text and graphical elements well balanced?
- Are tables and lists used effectively?
- Are the typefaces that are used for headings, text, and highlighting clear and helpful?
- Are graphical elements and typefaces used consistently?

The following sections describe what you should look for as you start to edit different types of information, regardless of the approach you take to editing.

**For embedded assistance**

Focus on retrievability by asking the following questions about the information:

- Is programmatic assistance available for some fields?
- Are input hints clear and easy to see?
- Is static text placed where users will read it?
- If fields include hover help, does it display appropriately?
- Are links from embedded assistance used judiciously?
- Is more information easy to get to?

Reviewing the user interface as a whole can also help you determine visual effectiveness, task orientation, and organization. Reviewing the user interface together with task topics can help you evaluate accuracy and completeness of procedures.

**For task information**

Focus on task orientation by asking the following questions about the information:

- Is the intended audience clearly stated?
- Does the navigation pane or table of contents clearly show the real tasks for the product?
- Are the tasks presented in the order of use?
- Do all necessary tasks appear to be included?
- Are optional and conditional steps handled appropriately?
- Can a user read the task help while doing the steps?
- Do task topics provide links or cross-references to necessary supporting conceptual and reference information?

Checking the navigation pane or the table of contents can help you determine the task orientation, completeness, and organization of the information as a whole. Looking for paths to the task information from the user interface helps you assess retrievability.

### For conceptual information

Focus on organization by asking the following questions about the information:

- Are major concepts presented as topics?
- Are concept topics related to appropriate task and reference topics?
- Are concept topics related to each other?
- Are graphics, examples, similes, and analogies used effectively in concept topics?
- Are minor concepts treated appropriately within other topics, such as with brief explanations and glossary definitions?
- Are illustrations used and placed appropriately?

### For reference information

Focus on retrievability and visual effectiveness by asking the following questions about the information:

- What retrievability aids does the information use? Does it use color or graphics to draw the eye to topics of importance that recur, such as restrictions or operating system-dependent information?
- Does the organization enhance retrievability? Will the subdivisions of information in the table of contents seem logical to someone who is looking up information?
- Are the headings appropriate for reference information? Are things such as command names followed by a common noun?
- Do the type style and format enhance retrievability? For example, do code examples stand out as examples? Is the explanation of a syntax element easy to pick out?
- Is the information retrievable by users who have impairments, such as blindness, color blindness, deafness, or mobility restrictions?

You can also consider the organization and completeness of reference information by asking:

- For items such as classes, methods, commands, or programming statements, are the explanations in alphabetic order by name?
- For messages, are the explanations ordered by message identifier?
- If the information contains syntax diagrams, does it also have an explanation of how to read the diagrams?
- Is the information presented in a pattern, such as purpose, syntax, restrictions, and usage notes?

## Reading and editing the information

Read and edit the information in detail. For embedded assistance, decide whether to read all information of the same type, such as all hover help, or all the embedded assistance for a panel. For information in book format, reading should be linear, from front to back. Defer more active testing, searching, and using the information until the next step.

After you get into the material, you will probably want to note items to add to the style sheet and items to watch out for that you suspect might be a problem. At some point, you can begin to list problems, perhaps under the quality characteristic where you think that they apply.

For a given scenario's task and concept information, keep these questions in mind:

- Can I understand the steps in a procedure?
- Do the procedures tell me what I need to do the task without repeating information in the user interface?
- Is the information easy to navigate? Is the sequence clear?
- Can I understand topics the first time that I read them?
- Are new terms defined?
- Do I see anything that is obviously inconsistent or incorrect?
- Are the transitions from subject to subject logical?

For reference information, keep in mind that users will not read it from cover to cover, or even section by section. They will enter it almost anywhere, in search of small bits of information. So as you read and edit, make a list of the kinds of information that you come across—syntax diagrams, keyword explanations, restrictions, operating-system considerations, examples, and so on. You will use this list in the next step when you check retrievability.

Use the checklists at the end of each chapter in this book to help ensure that you look for the problems related to each quality characteristic. If you find more problems, add them to the lists where appropriate. Here's a brief list of major areas to check:

- Factual consistency
- Organizational consistency
- Appropriate level of detail
- Appropriate visual elements
- Consistent style

## Check for factual consistency

Make sure that the information in examples, tables, and figures agrees with the surrounding text. Information about the same subject should be consistent.

Check that elements of a procedure are logical. For example, prerequisites should be stated before the procedure, and each step should logically follow the previous.

Verify product and component names, and ensure that any abbreviated names are legally approved.

Make sure that the keyword relationships shown in syntax diagrams agree with the explanations of the keywords. For example, a diagram might show that keyword X and keyword Y can coexist in the same command, but the explanations of X and Y might suggest that they are incompatible. Either the diagram or the explanations need to be fixed.

Also look at the examples and ensure that they can actually be derived from the syntax. Make sure that the examples and supporting text agree and that scenarios are consistent.

## Check for organizational consistency

Check that the same type of information is presented in the same way. For example, if details about a field are in hover help on one screen, the same type of information should not be presented in static text on another screen.

Compare the information to your pattern for progressive disclosure to ensure that the right type of content is in the right place. For example, does static descriptive text come before a field when your pattern specifies that it should come after?

For tasks, make sure that each topic is properly sectioned into prerequisites, steps, and postrequisites.

In reference topics that cover items such as command syntax, check that all necessary standard parts are included.

## Check for appropriate level of detail

The need for some details depends on the experience level of the audience. Other details must be consistently provided. As you read the embedded assistance for interface elements (fields, buttons, commands, statements, keywords), for example, make sure that their purpose is clearly stated. A similar requirement holds for examples: the purpose or effect of examples must be clear.

When you review embedded assistance, make sure that only the elements that need assistance have it. For example, a **Confirm Password** field should never have any assistance.

As you review topics, watch for any information about the internal workings of the product or the level of detail that might come from a technical specification.

### Check for appropriate visual elements

Looking at one type of visual element such as tables can reveal problems that you wouldn't otherwise notice. Consider, for example, whether the content of tables effectively supports or replaces text. If your audience might include vision-impaired users, check for one of the following methods as appropriate so that these users can still benefit from the information in graphics:

- The key content of graphics is summarized in surrounding text.
- For web information, the graphic is briefly described in alternative text, or a longer description is provided as text that most users would not see.

### Check for consistent style

Notice choices of style and whether they are consistent throughout the information. Examples of style choices to check are tone, spelling, punctuation, capitalization, terminology, and parallelism of headings, list items, and index entries.

If you edit source files, check for correct tagging, particularly that correct semantic elements are used in XML.

## Looking for specific information

Check links from the user interface to topics and links between topics to verify that information is linked appropriately. Look in the table of contents or index for topics that you have read; you should be able to find them again.

Use the native search to look for terms. Do the results match what you'd expect to see based on the existing information?

Check whether reference information has effective entry points to each of the kinds of information that you identified while reading the information.

## Summarizing your findings

Use Appendix A, "Quality checklist," on page 545 to summarize the kinds of problems that you found.

If a visual designer reviewed the information, combine the comments from that review with your own before you make your final assessment. See "Reviewing the visual elements" on page 536 for relevant guidelines.

Putting it all together

Your comments in the draft and the summary tell writers and planners about strengths that you find as well as changes that can improve the quality of the information. Your summary should identify the strengths and the weaknesses that, in your opinion, would have the most impact on how satisfied users are with the information.

### Assigning problems to the quality characteristics

Some problems affect more than one quality characteristic. In that case, consider the importance of the problem to the user. When a problem might affect any of two or more quality characteristics, give preference to the quality characteristic that is more important to the user. For example, a typo like *its* instead of *it's* might be a style problem in one situation but a clarity problem in another situation. Transposing letters in a file name has a definite impact on accuracy, but *form* instead of *from* might hardly be noticed.

However, also note when a problem affects another quality characteristic. For example, note when a problem that you report in concreteness also has an effect on retrievability. This information helps the writer understand the impact of a problem.

Style has a dependence on other quality characteristics, because some characteristics provide a basis for style decisions. For example, a style guideline might be made because following it promotes clarity, visual effectiveness, or retrievability. Style in itself does not have a rationale other than the force of convention.

You might choose to put a deviation from the chosen style guidelines in another quality characteristic when both of the following statements are true:

- Failure to follow the style has a major impact on the other quality characteristic.
- The error occurs frequently. The writer seems to have forgotten the rationale or seems not to understand it.

You might find other "rules" for assigning problems to quality characteristics as you become more familiar with thinking about problems in terms of the quality characteristics and their effect on users.

### Assigning quality ratings

After you categorize and summarize your comments, you can assign ratings to the quality characteristics. A quality rating is a numeric representation of the quality of information on a particular characteristic or on the whole set of quality characteristics. You can use this quality rating as an indicator of the progress of information either during its development cycle or from release to release.

The quality rating is the result of:

- Thoroughly editing the information
- Summarizing your editorial comments
- Rating each of the nine quality characteristics

You can use Appendix A, "Quality checklist," on page 545 to summarize your ratings. To say how satisfied you are with each quality characteristic, review the strengths and weaknesses that you classified in a given quality characteristic and then consider the meanings of the scale points.

As you work with quality ratings, you will probably find situations where one problem could be enough to bring a quality characteristic down a point and where many small problems would not change a score. Again, you will probably derive some rules to help you decide what ratings to give. You might decide, for example, that when in doubt about a rating for a quality characteristic, pick the better rating.

However, assigning a quality rating is not like grading a math test, where you subtract points for errors. Assigning a quality rating is more like comparing the information to what you consider average and then figuring out whether the strengths and problems that you found make the information better or worse than average.

You need to take into account where the information is in the development cycle. For example, an early draft might not have all of the information, but the writer should indicate what is missing and where it will go. The effect of the missing information on the completeness rating for the information would be much smaller than in a final draft.

Finally, when you determine the accuracy quality rating for information, take into account your level of knowledge about the subject. You probably need to consider whether the technical reviewers reported errors in the information and, if so, the number and scope of the errors and their impact.

## Conferring with the writer

Even though writers need reviewers (especially editors), few people have thick skins when their writing draws critical comments. If you meet with the writer to go over your comments, you can make sure that the writer understands your suggested changes and your reasons for them. You can also find out areas where your suggested changes might not be appropriate. After all, you and the writer are working together to create the best possible information for the users.

## Reviewing the visual elements

Visual designers review information and software interfaces for how well their visual elements attract and motivate users and convey the intended meaning. The following table shows the relevant quality characteristics.

Table 13.6  Guidelines for reviewing visual elements

| Quality characteristic | | Guidelines |
|---|---|---|
| Easy to use | Task orientation | <ul><li>Information is appropriate for the intended audience.</li><li>Information is presented from the users' point of view.</li><li>Tasks support user goals.</li><li>Steps are grouped for usability.</li><li>Unnecessary focus is not given to product features.</li></ul> |
| | Accuracy | <ul><li>Information is current.</li><li>Information about all subjects is consistent.</li></ul> |
| | Completeness | <ul><li>User interfaces are self-documenting.</li><li>Embedded assistance follows a predictable pattern.</li><li>Embedded assistance and topics support users' goals.</li><li>Each subject is covered in only as much detail as users need.</li><li>Information is repeated only when needed.</li></ul> |
| Easy to understand | Clarity | <ul><li>Sentences, lists, tables, and visual elements are free from wordiness.</li><li>Technical terms are used consistently and appropriately.</li></ul> |
| | Concreteness | <ul><li>Information is appropriate to the skill level and needs of users.</li><li>Concreteness elements are appropriate for the information type.</li><li>Code examples and samples are easy to use.</li></ul> |
| | Style | <ul><li>Highlighting is applied correctly and consistently</li><li>Templates are used correctly and applied consistently.</li><li>Markup language tagging is consistent.</li></ul> |

Table 13.6  Guidelines for reviewing visual elements *(continued)*

| Quality characteristic | | Guidelines |
|---|---|---|
| Easy to find | Organization | • Emphasis and subordination are appropriate. |
| | | • Users can see how the pieces fit together. |
| | | • Contextual information is in the appropriate type of embedded assistance |
| | Retrievability | • Entry points are helpful. |
| | | • The table of contents has an appropriate level of detail. |
| | Visual effectiveness | • Visual design practices are applied to textual elements. |
| | | • Graphics are meaningful and appropriate. |
| | | • A consistent visual style is applied. |
| | | • Visual elements help users find what they need. |
| | | • Visual elements are accessible to all users. |

When you review the visual elements, you can apply the quality characteristics throughout the following steps:

- Prepare to review.
- Get an overview.
- Review individual visual elements.
- Summarize your findings.
- Confer with the writer or editor or both.

## Preparing to review

Familiarize yourself with the plan for the information and with the user interface. Get copies of the product and information design specifications and style guidelines that the writer used.

Before you begin your review, you should know the following information:

- Type of information, such as embedded assistance or topics
- Delivery medium or media, such as embedded assistance, EPUB or other softcopy format, help, wizard, multimedia
- Target audience, their tasks and experience level
- Visual design specifications and guidelines
- What style guidelines apply
- What changes are to be made
- Plans for translation into other languages
- Ways for the user to interact with the product

Putting it all together

If your review is to happen along with an edit, schedule your review to follow the edit.

**Getting an overview**

For your first pass through the information, look at it without actually reading the text. Step through the information, noting both positive and negative impacts of its presentation. You can supplement the visual effectiveness checklist with the following list:

- Are pages visually balanced?
- Is white space used effectively? If the information is to be translated, can the white space accommodate the longer text that some languages require?
- Are task lists obvious? Is their sequence clear?
- Is the information free of visual clutter? Note, for example:
    - Irrelevant graphics such as excessive screen captures
    - Unnecessary lines and symbols
    - Overabundant highlighting
    - Distracting background patterns or animations
- Are the typeface and size that are used for text in each element legible and consistent?
- Are elements presented consistently and according to style guidelines? Check, for example, the following visual elements:
    - Lists
    - Tables
    - Programming syntax
    - Headings
    - Highlighting
    - Rules
    - Spacing
- Are all illustrations rendered in a consistent style? Do the illustrations look as if the same illustrator did them? Check, for example, the following visual elements:

- Line characteristics
- Fonts and font sizes
- Color
- Shading and contrast
- Fill patterns
* Are the visual metaphors in the information the same as those in the product's user interface?
* If color, shading, or a patterned background is used, does it enliven and enhance the information without being overpowering or distracting? Are color choices logical and consistent? If text is placed on top of these graphic elements, is the text legible?
* Is the information broken into manageable chunks?
* Are visual retrievability aids used effectively? Check, for example, the following visual elements:
  - Labels and headings
  - Rules and other separators
  - Captions
  - Icons
  - Color or shading
  - Running heads and running feet
* Is the organization and flow of the information clear at a glance?

When you review online information, consider also the following questions:

- Can visual elements be viewed in all supported resolutions?
- Are the visual elements usable in the supported resolutions?
- Are backgrounds simple and unobtrusive?
- Are links clearly identified and active?
- Are navigational aids consistent in design and placement?
- Are navigational aids obvious?
- Are graphics accompanied by alternate text that is meaningful for visually impaired users?

Putting it all together

When you review multimedia presentations, consider also the following questions:

- Are all windows laid out consistently?
- Is a consistent, clear color scheme used throughout?
- Do visual transitions all use the same one or two methods?
- Is text sufficient without graphics or audio?

## Reviewing individual visual elements

Go through the information again from the beginning, focusing on individual visual elements that you feel need attention. This time you might read the text that surrounds graphics and illustrations, especially if the text has been revised.

If an edit was done, review the editor's comments on visual elements and design. Note whether you agree and, if you disagree, why. At places where the editor has suggested adding or changing a visual element, read the surrounding text and make recommendations, if possible, on the design of the visual element.

Consider these issues as you focus on individual elements:

- Are illustrations meaningful and appropriate? Have the graphics been updated to match the revised text?
- Are the graphics legible and clear in all presentation media and formats?
- In icons:
  - Are the visual metaphors that are used in the icons unambiguous and meaningful? If the information will be translated, are the icons universal, so that people of other cultures will understand them?
  - If you have seen these icons before, is their meaning here consistent with their earlier use?
  - If the icons show different states (such as running and stalled), is the meaning of each state clear?
- In images and toolbars:
  - Are the images that are used for toolbar actions clear?
  - Are common images used for common actions (for example, scissors for "cut")?
  - Are the images universal and not offensive to other cultures?
  - Is the meaning of toolbar buttons intuitive? Are they supplemented with tooltips so that visually impaired users can understand them?

- In animations or video clips:
  - Are the animations smooth?
  - If the animations continuously loop, can users stop them? Can users easily figure out how to stop them?
  - Can users rerun animations that run once and stop? Can users easily figure out how to rerun them?
  - If the animations include text, can the text be read easily? Is the text presented long enough for users to read it? Is any text displayed intermittently, so that it blinks or flashes? Such text can be both annoying and unreadable.

You might find other visual elements that need a focused review, such as the icons in the user interface if you did not design them.

## Summarizing your findings

Gather your significant concerns and comments into a report. Note the quality characteristic that you think is most seriously affected by each problem. The ultimate impact of a visual effectiveness problem is often to one of the other quality characteristics described in this book, such as clarity, accuracy, or organization.

If you identified areas where you disagree with the editor, mention these in your report.

As much as possible, recommend solutions to problems that you identified. It helps to be familiar with the tools that were used to develop the information and its visual components. You can then recommend solutions that are within the capabilities of the tools.

## Conferring with the editor or writer or both

If you did your review as part of an edit, give your comments and summary to the editor to compile into a summary report.

Schedule a meeting with the writer or both the writer and the editor to discuss the results of your visual review. If an illustrator or another designer was involved in developing the visual elements, include that person in the meeting, too. At the meeting, reach agreement among all participants on the changes that will be made in the visual elements.

# PART 6

# Appendixes

| | |
|---|---|
| Appendix A. Quality checklist | 545 |
| Appendix B. Who checks which characteristics? | 549 |
| Glossary | 555 |
| Resources and references | 565 |
| Index | 573 |

# APPENDIX A

# Quality checklist

After you edit or review technical information, you can use this quality checklist to summarize your ratings for each quality characteristic. If you used the checklist at the end of each chapter to keep track of your findings, you can pull together your ratings here. You can then get an overall picture of the strengths and weaknesses of the information and make a plan for working on the weaknesses.

**Rating scale:**

1. Very dissatisfied: Among the worst.
2. Dissatisfied: Clearly subpar.
3. Neither satisfied or dissatisfied: OK—par for the course; overall, neither praiseworthy nor blameworthy.
4. Satisfied: Solid, professional work.
5. Very satisfied: Among the best; could be used as a model.

# Appendices

| Quality characteristic and guidelines | Quality rating |
|---|---|
| **Easy to use** | |
| **Task orientation** | **1 2 3 4 5** |
| Information is appropriate for the intended audience. | 1 2 3 4 5 |
| Information is presented from the users' point of view. | 1 2 3 4 5 |
| Tasks support users' goals. | 1 2 3 4 5 |
| Unnecessary focus is not given to product features. | 1 2 3 4 5 |
| A practical reason for the information is evident. | 1 2 3 4 5 |
| Step-by-step instructions are clear. | 1 2 3 4 5 |
| Steps are grouped for usability. | 1 2 3 4 5 |
| **Accuracy** | **1 2 3 4 5** |
| Subjects have been adequately researched. | 1 2 3 4 5 |
| Information has been verified. | 1 2 3 4 5 |
| Information is current. | 1 2 3 4 5 |
| Information about all subjects is consistent. | 1 2 3 4 5 |
| Tools have been used to check the accuracy of information. | 1 2 3 4 5 |
| **Completeness** | **1 2 3 4 5** |
| User interfaces are self-documenting. | 1 2 3 4 5 |
| Information is disclosed according to a pattern. | 1 2 3 4 5 |
| All subjects that support users' goals are covered, and only those subjects. | 1 2 3 4 5 |
| Each subject has just the detail that users need. | 1 2 3 4 5 |
| Information is repeated only when needed. | 1 2 3 4 5 |
| **Easy to understand** | |
| **Clarity** | **1 2 3 4 5** |
| The focus is on the meaning. | 1 2 3 4 5 |
| Text is free from wordiness. | 1 2 3 4 5 |
| Elements flow from one to another, and paragraphs stay on point. | 1 2 3 4 5 |
| The information is free from ambiguity. | 1 2 3 4 5 |
| Technical terms are used consistently and appropriately. | 1 2 3 4 5 |

# Quality checklist

| Quality characteristic and guidelines | Quality rating |
|---|---|
| **Concreteness** | 1 2 3 4 5 |
| Information is appropriate to the skill level and needs of users. | 1 2 3 4 5 |
| Concreteness elements are appropriate for the information type. | 1 2 3 4 5 |
| Concreteness elements are focused, realistic, and up to date. | 1 2 3 4 5 |
| Scenarios illustrate tasks and provide overviews. | 1 2 3 4 5 |
| Code examples and samples are easy to use. | 1 2 3 4 5 |
| Examples and scenarios are presented in context. | 1 2 3 4 5 |
| Similes and analogies are used to relate unfamiliar information to familiar information. | 1 2 3 4 5 |
| Language is specific. | 1 2 3 4 5 |
| **Style** | 1 2 3 4 5 |
| Passive and active voice is used appropriately. | 1 2 3 4 5 |
| The tone is appropriate for the audience. | 1 2 3 4 5 |
| The information is free from bias. | 1 2 3 4 5 |
| Terms are spelled correctly. | 1 2 3 4 5 |
| Terms are correctly capitalized. | 1 2 3 4 5 |
| Punctuation is consistent and appropriate. | 1 2 3 4 5 |
| Highlighting is consistent. | 1 2 3 4 5 |
| Templates are used consistently. | 1 2 3 4 5 |
| Common expressions are reused when possible. | 1 2 3 4 5 |
| Markup tagging is correct and consistent. | 1 2 3 4 5 |
| **Easy to find** | |
| **Organization** | 1 2 3 4 5 |
| Contextual information is separated from other types of information. | 1 2 3 4 5 |
| Contextual information is in the appropriate type of embedded assistance. | 1 2 3 4 5 |
| Noncontextual information is separated into discrete topics by type. | 1 2 3 4 5 |
| Elements are arranged to facilitate navigation. | 1 2 3 4 5 |
| Tasks are organized sequentially. | 1 2 3 4 5 |
| Elements are organized consistently. | 1 2 3 4 5 |

# Appendices

| Quality characteristic and guidelines | Quality rating |
|---|---|
| Users can see how the pieces fit together. | 1 2 3 4 5 |
| Emphasis and subordination are appropriate. | 1 2 3 4 5 |
| **Retrievability** | **1 2 3 4 5** |
| Content is optimized for searching and browsing. | 1 2 3 4 5 |
| Paths guide users through the information. | 1 2 3 4 5 |
| Links are appropriate. | 1 2 3 4 5 |
| Entry points are helpful. | 1 2 3 4 5 |
| **Visual effectiveness** | **1 2 3 4 5** |
| Visual design practices are applied to textual elements. | 1 2 3 4 5 |
| Document elements and structures are used appropriately. | 1 2 3 4 5 |
| Font and font style variations are limited. | 1 2 3 4 5 |
| Graphics are meaningful and appropriate. | 1 2 3 4 5 |
| Screen captures are used judiciously. | 1 2 3 4 5 |
| A consistent visual style is applied. | 1 2 3 4 5 |
| Visual elements help users find what they need. | 1 2 3 4 5 |
| Visual elements are accessible to all users. | 1 2 3 4 5 |

APPENDIX B

# Who checks which characteristics?

Writers have responsibility for checking all of the items that affect the quality of technical information. Technical editors have a similar responsibility except that they usually are not responsible for verifying the accuracy of the information. Peer editors (or copy editors in some situations) typically edit near the end of the development cycle; they need to check certain aspects of all of the quality characteristics.

The more limited responsibility of technical reviewers (people skilled in the content of the information) spans several quality characteristics—mainly accuracy, completeness, concreteness, and visual effectiveness. Sometimes technical reviewers are asked to check certain aspects of clarity, organization, and retrievability.

During usability testing, users (or people similar to users) provide information that mainly affects task orientation but also parts of all the other quality characteristics.

The work of testers mainly affects accuracy and concreteness.

In addition to checking all aspects of visual effectiveness, visual designers need to check certain aspects of clarity, concreteness, style, and retrievability.

# Appendices

| Quality characteristic and guidelines | Technical reviewers | Users | Testers | Technical editors | Visual designers | Peer editors |
|---|---|---|---|---|---|---|
| **Easy to use** | | | | | | |
| **Task orientation** | | | | | | |
| Information is appropriate for the intended audience. | X | X | | X | X | |
| Information is presented from the users' point of view. | | X | | X | X | |
| Tasks support users' goals. | X | X | X | X | X | |
| Unnecessary focus is not given to product features. | | X | | X | | X |
| A practical reason for the information is evident. | | X | | X | | |
| The step-by-step instructions are clear. | X | X | X | X | | X |
| Steps are grouped for usability. | | X | | X | | X |
| **Accuracy** | | | | | | |
| Subjects have been adequately researched. | X | | | X | X | |
| Information has been verified. | X | | | X | | |
| Information is current. | X | X | X | X | | X |
| The information about all subjects is consistent. | X | X | | X | | X |
| Tools have been used to check the accuracy of information. | | | X | X | | X |
| **Completeness** | | | | | | |
| User interfaces are self-documenting. | X | X | X | X | | X |
| Information is disclosed according to a pattern. | | X | | X | | X |

# Who checks which characteristics?

| Quality characteristic and guidelines | Technical reviewers | Users | Testers | Technical editors | Visual designers | Peer editors |
|---|---|---|---|---|---|---|
| All subjects that support users' goals are covered, and only those subjects. | | X | | X | | |
| Each subject has just the detail that users need. | X | X | X | X | | |
| Information is repeated only when needed. | | X | | X | | |
| **Easy to understand** | | | | | | |
| **Clarity** | | | | | | |
| The focus is on the meaning. | | | | X | | X |
| Text is free from wordiness. | | | X | X | | X |
| Elements flow from one to another, and paragraphs stay on point. | | X | | X | | X |
| The information is free from ambiguity. | | X | X | X | | X |
| Technical terms are used consistently and appropriately. | X | X | X | X | | X |
| **Concreteness** | | | | | | |
| Information is appropriate to the skill level and needs of users. | X | X | | X | X | |
| Concreteness elements are appropriate for the information type. | | | X | X | | X |
| Concreteness elements are focused, realistic, and up to date. | X | | X | X | | X |
| Code examples and samples are easy to use. | X | X | | X | | X |
| Scenarios illustrate tasks and provide overviews. | | | | X | | X |

# Appendices

| Quality characteristic and guidelines | Technical reviewers | Users | Testers | Technical editors | Visual designers | Peer editors |
|---|---|---|---|---|---|---|
| Examples and scenarios are presented in context. | | | X | X | | X |
| Similes and analogies are used to relate unfamiliar information to familiar information. | X | X | | X | | |
| Language is specific. | | X | | X | | |
| **Style** | | | | | | |
| Passive and active voice is used appropriately. | | | | X | | X |
| The tone is appropriate for the audience. | | X | | X | | X |
| The information is free from bias. | | | | X | | X |
| Terms are spelled correctly. | | | | X | | X |
| Terms are capitalized correctly. | | | | X | | X |
| Punctuation is consistent and appropriate. | | | | X | | X |
| Highlighting is consistent. | | | | X | | X |
| Templates are used consistently. | | | | X | X | X |
| Common expressions are reused when possible. | | | | X | | X |
| Markup tagging is correct and consistent. | | | | X | X | X |
| **Easy to find** | | | | | | |
| **Organization** | | | | | | |
| Contextual information is separated from other types of information. | | | | X | | |
| Contextual information is in the appropriate type of embedded assistance. | X | X | X | X | | |

# Who checks which characteristics?

| Quality characteristic and guidelines | Technical reviewers | Users | Testers | Technical editors | Visual designers | Peer editors |
|---|---|---|---|---|---|---|
| Noncontextual information is separated into discrete topics by type. | | X | | X | | |
| Elements are arranged to facilitate navigation. | | X | | X | X | X |
| Tasks are organized sequentially. | X | X | X | X | | |
| Elements are organized consistently. | | X | | X | X | X |
| Users can see how the pieces fit together. | | X | | X | X | |
| Emphasis and subordination are appropriate. | X | | | X | X | |
| **Retrievability** | | | | | | |
| Titles are clear and descriptive. | X | X | X | X | | |
| Keywords are used effectively. | | | X | X | | X |
| The table of contents is optimized for scanning. | | X | X | X | X | X |
| Users are guided through the information. | | X | X | X | | X |
| Links are appropriate. | X | X | X | X | | X |
| Entry points are helpful. | | X | | X | X | X |
| **Visual effectiveness** | | | | | | |
| Visual design practices are applied to textual elements. | | | | X | X | |
| Document elements and structures are used appropriately. | | | | X | X | X |
| Font and font style variations are limited. | | | X | X | X | |
| Graphics are meaningful and appropriate. | X | X | | X | X | |

Appendices

| Quality characteristic and guidelines | Technical reviewers | Users | Testers | Technical editors | Visual designers | Peer editors |
|---|---|---|---|---|---|---|
| Screen captures are used judiciously. | | | | X | X | X |
| A consistent visual style is applied. | | | | X | X | X |
| Visual elements help users find what they need. | X | X | | X | X | X |
| Visual elements are accessible to all users. | | X | X | X | X | X |

# Glossary

**accessibility.** In information technology, accommodations to enable people with disabilities (such as those that affect vision, hearing, mobility, and attention) to use computer software and hardware.

**accuracy.** Freedom from mistake or error; adherence to fact or truth.

**active voice.** A grammatical form in which the doer of the action is expressed as the subject rather than as an agent or not expressed at all. For example, the following sentence uses active voice: "Administrators can update the files." Contrast with *passive voice*.

**agile development.** An iterative approach to product development in which the product evolves and improves as a result of continuous analysis and testing throughout multiple development cycles. Team members from all involved disciplines participate in the entire development process and target users are consulted regularly. Contrast with *waterfall development*.

**alternative text.** A textual description of an image or icon that conveys the same essential information as the image. In web pages, alternative text is often inserted in the alt attribute of an <image> element. Alternative text is typically added to meet accessibility requirements.

**analogy.** An explicit comparison based on a resemblance between things that are otherwise not alike. See also *metaphor*, *simile*.

**animation.** The motion that is simulated by displaying a series of images. Multimedia presentations often include animation.

**antecedent.** A word, phrase, or clause that a pronoun refers to. See also *vague referent*.

**audience.** A group of users who have similar tasks and a similar background; they might have differing expertise in doing these tasks.

**audience analysis.** An investigation of user characteristics such as education and training,

# Glossary

level of experience with the subject or with related subjects, and work environment.

**balance.** A pleasing arrangement of visual elements, or the act of integrating visual elements in an effective way. See also *visual element*.

**boilerplate.** Any standardized text that must be used consistently across documents and products. Legal notices are an example of boilerplate.

**breadcrumb trail.** A set of links that shows the hierarchy of the current page or topic. Such links facilitate navigation to other pages in the hierarchy.

**bulleted list.** See *unordered list*.

**cascading style sheet (CSS).** A style sheet language that is used for describing the appearance and formatting of a document that is written in a markup language, such as HTML or XML. See also *style sheet*.

**circularity.** A logic error in which the evidence that supports a statement is basically the same as the statement itself.

**clarity.** Freedom from ambiguity or obscurity; the presentation of information in a way that helps users understand it the first time.

**code example.** One or more lines of code, in any programming language, that users can use as is or can adapt to their own situation.

**coherence.** Characterized by a readily understandable flow and logic of information from sentence to sentence, from paragraph to paragraph, and from text to a graphic or table and back; adherence to a single idea in a paragraph or topic.

**colloquialism.** An informal expression that is more often used in casual conversation than in formal speech or writing. For example, the phrase *a can of worms* to mean a source of trouble or complexity is a colloquialism. See also *idiom*, *jargon*, *slang*.

**completeness.** The inclusion of all necessary parts—and only those parts.

**conceptual information.** Information that pertains to design, ideas, relationships, and definitions. See also *reference information*, *task information*.

**concreteness.** The inclusion of appropriate examples, scenarios, similes, analogies, specific words, and graphics.

**conditional step.** In a procedure, a step that users follow only if certain criteria are met, such as having a certain fix pack or service agreement. Conditional steps usually begin with the word *If*, as in, "If you are using Linux, ..." Contrast with *optional step*.

**conditional text.** Designated text that is included or excluded based on conditions that the writer specifies. This type of text helps enable the reuse of text in multiple contexts.

**container.** In a topic hierarchy, a topic that gathers together related subtopics. A container can have substantive information itself (often an overview of the subtopics) or can serve only as a heading in the hierarchy.

**contextual information.** Information that explains the piece of the product that users are working with. Contextual information makes sense only when users are interacting with the product. For software products, contextual information is provided in the user interface and might explain product windows, fields, and controls. Contrast with *noncontextual information*.

**control.** See *interface control*.

**control-level assistance.** A type of embedded assistance that provides information about how to work with a specific user interface control such as a field or radio button. One common method for delivering control-level assistance is hover help.

**coordination.** The grammatical connection of two or more words, phrases, or clauses that gives each equal emphasis and importance. For example, the following sentence uses two coordinated clauses joined by *or*: "You can reuse the original metal gasket, or you can replace it with a rubber gasket."

**cross-reference.** See *link*.

**dialog.** In a software graphical user interface, a type of window that is used to enable reciprocal communication or interaction between a computer and its user. Also referred to as *dialog box*.

**DITA.** Darwin Information Typing Architecture (DITA), a standard for authoring topic-based information in individual XML files. Each topic covers a discrete subject, for example, a task topic for a user task or a conceptual topic for an overview or technical concept.

**domain expertise.** Accumulated practical knowledge about a subject area, for example, international banking, nursing, or software security.

**elliptical style.** A style of writing that eliminates words or phrases that are implied by context. For example, this sentence omits the definite article *the* before both noun phrases: "Connect fuel hose to fuel tank petcock."

**embedded assistance.** Labels, messages, and other text in a hardware or software user interface. In software user interfaces, embedded assistance can also include programmatic assistance.

**entry point.** An element that marks the presence and location of information. An entry point can be, for example, a heading, field label, link, or table of contents.

**evaluation test.** A type of usability test that is done early in a development cycle to determine whether a design is likely to meet usability criteria if fully implemented. Contrast with *usability walkthrough, validation test*.

**example.** A representative of a set of related things. In technical writing, an example can be a piece of code, a graphic, or text. See also *sample*.

**experience level.** A characteristic of users that deals with their first-hand knowledge of the subject as opposed to their education and training.

**expert user.** A person whose experience and training makes him or her comfortable using a tool and able to handle most tasks. Contrast with *novice user*. See also *user*.

**field test.** A test of a product in an actual or simulated work environment before the product is released.

**figure.** An illustration or graphic that is inserted into a block of textual information to complement or clarify the information. Figures usually, but not always, include captions. See also *illustration*.

**follow-on task.** See *postrequisite*.

**front-loaded.** Pertaining to the placement of keywords at the start of a title. When keywords and other important terms come at the start of a title, the title is front-loaded.

**function-oriented task.** A task that is presented in terms of using a product component, service, or technology. For example, *using the CNTREC utility* is a function-oriented task, but *counting records* is not. Contrast with *user task*.

**gender-neutral language.** Words, such as the pronoun *they* and the noun *person*, that do not specify a gender.

**gerund.** A word that ends in *-ing* and functions as a noun. A gerund can have objects, complements, or modifiers.

557

# Glossary

**globalization.** The process of creating a product that can be used in many cultural contexts without modification. Such a product can process multilingual data and present culturally correct information (such as collation, date format, and number format). See also *internationalization*, *localization*.

**graphical element.** Any nontextual visual element that is intentionally added to text to enhance its visual impact. This term applies to graphics, illustrations, or visual organizers such as rules to separate running heads from text and bullets to present parallel items. See also *visual element*.

**graphical user interface (GUI).** A type of software user interface that takes advantage of high-resolution graphics. A typical graphical user interface combines graphics, object-action paradigm, pointing device, menu bars, menus, overlapping windows, and icons. See also *user interface*.

**heading.** A prominent phrase that describes the information after it. Contrast with *label*.

**help.** Discrete pieces of online information about elements of a product such as a window, field, message, or task. Also, the set of online information as a whole. See also *contextual information*.

**high-level step.** In a procedure, a step for a major action with a clear purpose. A high-level step typically requires lower-level actions. For example, "Pick up Joe at the airport" is a high-level step, but "Open the car door" is not. Contrast with *low-level step*.

**hot spot.** An area in an illustration that a user can hover over or click to display hover help or perform an action. See also *image map*.

**hover help.** A mechanism for displaying hover text. To view the help, users hover over a user interface control or an icon that specifically denotes that there is help available, for example, a ❓ icon. See also *hover text*.

**hover text.** A short description for user interface controls such as fields, check boxes, radio buttons, lists, and grids. Hover text is displayed in *hover help*.

**human factors engineer.** See *usability engineer*.

**icon.** A small graphic symbol on a computer screen that represents a tool, task, or thing, such as a file or a folder.

**idiom.** An expression whose meanings cannot be inferred from the meanings of the words that make it up. For example, the sentence "You should keep an eye out for performance problems" contains the idiom *keep an eye out for*. See also *colloquialism*, *jargon*, *slang*.

**illustration.** A created piece of artwork such as a drawing, chart, or diagram that is used to explain, depict, or complement text. See also *figure*.

**image map.** A graphic that contains more than one hot spot. See also *hot spot*.

**imperative mood.** A verb form that indicates a command, such as "Press Enter."

**information architecture.** The practice of applying principles of design and structure to support the usability and findability of information. Information architecture applies to the organization of an information environment, such as a website or online information system, to support navigation as well as to the use of metadata and indexing to support searching and browsing. See also *metadata*, *organization*, *usability*.

**information center.** A set of online topics presented with a navigation pane on one side and a content pane on the other side. Often, an information center includes a search engine and an index.

# Glossary

**information set.** All of the documentation for a product or set of products. An information set can include books, topics, videos, embedded assistance, articles, and more.

**information type.** A category of information that is based on its content and use. The major types that are discussed in this book are task, concept, reference, contextual, and messages. Other types include troubleshooting, tutorial, and wizard. Each type has a set of common qualities that can allow for the creation of templates or standards. See also *topic*.

**interaction design.** See *user interaction design*.

**interface.** The means by which users interact with hardware and software.

**interface control.** In an interface, a device such as a button, menu, or lever that a person uses to do tasks with the product.

**internationalization.** The design and development of information or of an application so that it can easily be modified to support different national languages and cultural conventions. See also *globalization*, *localization*.

**jargon.** Technical terminology or a characteristic idiom of a science, trade, profession, or similar group. See also *colloquialism*, *idiom*, *slang*.

**keyword.** A term that represents an important concept in a document or topic. Users also use keywords to search for content. Where and how keywords are applied in the information can greatly affect users' ability to find what they need.

**label.** (1) A heading or other textual name for a user interface control or section of a user interface. (2) A heading (usually one word such as Tip, Restriction, or Important) that governs a paragraph within a larger heading structure. Labels do not show up in a table of contents. See also *heading*.

**link.** A pointer to related information. Links provide the connections among pieces of information. When information is hypertext-enabled, users can click a link to go directly to the related information.

**localization.** The process of translating text and creating culturally specific settings for a software product. The process can include packaging the product into a national language version. See also *globalization*, *internationalization*.

**low-level step.** In a procedure, a step for a minor action, such as for clicking a button or filling in a field. Contrast with *high-level step*.

**machine translation.** The use of computer software to translate text from one human language to another. See also *translation memory*.

**message.** Information that is presented to the user that provides status, warns the user about a potential problem, or indicates that an error has occurred.

**metadata.** Information about information, which can be used to manipulate, filter, sort, or otherwise manage a set of information. In the context of retrievability, metadata consists of the keywords and synonyms that are included in the meta tag to help users find what they are searching for.

**metaphor.** A word, phrase, or visual representation that denotes or depicts one object or idea but suggests a likeness to or analogy with another object or idea. See also *analogy*, *simile*, *visual metaphor*.

**meta tag.** An HTML tag that provides information about an HTML page. The information is not displayed on the page but can be read by search engines. Some information from the meta tag, such as the description, might appear in search results.

**minimalism.** A set of principles for presenting instructional material that supports exploration,

## Glossary

uses real tasks, supports recovering from errors, and provides concise information.

**multimedia.** An application that uses any combination of text, image, animation, video, and audio effects to present information to the user.

**navigate.** To find the way to a destination. Mechanisms, such as links, scroll bars, tables of contents, and breadcrumbs help users navigate through user interfaces, tasks, or portions of information sets to complete tasks or find information.

**navigation pane.** A section of a window that contains links to content in an application or information set. See also *pane, table of contents*.

**negative expression.** A phrase that includes negative words such as *no, not, none, never,* or *nothing*.

**nominalization.** A word formation in which a verb or an adjective or other part of speech is used as or transformed into a noun. For example, *make a decision* is the nominalization of *decide*.

**noncontextual information.** Information that is delivered outside the product interface. Noncontextual information might include product overview information, installation information, and information about high-level tasks. Contrast with *contextual information*.

**nonrestrictive clause.** A descriptive clause that adds nonessential information about the noun that it modifies. For example, the following sentence includes a nonrestrictive relative clause that starts with *which*: "Hybrid segmentation, which supports dictionary-based segmentation, can be used for all bi-directional languages." Contrast with *restrictive clause*.

**noun string.** A set of two or more nouns that together identify one thing, for example, *torque converter* or *application programming interface*.

**novice user.** Someone who has little or no experience in a particular area and can require more help to complete a task. Contrast with *expert user*. See also *user*.

**numbered list.** See *ordered list*.

**optional step.** In a procedure, a step that users can choose to skip and still complete the task successfully. Optional steps usually begin with the word *Optional*. Contrast with *conditional step*.

**ordered list.** A set of items that are in sequential (numerical or alphabetical) order. Nested ordered lists usually use an alphabetical format within a numerical format, as in an outline. Sometimes called *numbered list*. Contrast with *unordered list*.

**organization.** A coherent arrangement of parts that makes sense to the user.

**pane.** A segment of a window that can be scrolled separately and that might have its own title. Selection of items in a pane can cause items in another pane to change. See also *navigation pane*.

**parallelism.** The use of the same grammatical structure for similar elements such as items in a list, nouns or adjectives in a series, and compound clauses.

**passive voice.** A grammatical form in which the doer of the action is expressed as an agent or not at all. For example, the following sentence uses passive voice: "Files can be updated by administrators." Contrast with *active voice*.

**pattern.** A form or model for including information in the appropriate elements. A pattern might identify the type of information to provide in hover help, in static text, in a user interface, or in a separate help topic. See also *template*.

**point.** A standard unit of type size. The British and American standard is 72 points to an inch.

**pop-up help.** See *hover help, tooltip*.

**postrequisite.** A task that users need to do after they do a different task. Sometimes called *follow-on task*. Contrast with *prerequisite*.

**prerequisite.** A requirement that is needed before a user can do a task. Contrast with *postrequisite*.

**primary audience.** The group of people who are expected to use a particular part of the documentation most often. An audience is grouped by a set of defining characteristics such as occupation or experience level. See also *secondary audience*.

**procedure.** A set of steps in a definite order (though some steps can be conditional or optional) for a user to do a task that is well defined, such as creating index entries or replacing a piece of hardware. See also *process, task*.

**process.** A series of events, stages, or phases that lead to a particular output or result. The parts of a process must usually be in a definite order and often involve more than one person. Some processes are carried out only by a machine or computer. See also *procedure*.

**programmatic assistance.** A type of embedded assistance that programmatically does a step or assists a user. Examples of programmatic assistance include default, detected, or autocompleted values. See also *embedded assistance*.

**progressive disclosure.** An interaction design technique to defer complexity and provide it on user request. When applied to documentation, this technique begins with information that is embedded in the user interface. It then provides links to additional information either within or outside of the product.

**prototype.** A mock-up that helps developers experiment with a design for a product, a system, or an interface. A low-tech (or low-fidelity) prototype is made quickly out of temporary materials or is sketched quickly.

**quality.** Excellence or superiority in kind.

**quality characteristic.** An attribute that is essential to describing quality.

**real task.** See *user task*.

**redundancy.** Repetition in a phrase where two or more words are synonymous or almost synonymous, as in *repeat again*; content that is repeated and therefore unnecessary.

**reference information.** A collection of terms, commands, syntax elements, rules, messages, parts, components, or conventions that is organized for quick retrieval. See also *conceptual information, task information*.

**relative clause.** A clause that is introduced by a relative pronoun such as *who, that*, or *which*. For example, in the sentence "You can upgrade the database that stores personnel files," *that stores personnel files* is a relative clause.

**resolution.** The dimensions in pixels of the display area of a computer monitor.

**restrictive clause.** A descriptive clause that is essential to the meaning of the noun that it modifies. Contrast with *nonrestrictive clause*.

**retrievability.** The presentation of information in a way that enables users to quickly and easily find specific items.

**reuse.** To use the same piece of information repeatedly or in multiple places with few or no changes, perhaps for a different document or a product. See also *single source*.

**roundabout expression.** A phrase that uses more words than necessary, such as *in the course of* instead of *during*.

# Glossary

**rule.** A line of any thickness, or weight, that serves as a nontextual visual element. Such a line can, for example, set off rows in a table, surround a figure, or separate text from a running head.

**sample.** A set of code or data that is provided with a software product to help users try out the product or parts of it. See also *example*.

**scenario.** A series of events over time, usually involving a fictitious but realistic set of circumstances.

**scenario-based information.** Information that is designed and written to support the paths that users take to accomplish their key goals.

**screen capture.** A graphic likeness of a screen or a portion of the screen for inclusion in technical information. Sometimes called *screen shot*.

**screen reader.** An assistive technology device that provides speech output (and sometimes Braille output) while a visually impaired person uses a software program.

**search engine.** Software that acts on user input to find likely matches for what the user is seeking. The space being searched can be a page, a document, a large set of items, or the entire web. Each search engine uses its own algorithm to create and return the results of a query.

**secondary audience.** The group of people who are expected to use a particular part of the documentation less often than the primary audience does. An audience is grouped by a set of defining characteristics such as occupation or experience level. See also *primary audience*.

**simile.** An explicit comparison (often introduced by *like* or *as*) of two things to show a similarity that they share. See also *analogy*, *metaphor*.

**single source.** An approach to producing information in which the same content is used in more than one document or media type. For example, the same content can be used in two user's guides: one for the Linux version of a product and one for the UNIX version of the same product. See also *reuse*.

**site map.** A hierarchical rendering of the pages of a website as links to help users find information. See also *table of contents*.

**slang.** Words and phrases that are regarded as very informal, are more common in speech or social media than writing, and are typically restricted to a particular context, point in time, or group of people. See also *colloquialism, idiom, jargon*.

**static text.** In a user interface, words or sentences that provide information to help users understand what to do on a window or in a field. Unlike hover help, static text is always displayed. Static text is often placed at the top of windows and dialogs or near fields and controls. See also *embedded assistance*.

**style.** (1) Correctness and appropriateness of writing conventions and of words and phrases. (2) A convention with respect to spelling, punctuation, capitalization, or typographic arrangement and display.

**style guide.** A collection of rules and guidelines to establish what usage is correct and incorrect or just preferred for the sake of consistency. Usually, a style guide contains many style guidelines and examples. Sometimes called *style manual*. See also *style guideline, style sheet*.

**style guideline.** A directive about what to do in a specific situation that affects the appearance, consistency, or structure of information. See also *style guide, style sheet*.

**style sheet.** A definition of the appearance of a document in terms of typefaces, fonts, spacing,

# Glossary

tagging, and other visual elements. See also *cascading style sheet, style guide, style guideline.*

**subject.** (1) The part of a clause on which the rest of the clause is predicated. (2) The area of concern in a topic or document. See also *topic.*

**subtask.** A portion of a larger task that is broken into a discrete subsection. A subtask has enough information (such as rationale and steps) that it makes sense as a discrete topic. See also *procedure, task, user task.*

**syntactic cue.** A grammatical element of language (such as word order, verb endings, articles, and conjunctions) that helps users correctly understand the structure of a sentence.

**table of contents.** A structured list of headings in an information set to help users find information. A table of contents has a link (or page number) for each heading. See also *navigation pane, site map.*

**tag cloud.** For online content, a visual presentation of keywords entered as tags by users, authors, or both.

**target.** The destination for a link or cross-reference.

**task.** A physical or mental activity. See also *procedure, subtask, user task.*

**task analysis.** The process of determining primarily who does a task and how, when, where, and why.

**task information.** Information that makes clear how to do a task. It includes procedures and can also include items such as examples and tips. See also *conceptual information, reference information.*

**task orientation.** A focus on helping users do tasks that support their goals.

**template.** A document, file, or other resource that serves as a pattern for creating a similar thing. See also *pattern.*

**tooltip.** One- to two-word names for tools that do not have permanent labels in the interface and that appear when users hover over the icon for the tool. Contrast with *hover text* and *hover help.* See also *interface control.*

**tone.** A manner of expression in writing that conveys the viewpoint of the writer. In technical writing, a tone is more likely to be serious and authoritative than humorous or condescending.

**topic.** A unit of information about a particular subject. Sometimes called *article* or *module.* See also *information type, subject.*

**topic model.** A type of outline that shows the structure of topic-based information. See also *topic set.*

**topic set.** A structured group of related topics that are delivered together and usually cover a user task or user goal. For example, a product might have several topic sets: one for installation instructions, one for troubleshooting content, one for each main user goal, and one for command reference information.

**translation memory.** The software that stores matching segments of a source language and a target language as translated by a human translator. See also *machine translation.*

**typeface.** A complete set of letters, numbers, and punctuation of a particular design, such as Helvetica or Times Roman.

**UI control.** See *interface control.*

**unordered list.** A set of items that are arranged in an order that does not require a sequential (numerical or alphabetical) format. Sometimes called *bulleted list.* Contrast with *ordered list.*

# Glossary

**usability.** The ease with which people in a defined group can learn and use a product (including the information) to accomplish certain tasks.

**usability engineer.** A person who specializes in researching, designing, developing, and validating the usability of a product. Sometimes called *human factors engineer* or *user experience designer*. See also *usability, usability test, validation test*.

**usability test.** A test of the interaction between the user and the product or information for the product. See also *evaluation test, usability walkthrough, validation test*.

**usability walkthrough.** A usability inspection approach in which users follow a typical scenario to determine the ease of learning and using a product interface. See also *usability*. Contrast with *evaluation test, validation test*.

**user.** A person who does any of the following tasks with a product: evaluating, buying, planning, learning, installing, migrating, using, operating, administering, developing applications, customizing, diagnosing, or maintaining. See also *expert user, novice user*.

**user experience designer.** See *usability engineer*.

**user goal.** An outcome that users want, such as a working bicycle or an optimally performing database. Achieving a goal generally involves doing multiple tasks. See also *user task*.

**user interaction design.** The professional discipline of designing product interfaces that are easy to use. See also *user interface design*.

**user interface.** The area where a user and an object come together to interact; as applied to computers, the ensemble of hardware and software that enables a user to interact with a computer program. See also *graphical user interface (GUI)*.

**user task.** A task that users want to perform regardless of the product or method they use to do it. For example, setting up a bank account is a user task, but filling out the web form to do so is not. Sometimes called *real task, user-oriented task*. See also *user goal*. Contrast with *function-oriented task*.

**vague referent.** The noun, phrase, or clause that is referred to by a pronoun in such a way that it is difficult to determine which noun, phrase, or clause is being referred to. See also *antecedent*.

**validation test.** A type of usability test that is done late in a development cycle to determine whether an implementation meets usability criteria. Contrast with *evaluation test, usability walkthrough*.

**visual effectiveness.** Attractiveness and enhanced meaning of information through use of layout, illustrations, color, type, icons, and other graphical devices.

**visual element.** Any component that affects the visual impact of the information. A visual element could be, for example, highlighting a heading, table, or illustration. See also *graphical element*.

**visual metaphor.** An implied likeness between an image or images and familiar objects or concepts in a particular domain. See also *metaphor*.

**walkthrough.** See *usability walkthrough*.

**waterfall development.** A development methodology in which products are developed to meet established specifications, and the specifications ideally do not deviate throughout the development process. Waterfall development follows a sequential path in which each participating team completes their work and hands it off to the next team. Contrast with *agile development*.

**white space.** The portion of a printed page or online window that is blank.

**wizard.** A type of interface that leads users through well-defined steps to produce a result, such as creating a chart from information in a database.

# Resources and references

## Easy to use

Android Open Source Project. *Android Design.* Available at developer.android.com.

> The design guidelines include a topic on writing style, which advocates a concise, simple, and friendly style.

Apple Inc. *iOS Human Interface Guidelines.* Available in the iOS Developers Library.

> The "Terminology and Wording" section of these guidelines says that every word in an app is part of a conversation.

Carroll, John M. *Minimalism beyond the Nurnberg Funnel.* Cambridge, MA: The MIT Press, 1998.

> Both practitioners and academicians wrote papers for this book to take stock of minimalism. Three papers are authored or coauthored by John Carroll. In one article, he presents four minimalist principles: choose an action-oriented approach; anchor the tool in the task domain; support error recognition and recovery; and support reading to do, study, and locate.

Carroll, John M. *The Nurnberg Funnel: Designing Minimalist Instruction for Practical Computer Skill.* Cambridge, MA: The MIT Press, 1990.

> Carroll, whose usability research at IBM in the early 1980s first showed the efficacy of minimalism for training materials, presents an approach that supports the natural tendency of many new users to learn through trial and error.

Resources and references

Hackos, JoAnn T., and Janice C. Redish. *User and Task Analysis for Interface Design*. New York: John Wiley & Sons, 1998.

> This in-depth treatment of user and task analysis includes how to apply the analysis results to documentation and training.

Mike Beedle et. al. "Manifesto for Agile Software Development." www.agilemanifesto.org/.

> This website summarizes the 12 principles for agile software development.

McKay, Everett N. *UI is Communication: How to Design Intuitive, User Centered Interfaces by Focusing on Effective Communication*. Boston, MA: Morgan Kaufmann Publishers, 2013.

> This book offers a simple perspective to user interface design: a user interface is essentially a conversation between a user and the product. McKay's approach evaluates a user interface by how easily it communicates with its users to help them complete tasks. Scenarios and human communication rather than features and requirements should drive design.

Microsoft Corporation. *Windows User Experience Interaction Guidelines for Windows Desktop apps*. Available on microsoft.com.

> Microsoft provides detailed guidelines on style and tone.

Snyder, Carolyn. "Docs in the Real World." User Interface Engineering. Last modified November 1, 1998. www.uie.com/articles/documentation/.

> This article describes the findings of usability consultants at User Interface Engineering when they observed real users as they used technical information in their jobs.

# Easy to understand

Baugh, Sue L. *Essentials of English Grammar*. Third Edition. New York, NY: McGraw-Hill, 2005.

> This book is a quick reference for basic grammar, punctuation, capitalization, and tips for writing clear and concise information.

Bellamy, Laura, Michelle Carey, and Jenifer Schlotfeldt. *DITA Best Practices: A Roadmap for Writing, Editing, and Architecting in DITA*. Upper Saddle River, NJ: IBM Press, 2012.

> This book provides tips and guidelines for creating topic-based information in Darwin Information Typing Architecture (DITA). The authors explain the most commonly used DITA XML elements and show how to create consistent, semantically valid technical information. The book also covers linking, conditional coding, DITA maps, metadata, and content reuse.

Brogan, John A. *Clear Technical Writing*. New York, NY: McGraw-Hill, 1973.

> This book is out of print but well worth tracking down. It's packed with examples and exercises for learning how to eliminate "gobbledygook" from your writing. It focuses on major sources of unclear writing, such as redundancies, weak verbs, abstract nouns, showy writing, and improper subordination.

*The Chicago Manual of Style: The Essential Guide for Writers, Editors, and Publishers*. 16th Edition. Chicago, IL: University of Chicago Press, 2010.

> This style guide is a widely used general reference on matters of academic and technical style. It has chapters devoted to topics such as punctuation, numbers, and indexes. The latest edition includes information about electronic publishing and a primer on XML. You can consult the Chicago Manual FAQ at www.press.uchicago.edu/Misc/Chicago/cmosfaq/cmosfaq.html. Goldstein, Norm, editor.

Crivello, Roberto. Concision in Technical Translations from English into Italian. *ATA Chronicle*, October 2000. robertocrivello.com/concision.

> Crivello's article describes the problems in English-to-Italian translations when the English text contains redundancies, unnecessary repetition, and wordiness. His article provides tips for translators to create more effective and less wordy Italian translations. Such tips should also be heeded by English content authors.

DeRespinis, Francis, Peter Hayward, Jana Jenkins, Amy Laird, Leslie McDonald, Eric Radzinski. *The IBM Style Guide: Conventions for Writers and Editors*. Upper Saddle River, NJ: IBM Press, 2012.

> Available only within IBM for over 30 years, the recent publication of these extensive guidelines for hardware and software information is a comprehensive guide to writing content that is consistent, concise, and easy to translate. The book demonstrates each guideline with correct and incorrect examples. In addition to the usual style topics, it includes chapters on indexing and writing for diverse audiences, as well as an appendix showing which DITA elements to use for highlighting.

Hacker, Diana, and Nancy Sommers. *A Writer's Reference*. Seventh Edition. Boston, MA: Bedford/St. Martin's, 2011.

> This handbook has chapters on composition and style, correctness, and format (including basic grammar).

Hughes, Mike. "User Assistance in the Role of Domain Expert." UXmatters. Last modified January 8, 2007. www.uxmatters.com/index.php.

> This article shows how usability testing identifies the information gap for domain expertise. Other articles on this website highlight ways that information and usability intersect.

Kohl, John R. *The Global English Style Guide: Writing Clear, Translatable Documentation for a Global Market*. Cary, NC: SAS Publishing, 2008.

> Packed with examples, this book provides excellent guidance on writing content that is clear for all audiences. Each guideline is accompanied by a ranking for its importance to human translation, machine translation, and nonnative readers of English.

Lynch, Patrick J., and Sarah Horton. *Web Style Guide: Basic Design Principles for Creating Web Sites*. Third Edition. New Haven, CT: Yale University Press, 2008.

> Based on the authors' experience with the Yale and Dartmouth websites, this style guide aims at improving the design of websites, including web pages, graphics, and multimedia. This guide includes chapters on information architecture and universal usability.

Microsoft Corporation. *Microsoft Manual of Style for Technical Publications*. Fourth Edition. Redmond, WA: Microsoft Press, 2012.

> This style guide for writers who document Microsoft products and technologies is arranged alphabetically. It has entries from "abbreviations and acronyms" to "zoom in, zoom out." This style guide often provides a rationale for its guidelines, emphasizing clarity and consistency, and can apply more broadly to the computer industry. Non-English versions are available to download at microsoft.com.

Tarutz, Judith A. *Technical Editing: The Practical Guide for Editors and Writers*. Reading, MA: Addison-Wesley, 1992.

> This book includes guidelines for creating a style guide and for editing with various purposes: editing for accuracy, editing for style, editing for usability, and editing for an international audience. One chapter deals with the 100 most common errors, grouped in the following categories: fuzzy thinking; style and usage; grammar and syntax; technical accuracy; judgment, taste, and sensitivity; typography and graphics; legal and ethical concerns; and proofreading oversights.

Williams, Joseph M., and Joseph Bizup. *Style: Lessons in Clarity and Grace*. Eleventh Edition. Boston, MA: Pearson.

> This book has a broader focus than just technical information, but its lessons about coherence, emphasis, and conciseness are valuable for technical writers. The book has many examples and an in-depth analysis of them.

World Wide Web Consortium. www.w3c.org.

> This website includes many standards, among them XML, XSL, and accessibility.

Zinsser, William. *On Writing Well: The Classic Guide to Writing Nonfiction*. 30th Anniversary Edition. New York, NY: HarperCollins, 2006.

> While focusing on how to write effective English, Zinsser offers principles (the transaction, simplicity, clutter, style, the audience, words, and usage) and methods (unity, the lead and the ending, bits and pieces). Primarily a journalist and a teacher of writing, Zinsser includes a chapter about writing for science and technology and several chapters about writing nonfiction.

# Easy to find

Lopuck, Lisa. *Web Design for Dummies*. Third Edition. Indianapolis, IN: John Wiley & Sons, 2012.

> This book includes practical advice for planning a website, designing the graphics, and testing the interface.

Mathewson, James, Frank Donatone, and Cynthia Fishel. *Audience, Relevance, and Search: Targeting Web Audiences with Relevant Content*. Upper Saddle River, NJ: IBM Press, 2010.

> This book is intended to help people such as marketers, copywriters, and information architects drive their target audiences to the website and keep them engaged. Its excellent description of how search engines work may become outdated, but the message about content quality is timeless.

Morville, Peter, and Jeffrey Callender. *Search Patterns: Design for Discovery*. Sebastopol, CA: O'Reilly Media, 2010.

> Filled with lively illustrations and plenty of food for thought, this book covers the design of user interfaces for search. It presents design patterns for everything from autocomplete to faceted navigation and unified discovery. The final chapter is dedicated to a discussion of "tangible futures."

Nielsen, Jakob, Nielsen Norman Group. www.nngroup.com/articles/.

> Since 1995, this site has offered monthly or bimonthly articles about web topics, most of which are based on Nielsen's usability studies. His articles on mobile intranet design (19 August 2013), progressive disclosure (4 December 2006), usability metrics (21 January 2001), and writing for mobile users (1 August 2011) are especially useful for technical writers.

Pfeiffer, William S., and Kaye E. Adkins. *Technical Writing: A Practical Approach*. Eighth Edition. Boston, MA: Pearson, 2013.

> This book has chapters about methods of organizing information, including induction and deduction.

## Resources and references

U.S. Department of Health and Human Services. "Research-Based Web Design and Usability Guidelines." www.guidelines.usability.gov.

> This website, updated in 2013, offers guidelines in areas such as navigation, search, and organization and cites the research to support each guideline.

Rosenfeld, Louis, and Peter Morville. *Information Architecture for the World Wide Web: Designing Large Scale Web Sites*. Third Edition. Sebastopol, CA: O'Reilly Media, 2006.

> This book emphasizes designing information from the user's perspective. It includes chapters about organizing complex content for delivery on the web, and designing systems for navigation, labeling, and search.

Smiciklas, Mark. *The Power of Infographics: Using Pictures to Communicate and Connect with Your Audience*. Indianapolis, IN: Que Publishing, 2012.

> This book describes how to transform complex data or concepts into information graphics that deliver ideas succinctly and help users take action quickly.

Weinman, Lynda. *Designing Web Graphics.4*. Fourth Edition. Indianapolis, IN: New Riders Publishing, 2003.

> This book provides a wealth of information for creators of graphics for the web, including motion graphics. It provides technical guidance about how to work within the constraints of varying web browsers, color palettes, and operating systems. Some guidance about good design of graphics, type, and page layout is included.

Welinske, Joe. *Developing User Assistance for Mobile Apps*. WritersUA, 2011.

> Many technical communication professionals will soon be developing assistance for mobile apps, if they aren't already. This book provides an introduction to mobile app user assistance and includes advice for iOS, Android, and Windows Phone.

## Putting it all together

Ames, Andrea, Alyson Riley, Deirdre Longo. "Mobile: The Crucible for Information Strategy." *Intercom: The Magazine of the Society for Technical Communication*, April 2012.

> This article introduces the critical methods for delivery of information in mobile environments.

Dumas, Joseph S., and Janice C. Redish. *A Practical Guide to Usability Testing*. Rev. Edition. Intellect Books, 1999.

> This book covers several methodologies for integrating usability testing into the development process. The focus is on overall product usability, including documentation.

Hackos, JoAnn T. *Content Management for Dynamic Web Delivery*. New York, NY: John Wiley & Sons, 2002.

> This book includes information about how to break linear content into more flexible modules, such as you would put into a content management system.

Hackos, JoAnn T., and Dawn M. Stevens. *Standards for Online Communication: Publishing Information for the Internet/World Wide Web/Help Systems/Corporate Intranets*. New York, NY: John Wiley & Sons, 1997.

> Emphasizing information design, this book focuses on analyzing information needs (in terms of procedural, conceptual, reference, and instructional information), designing online systems, and implementing the design. It includes information about organizing topics; using retrievability techniques such as links, tables of contents, indexes, and search; writing for readability; and adding graphics.

Krug, Steve. *Don't Make Me Think: A Common Sense Approach to Web Usability*. Third Edition. Indianapolis, IN: New Riders, 2014.

> "Test early and often" is the mantra of this book, which includes techniques to avoid spending lots of money and time on usability testing and yet get lots of value.

Mayhew, Deborah J. *The Usability Engineering Lifecycle: A Practitioner's Handbook for User Interface Design*. San Francisco, CA: Morgan Kaufmann Publishers, 1999.

> This book shows how to integrate usability techniques throughout the software development process, including development for the web.

Nielsen, Jakob. *Designing Web Usability*. Indianapolis, IN: New Riders, 2000.

> In addition to a major focus on providing quality content on the web and helping users find what they need, Nielsen devotes a chapter each to accessibility and to international use of the web.

Nielsen, Jakob, and Raluca Budiu. *Mobile Usability*. Berkeley, CA: New Riders, 2013.

> In addition to discussing strategy and design, this book devotes a full chapter to writing for mobile.

Price, Jonathan, and Lisa Price. *Hot Text: Web Writing That Works*. Indianapolis, IN: New Riders, 2002.

> This book offers advice for writing for the web such as "make text scannable" and "reduce cognitive burdens."

Rubin, Jeffrey, and Dana Chisnell. *Handbook of Usability Testing: How to Plan, Design, and Conduct Effective Tests*. Second Edition. Indianapolis, IN: John Wiley & Sons, 2008.

> This book is aimed at helping people who have not been trained as usability engineers and yet need to improve the usability of what they work on. This book provides practical guidance from developing a test plan through conducting the tests to analyzing the results. It includes information about designing the user experience.

# Index

## A

abstract language 256
accessibility
    screen captures 458
    alternative text for graphics 478
    color and contrast 481
    tables 482
accuracy
    automated tools for checking 93
    checklist 96
    definition 14
    example of in task topics 501
    grammar-checking tools 94
    guidelines 68
    information consistency
        types of inconsistencies 88
        information reuse 86
    information currency
        examples and samples 84, 242
        messages 81
        increasing content lifespan 82
        technical changes 79
    link-checking tools 95
    overview 67
    research
        direct observation 69
        hands-on experience 69
        information plans 71
        interviews 73
        outlines 71
        rough drafts 71
        topic models 71
    spell-checking tools 93
    technical reviews
        choosing reviewers 76
        exit criteria 77
        focus for 76
        interface testing 74
        quality control tests 78
        technical ownership 77
    user interfaces 80
    verifying
        hands-on testing 74
        quality control tests 78
        technical reviews 76
active voice 263
adverbial conjunctions 286

# Index

agile development process  12
ambiguity
    coordination  204
    empty words  183
    fragments  198
    keywords as plain text  182
    long noun phrases  204
    modifiers
        dangling  201
        misplaced  202
        squinting  203
    negative expressions  191
    noun phrases  204
    overview  180
    positive writing  189
    pronouns  199
        *that* and *which*  201
        vague referents  199
    syntax  194
    translation problems  188
    word as more than one part of speech  180
analogies  218, 253
animation  443
audience
    *See also* users
    cultural backgrounds  222
    international  494
    mixed  129
    primary  123
    secondary  123
    skill levels  222
    writing for intended  25
autocompled values  9, 104

# B

beta test  523
bias in writing  273
breadcrumb trail  394
browsing, optimizing for  381
bulleted lists in steps  55
business scenarios  243, 245

# C

callouts in illustrations  473
*can* vs. *may*  188
capitalization
    technical terms  280
    translation problems  280
    user interfaces  281
characteristics, quality  13
checklists
    accuracy  96
    all quality characteristics  545
    clarity  212
    completeness  148
    concreteness  259
    organization  366, 376
    retrievability  420
    style  314
    task orientation  64
    visual effectiveness  483
circular statements  166
clarity
    ambiguity
        coordination  204
        empty words  183
        fragments  198
        guidelines  180
        keywords as plain text  182
        modifiers  201
        negative expressions  191
        noun phrases  204
        positive writing  189
        pronoun use  199
        syntax  194
        *that* and *which*  201
        vague referents  199
        word used as more than one part of speech  180
    checklist  212
    coherence
        coordinating ideas  174
        digression  174
        subordinating ideas  175
        transition words  174

# Index

conceptual information 508
coordinated phrases and clauses 157
definition 14
gerund phrases 156
guidelines 154
modifying phrases and clauses 159
reference information 513
rewriting 153
subject-verb proximity 155
task topics 506
technical terms
    appropriateness 205
    consistency 205
    defining 210
testing 524
usability testing 518
wordiness
    expletives 162
    imprecise verbs 169
    Latinate verbs 172
    nominalizations 170
    passive verbs 170
    phrasal verbs 173
    redundancies 163
    roundabout expressions 161
    unnecessary modifiers 167
code samples 247
coherence
    coordinating ideas 174
    subordinating ideas 174
    transition words 174
colloquialisms, tone 268
colons
    explaining ideas 290
    lists 289
color and contrast 481
commands, as plain text 182
commas
    comma splices 287
    essential phrases 284
    items in series 288
    nonessential phrases 284
    run-on sentences 287
    serial 94
    with conjunctions 286
    with quotation marks 294
common expressions, reusing
    error messages 309
    reference topics 309
    videos 307
completeness
    checklist 148
    definition 14
    embedded assistance 493
    guidelines 100
    instructions 133
    outlines 115
    overly complete 136
    progressive disclosure
        hover help 113
        messages 111
        pattern 107
    testing 524
    usability testing 518
    user goals 118, 133
    user interfaces 101
concept topic type 339
conceptual information
    applying quality characteristics 506
    concreteness elements in 229
    editing 530
    examples in 229
    scenarios in 232, 243
    similes in 230
concreteness
    analogies 218, 253
    audience characteristics 222
    challenges to 215
    checklist 259
    conceptual information 229, 508
    control-level assistance 226
    definition 14
    domain expertise 220
    elements
        analogies 218, 253
        current 216, 242
        examples 218
        focused 240

# Index

organization 320
precise language 219
programmatic assistance 217, 223
realistic 241
samples 218
scenarios 217, 243, 251
similes 218
visual 218
embedded assistance 223
error messages 225, 237
examples
    code 247
    highlighting in 250
    level of detail 221
    overview 218
    realistic 241
guidelines 219
information currency 216
labels 224
nonnative users 494
overview 215
programmatic assistance 217, 223
reference information 233, 513
samples
    code 247
    highlighting in 250
    overview 218
    realistic 241
scenarios
    business scenarios 243
    overview 217
    realistic 241
    task scenarios 243
similes 218, 253
specific language 256
task information 227
testing 524
troubleshooting information 235
usability testing 518
conditional steps 60
conjunctions 286
consistency
    common expressions 306

factual 532
information reuse 86, 306
organizational 354, 532
spelling 276
stylistic 533
visual
    design guidelines 465
    layout 463
    style 460
contextual information
    definition 323
    embedded assistance 332
    organization 352
    pattern 334
    separating 324
    window-level assistance 352
contrast and color 481
control-level assistance
    concreteness elements 226
    definition 107
    planning for 107
    unnecessary repetition 145
coordinate conjunctions 286
coordinated clauses 157
cross-references
    *See* links
cueing graphics 475
cultural bias 273
currency of information
    examples and samples 242
    increasing content lifespan 82
    technical changes 79
customized documentation 130

## D

dangling modifiers 201
dashes
    *See* em dashes
    *See also* hyphens
default values
    definition 104
    programmatic assistance 9, 223

design review meetings  70
detected values
    definition  104
    programmatic assistance  9
digression, coherence  174
disabled UI control  105
DITA
    definition  305
    tagging style  312
domain expertise
    concreteness  220
    definition  13
    example  130
drafts, rough  71

# E

editing
    conceptual information  530
    embedded assistance  529
    organization  530
    overview  528
    preparation  528
    reference information  530
    summarizing findings  534
    task information  529
elliptical style  194
embedded assistance
    API names  7
    capabilities  107
    color  8
    command names  7
    concreteness  223
    control-level  107, 226
    definition  4
    editing for  529
    hardware  8
    help pane  6, 109
    hover help  6
    icon assistance  107
    illustration  6
    inconsistent  358
    input hints  6
    interface labels  6
    keyword names  7
    messages  6, 7, 108, 225
    multiple quality characteristics applied to  493
    navigation  395
    parameter names  7
    programmatic assistance  9, 104, 223
    redundancy  145
    retrievability  493
    task orientation  43
    text labels  224
    tool names  7
    tooltips  6, 107
    types  332
    user interface text, static  107
    utility names  7
    wizards  6
embedded help pane  6
em dashes
    guidelines  292
    using for emphasis  290
empty words  183
entry points
    definition  413
    highlighting  414
    tables  416, 420
    user interface  417
error messages
    appropriate wording  192
    common expressions  308
    concreteness  237
    troubleshooting information  237
    users' perspective  30
evaluation tests  520
examples
    code  247
    level of detail  221
    overview  218
    realistic  241
experience levels
    experienced  128
    mixed  129
    novice users  126, 128

**577**

# Index

expertise, domain 13
expletive constructions 163

## F

factual consistency 532
feature-focused content 41, 45
field observation 523
fonts 429
fragments
    ambiguity 198
    list introductions 300
front-loading 382
function-oriented tasks 35

## G

gender bias 273
gerund phrases 156
goals, user
    See user goals
grammar-checking tools 94
graphics
    accessibility 478
    cueing 475
    illustrations
        big-picture 438
        consistency 460
        creating 439
        guidelines 439
        interactive 441
        scenarios 434
        spatial relationships 437
        task flows and processes 431
        visual effectiveness 431
    screen captures
        appropriate uses 449
        design tips 458
        drawbacks 458
        example 451
        unnecessary 453
    videos
        appropriate uses 444

        common expressions 307
        definition 444
        design tips 447
        tours 445
        tutorials 447
    white space 426

## H

hands-on experience 69
headings
    See titles
help
    See contextual information
help pane
    capability 109
    embedded 6
high-level steps 49
highlighting
    examples 250
    entry points 414
    text 296
    user interfaces 296
hints
    See input hints
hover help
    focus on user tasks 43
    illustration 6
    inconsistent 88
    navigation 396
    redundant 145
    sample content 11
hyphens
    ambiguous noun phrases 204
    guidelines 292
    spelling 293

## I

icons
    legibility 481
    user assistance 107

illustrations
    big-picture  438
    consistency  460
    creating  439
    guidelines  439
    interactive  441
    scenarios  434
    spatial relationships  437
    task flows and processes  431
    visual effectiveness  431
image maps  441
imprecise verbs  169
informal tone  268
information, unnecessary  136
information verification
    interface testing  74
    quality control tests  78
    review exit criteria  77
    technical ownership  77
    technical reviews  76
indexes  388
information plans  71
inline (in-sentence) links  403
input hints
    examples  258
    illustration  6
    structure  334
in-sentence (inline) links  403
instructions  133
interactive illustrations  441
interviews  73
introductions, organization  366
items in series, commas  288

## J

jargon
    terminology  209
    tone  268

## K

keywords
    clarity  182

    density  387
    location  385
    names  7
    proximity  387
    retrievability  384
    stuffing  387

## L

labels
    as plain text  182
    concreteness  224
    embedded assistance  224
    illustration  6
    inconsistent  88
    notes  374
    sample content  11
layout, consistent style  463
legal boilerplate  88
link-checking tools  95
links
    appropriateness  399
    definition  399
    essential  400
    in-sentence (inline)  403
    navigation  394
    redundant  405
    search engine optimization  400
    strategy  396
    wording  409
lists
    bulleted  55
    instructions  55
    length  300
    nesting  300
    parallelism  302
    punctuation  300
    unordered  55
low-level steps  49

## M

machine translation
    clarity  188

**579**

# Index

syntactic cues 195
main points
    emphasizing 366
    websites 369
markup languages 311
messages
    accuracy 81
    capability 108
    concreteness 225
    embedded assistance 6
    error 237
    illustration 108
    inconsistent 88
    logged 7
    reusing 88
    sample content 11
meta tags 388
minimalism 136
misplaced modifiers 202
model, topic
    *See* topic models
modifiers
    dangling 201
    misplaced 202
    squinting 203
    unnecessary 167
modifying clauses 159

# N

navigation
    breadcrumb trail 394
    consistent 355
    embedded assistance 395
    hover help 396
    links 394
    optimizing 389
    organization 345, 364
    paths 394
    retrievability 394
negative expressions 190
negative prefixes 192
negative space 467
    *See* white space

nominalizations 170
noncontextual information
    definition 323
    separating 324
    topics 337
    topic sets 343
    topic types 339
nonnative users
    ambiguity 180
    clarity 494
    concreteness 494
    multiple quality characteristics applied to 500
    style 494
notes
    in examples 250
    labels 374
    misusing 374
noun phrases, ambiguous 204
novice users 126, 129

# O

observation, direct 69
*only*, ambiguous use of 202
optional steps 58
organization
    challenges 327
    checklist 376
    consistency 354, 532
    contextual information
        definition 323
        embedded assistance 332, 352
    definition 14, 319
    editing 530
    embedded assistance 332, 334
    guidelines 321
    information elements 320
    information delivery mechanisms 322
    main points 366
    navigation 345
    noncontextual information 323
    notes 374
    overview 319
    progressive disclosure pattern 334

redundancy 326
relationship of elements 360
retrievability 357
secondary points 371
separation of information 322
sequential 350
step-by-step instructions 353
textual elements 424
topics 337
topic sets 343
topic types 339
outdated content 82
outlines
  completeness 115
  research 71
overcompleteness 136

# P

paragraphs, coherent 174
parallelism
  lists 302
  sentences 302
  user interface elements 302
parameters
  names as embedded assistance 7
  reusing 88
parentheses with periods 295
passive voice
  appropriateness 263
  error messages 264
  imprecise verbs 170
  weak verbs 170
patterns
  application of 333
  definition 107
  progressive disclosure 9, 334
periods with parentheses 295
phrases
  coordinated 157
  modifying 159
  prepositional 160
placement
  main points 366

secondary points 371
positive writing 189
precise language 219
prefixes, negative 192
prepositional phrases 160
pretentious tone 270
primary audience, definition 123
procedures 133
process, writing 15
products
  features, avoid focus on 41
  scenarios 33
programmatic assistance
  autocompleted values 9
  concreteness 217, 223
  default values 9
  definition 9
  detected values 9
  embedded 223
  examples 104
  illustration 9
  types 104
progressive disclosure
  applying a pattern 107
  definition 10
  embedded assistance capabilities 107
  hover help example 113
  messages example 111
  multiple quality characteristics applied 488
  navigation aids 394
  sample pattern 11, 109, 334
pronouns
  *that* and *which* 201
  vague referents 199
prototypes 70
punctuation
  colons
    explaining ideas 290
    lists 289
  combinations of 294
  commas
    comma splices 287
    essential phrases 284
    grammar tools 94

**581**

items in series  288
nonessential phrases  284
run-on sentences  287
serial  288
with conjunctions  286
with quotation marks  294
correct  284
em dashes  292
hyphens
ambiguous noun phrases  204
guidelines  293
compound words  293
lists  300
semicolons  291

# Q

quality characteristics
checklist  545
definition  13
determining  534
groupings  14
international audience  494
progressively disclosed information  488
reviews  516
topic-based information  501
usability testing  518
verification  549
writing cycle  17
quality
control tests  78
ratings  534

# R

redundancy
embedded assistance  145
helpful  141
unnecessary  145
user interfaces  164
words  163
reference information
multiple quality characteristics applied  509

organization  339
reusing common expressions  309
samples in  233
titles  382
topic types  339
referents, vague  199
relationships of topics, revealing  360
relative clauses, long sentences  159
repetition
helpful  141
main points  366
reasons for  141
unnecessary  145
research
observation  69
hands-on  69, 74
interviews  73
retrievability
checklist  420
conceptual information  508
definition  14
description  379
editing for reference information  530
embedded assistance  493
entry points
definition  413
highlighting  414
tables  416
user interface  417, 421
guidelines  380
links
cross-references  411
essential  400
in-sentence (inline)  403
redundant  405
wording  409
navigation aids  394
optimizing
browsing  381
indexes  388
keywords  384
searching  381
tables of contents  389
topic titles  381

Index

organization 357
reference information 513
task topics 506
testing 525
usability testing 519
user interfaces 397
reusing information 86
reviews
    conferring with writer 535
    design 70
    exit criteria 77
    guidelines 517
    problem reporting 517
    quality characteristics 516
    technical 76
    visual elements
        conferring with editor and writer 541
        description 538
        guidelines 536
        individual 540
        preparation 537
        summarizing findings 541
rewriting for clarity 153
roundabout expressions 161

# S

samples
    code 247
    concreteness element 218
    currency 242
    realistic 241
scenario-based information 32, 217
scenarios
    business
        definition 243
        effective 245
    conceptual information 232
    context 251
    definition 32, 243
    description 217
    illustrations 434
    integration 33
    product design 33

realistic 241
tasks 243
screen captures
    accessibility 458
    appropriate uses 449
    currency 458
    design tips 458
    example 452
    translation 458
    unnecessary 453
    usability testing 450
search engine optimization
    keywords 387
    links 400
    titles 381
secondary audience, definition 123
secondary points
    de-emphasize 371
    user interfaces 373
semicolons, guidelines 291
sentences
    fragments 198
    negative 190
    passive 170
sequential organization 350
serial commas 288
sexism 273
similes
    conceptual information 230
    description 218, 253
    example 253
    purpose 253
single sourcing 86
slang, tone 268
spatial relationships, illustrations 437
spelling
    consistency 276
    correctness 276
    hyphens 293
    technical terms 276
    tools for checking 93
squinting modifiers 203
static UI text
    affordance 107

583

inconsistent 88
sample content 11
step-by-step instructions
    lists 55
    organization 353
    progression 49
    reusing 88
    steps
        clear action 51
        conditional 58, 60
        grouping 53
        step levels 49
        optional 58
        passive voice 264
style
    active voice 263
    capitalization 280
    checklist 314
    common expressions 305, 306
    conceptual information 508
    consistency 460, 533
    cultural bias 273
    definition 14
    gender bias 273
    guidelines 262
    highlighting text 296
    nonnative users 494
    overview 261
    passive voice 263
    punctuation
        colons 289
        combinations of 294
        commas 284
        consistency 284
        em dashes 292
        hyphens 293
        semicolons 291
    sexism 273
    spelling 276
    tagging 311
    task topics 506
    templates 305
    testing 524
    tone 267

subject-verb proximity 155
subordinate
    clauses 175
    conjunctions 287
surveys 523
syntactic cues
    ambiguity 194
    *that* 196
syntax, inconsistent 88

# T

table of contents, optimizing 389
tables
    accessibility 482
    entry points 416, 420
    guidelines 473
    styles 470
    visual effectiveness 468
tagging
    DITA 312
    markup languages 311
    style 311
task information
    concreteness 227
    editing 529
    examples in 227
    illustrated 431
    multiple quality characteristics applied to 501
task orientation
    checklist 64
    definition 14, 23
    embedded assistance 43
    guidelines 24
    intended audience 25
    overview 23
    product features 41
    purpose of information 46
    step-by-step instructions
        clear actions 51
        conditional steps 58, 60
        grouping 53
        levels of steps 49
        optional steps 58

# Index

users' goals 32
users' point of view 27
task scenarios 243
task topics
    examples 35
    multiple quality characteristics applied to 501
    titles 382
    topic type 339
    user-oriented 35
technical ownership 77
technical reviews 76
technical terms
    consistency 207
    defining new terms 210
    jargon 209
    spelling 275
technical writers
    conferring with 535
    responsibilities 12
    role 11
templates
    DITA tagging 305
    role in visual effectiveness 421
    style 305
terminology management system 205
testing
    guidelines 524
    hands on on page 74
    information in test scripts 78
    laboratory 519
    prototyping 519
    style 524
    task orientation 524
    test cases 525
    tools 525
    usability
        evaluation tests 520
        outside laboratory 522
        organization 519
        quality characteristics 518
        remote 523
        retrievability 519
        screen captures 450
        validation tests 521

user interface 526
visual effectiveness 525
text, highlighting 296
text labels
    concreteness 224
    entry points 417
textual elements
    fonts 429
    highlighting 296
    organization 425
    user interface 426
*that*
    essential clauses 201
    syntactic cue 196
titles
    inconsistent 88
    for topic types 382
tone
    colloquialisms 268
    formal 268
    humor 268
    idioms 260
    informal 268
    neutral 271
    pretentious 270
    slang 268
tools
    embedded assistance 7
    grammar-checking 94
    link-checking 95
    spell-checking 93
    testing 525
tooltips
    embedded assistance 6
    inconsistent 88
topic models
    example 115, 120, 135
    publishing 135
    research 71
topics
    definition 337
    illustration 338
    keywords for 385
    multiple quality characteristics applied to 501

## Index

navigation 354
ordering 348
relationships 360
screen captures 449
types 339
topic sets
    definition 343
    hierarchy 344
    navigation 355
topic titles
    examples 383
    guidelines 382
    questions as 384
trademarked terms 183
transition words 174
translation
    machine
        capitalization 280
        clarity 188
        phrasal verbs 173
        spelling 275
        syntactic cues 195
    screen captures 458
troubleshooting information
    concreteness 235
    error messages 237
    procedures 236
troubleshooting topic titles 382

## U

unnecessary modifiers 170
usability testing
    laboratory
        evaluation tests 520
        validation tests 521
    organization 519
    quality characteristics 518
    remote 523
    retrievability 519
    screen captures 450
    outside testing 522
    visual effectiveness 519
    walkthroughs 70

user experience 123
user goals
    completeness 115, 133
    definition 32
    example 17
    investigating 16
    scenario-based 32
    task models 33
    task orientation 32
    understanding 16
user interfaces
    accuracy 80
    entry points 417
    examples in 225
    hands-on testing 74
    highlighting in text 296
    inline text 417
    mapping 364
    redundancies 165
    retrievability 397
    screen captures 449
    secondary points 373
    self-documenting 101
    specific language in 256
    testing 526
    textual elements 426
    verifying 74
    video tours 445
user-oriented tasks, definition 35
users
    *See also* audience
    edits or reviews 523
    experienced 129
    needs 220
    novice 126, 129
    point of view 27
    skill level 220

## V

vague referents 199
validation tests 521
variables, reusing 88
verbs

active  169
nominalizations  170
passive  170
phrasal  173
imprecise  169
with syntactic cues  196
weak  169
verb-subject proximity  155
videos
　appropriate uses  444
　commonly used expressions  307
　design tips  447
　tours  445
　tutorials  447
viewlets  444
visual effectiveness
　accessibility
　　alternative text for graphics  478
　　color  481
　　contrast  481
　　tables  482
　checklist  483
　description  421
　elements
　　callouts  473
　　cueing graphics  475
　　tables  468
　　white space  467
　graphics
　　illustrations  431
　　screen captures  448
　　videos  444
　guidelines  423
　impact on other quality characteristics  423
　style consistency
　　design guidelines  465
　　illustration  460
　　layout  463

textual elements
　fonts  429
　organization  425
　user interface  426
usability testing  519
visual elements
　appropriateness  533
　concreteness  218
　reviewing
　　guidelines  536
　　overview  538
　types  422

# W

waterfall development process, definition  11
websites
　main points  369
　secondary points  373
*which*  201
white space  467
window-level assistance  352
wizards
　navigation  365
　self-documenting user interface  101
wordiness
　expletive constructions  162
　imprecise verbs  169
　redundancies  163
　roundabout expressions  161
　unnecessary modifiers  167
words
　clear  187
　empty  186
　transition  174
writer
　*See* technical writer

587